T0074100

Information Security and Cryptography
Texts and Monographs

For further volumes:
http://www.springer.com/series/4752

Ahmad-Reza Sadeghi · David Naccache

Editors

Towards Hardware-Intrinsic Security

Foundations and Practice

Foreword by Pim Tuyls

 Springer

Editors
Prof. Ahmad-Reza Sadeghi
Universität Bochum
Horst Görtz Institut für
Sicherheit in der
Informationstechnik
Universitätsstr. 150
44780 Bochum
Germany
ahmad.sadeghi@trust.rub.de

Prof. David Naccache
École Normale Supérieure
Dépt. Informatique
rue d'Ulm 45
75230 Paris CX 05
France
david.naccache@ens.fr

Series Editors

Prof. Dr. David Basin
Prof. Dr. Ueli Maurer
ETH Zürich
Switzerland
basin@inf.ethz.ch
maurer@inf.ethz.ch

ISSN 1619-7100
ISBN 978-3-642-14451-6   e-ISBN 978-3-642-14452-3
DOI 10.1007/978-3-642-14452-3
Springer Heidelberg Dordrecht London New York

Library of Congress Control Number: 2010938038

ACM Computing Classification (1998): E.3, K.6.5, B.7, C.2

Cover design: KuenkelLopka GmbH, Heidelberg

Printed on acid-free paper

Springer is part of Springer Science+Business Media (www.springer.com)

Foreword

Nowadays, computing devices are omnipresent. The vision of the Ambient Intelligent society is becoming a reality very rapidly. Information is exchanged at the speed of light, everybody is connected anytime and anywhere, and new technological developments in the world are taking place faster than ever before. This evolution is the result of the progress in semi-conductor manufacturing processes and technologies which make ICs every year smaller, faster, and more powerful. Mobile devices, (smart) phones, PCs, laptops, smart cards, RFID-tags, personal secure tokens, sensors, etc., are typical products enabling ambient intelligence. Within this environment, information has become one of the most valuable goods and its early availability often means a competitive advantage or a guarantee to our overall security and safety. Human beings on the one hand and industrial as well as governmental organizations on the other hand have become highly dependent on the availability, accessibility, and the flow of correct information for their everyday operations.

Without proper protection, however, information is at the same time the Achilles heel of such a society. When a malicious person or organization can obtain or tamper with sensitive information, the most unexpected and severe consequences may arise. The competitive advantage of a company might disappear, the privacy of individuals and even the security of a whole nation can be compromised. In order to deal with the confidentiality and authenticity of information, cryptographic algorithms are implemented in modern computing devices to protect the link between endpoints. The fact that state-of-the-art cryptographic algorithms are very strong implies that not the links but the physical devices and implementation of the algorithms in those devices have become the weak link in the chain. In particular the secure storage of secret keys and the secure implementation of algorithms and architectures withstanding physical attacks represent some of the major challenges for the security community. The main problem stems from three facts. First, computations are physical processes that leak information on the data being processed through physical side-channels. Second, memories leak information on the stored data to attackers having the availability of "sophisticated" devices such as laser cutters, focused ion beams, and electron microscopes. Unfortunately such tools are readily available for rent nowadays. Third, security measures have to be based on and implemented in a low-cost manner to be economically viable while attackers have high to almost unlimited budgets available.

A particular field of applications where physical attacks pose an important threat is that of counterfeiting of goods. The terminology "goods" has to be understood here in its most general sense, i.e., physical goods as well as digital goods such as (embedded) software programs, music, video, and designs. Examples of physical goods being counterfeited are automotive and avionic parts, pharmaceuticals, bank passes, smart cards, routers, etc. The total annual value of the trade in fake goods has risen from $200 billion in 2002 to as much as $450 billion in 2006 and the number is expected to have risen to $600 billion in 2009. From these numbers it follows that counterfeiting has a huge economic impact. However, since those products have often lower quality they might additionally lead to brand damage for the legitimate company as well. When counterfeit components are used within critical infrastructures, it is important to realize that the quality level might not only cause damage but contain hidden components whose functionality is not specified. Without doubt this is a threat to the national security of a country.

Recently, a new field of security research dealing with the problem of "physical attacks" and "physical leakage of information" started to develop. Many research groups started to investigate algorithmic as well as physical countermeasures to these threats. Although no general theory dealing with this problem is available, several sub-fields are well developed. The general theory of side-channel secure cryptography has made big progress and goes under the name of physical observable cryptography. Apart from general theoretic developments various practical and efficient countermeasures have been developed as well. Hardware Intrinsic Security on the other hand is a much younger field dealing with secure secret key storage. By generating the secret keys from the intrinsic properties of the silicon, e.g., from intrinsic physical unclonable functions (PUFs), no permanent secret key storage is required anymore and the key is only present in the device for a minimal amount of time. The field of Hardware Intrinsic Security is extending to hardware-based security primitives and protocols such as block ciphers and stream ciphers entangled with hardware. When successful, this will raise the bar of IC security even further. Finally, at the application level there is a growing interest in hardware security for RFID systems and the necessary accompanying system architectures.

It is a pleasure for me to write the foreword of this book. The fields of Hardware Security in general and Hardware Intrinsic Security in particular are very challenging fields with many open problems of high practical relevance. It brings together researchers and practitioners from academia and industry from collaborating and competing groups. The field is highly interdisciplinary by nature. Here, expertises and results from different fields such as physics, mathematics, cryptography, coding theory, and processor theory meet and find new applications. The meeting at Dagstuhl in the summer of 2009, from which this book is the result, brought together many experts from all over the world to discuss these topics in an open and stimulating atmosphere. Personally, I am convinced that this book will serve as an important background material for students, practitioners, and experts and stimulates much further research and developments in hardware security all over the world. Without doubt the material covered here will lay the foundations of the future security devices guaranteeing the necessary privacy, confidentiality, and authenticity of information for our modern society.

January 2010   Pim Tuyls

Contents

Part III   Hardware Attacks

Part IV   Hardware-Based Policy Enforcement

Part V   Hardware Security in Contactless Tokens

Part VI   Hardware-Based Security Architectures and Applications

List of Contributors

Frederik Armknecht Horst Görtz Institute for IT Security, Ruhr-University Bochum, Bochum, Germany; Technische Universität Darmstadt, Darmstadt, Germany, Frederik.Armknecht@trust.rub.de

Muhammad Asim Philips Research Eindhoven, Information and System Security Group, The Netherlands, muhammad.asim@philips.com

Yoo-Jin Baek System LSI Division, Samsung Electronics Co., Ltd., Suwon, Korea, yoojin.baek@samsung.com

Lejla Batina Computing Science Department/DS group, Radboud University Nijmegen, 6525 AJ Nijmegen, The Netherlands, lejla@cs.ru.nl

Heike Busch Technische Universität Darmstadt, Darmstadt, Germany, busch@seceng.informatik.tu-darmstadt.de

Andrew H. Chan University of California, Berkeley, CA, USA, andrewhc@eecs.berkeley.edu

Aykutlu Dana UNAM, Institute of Materials Science and Nanotechnology, Bilkent University, Ankara, Turkey, aykutlu@unam.bilkent.edu.tr

Daniel Y. Deng Cornell University, Ithaca, NY, USA, dyd2@cornell.edu

Loic Duflot French Network and Information Security Agency (ANSSI), Paris, France, loic.duflot@ssi.gouv.fr

G. Edward Suh Cornell University, Ithaca, NY, USA, suh@csl.cornell.edu

Vanessa Gratzer Centre de recherche en informatique, Université Paris I, Panthéon-Sorbonne, Paris, France, vanessa@gratzer.fr

Olivier Grumelard French Network and Information Security Agency (ANSSI), Paris, France, olivier.grumelard@ssi.gouv.fr

Jorge Guajardo Philips Research Eindhoven, Information and System Security Group, The Netherlands, jorge.guajardo@philips.com

Ghaith Hammouri CRIS Lab, Worcester Polytechnic Institute, Worcester, MA 01609-2280, USA, hammouri@wpi.edu

Helena Handschuh Intrinsic-ID, San Jose, CA 95110, USA; ESAT-COSIC, Katholieke Universiteit Leuven, Leuven, Belgium, helena.handschuh@intrinsic-ID.com

Kimmo Järvinen Department of Information and Computer Science, Aalto University, Aalto, Finland, kimmo.jarvinen@tkk.fi

Stefan Katzenbeisser Technische Universität Darmstadt, Darmstadt, Germany, katzenbeisser@seceng.informatik.tu-darmstadt.de

Inyoung Kim Statistics Department, Virginia Tech, Blacksburg, VA 24061, USA, inyoungk@vt.edu

Sung-Hyun Kim System LSI Division, Samsung Electronics Co., Ltd., Suwon, Korea, teri_kim@samsung.com

Darko Kirovski Microsoft Research, Redmond, WA 98052, USA, darkok@microsoft.com

Vladimir Kolesnikov Alcatel-Lucent Bell Laboratories, Murray Hill, NJ 07974, USA, kolesnikov@research.bell-labs.com

Markus G. Kuhn Computer Laboratory, University of Cambridge, Cambridge CB3 0FD, UK, Markus.Kuhn@cl.cam.ac.uk

Yong Ki Lee Department of Electrical Engineering, University of California, Los Angeles, CA 90095-1594, USA, jfirst@ee.ucla.edu

Olivier Levillain French Network and Information Security Agency (ANSSI), Paris, France, olivier.levillain@ssi.gouv.fr

Roel Maes ESAT/COSIC and IBBT, Catholic University of Leuven, Leuven, Belgium, roel.maes@esat.kuleuven.be

Abhranil Maiti ECE Department, Virginia Tech, Blacksburg, VA 24061, USA, abhranil@vt.edu

Benjamin Morin French Network and Information Security Agency (ANSSI), Paris, France, benjamin.morin@ssi.gouv.fr

David Naccache Département d'informatique, École normale supérieure, Paris, France, david.naccache@ens.fr

Leyla Nazhandali ECE Department, Virginia Tech, Blacksburg, VA 24061, USA, leyla@vt.edu

Elisabeth Oswald Department of Computer Science, University of Bristol, Department of Computer Science, Bristol, UK, Elisabeth.Oswald@bristol.ac.uk

Olivier Pereira Crypto Group, Université catholique de Louvain, Louvain-la-Neuve, Belgium, olivier.pereira@uclouvain.be

Milan Petković Philips Research Eindhoven/Eindhoven University of Technology, Eindhoven, The Netherlands, milan.petkovic@philips.com, m.petkovic@tue.nl

Bart Preneel ESAT-COSIC, Katholieke Universiteit Leuven, Leuven, Belgium, bart.preneel@esat.kuleuven.be

Jean-Jacques Quisquater Crypto Group, Université catholique de Louvain, Louvain-la-Neuve, Belgium, quisquater@dice.ucl.ac.be

Ulrich Rührmair Technische Universität München, München, Germany, ruehrmai@in.tum.de

Ahmad-Reza Sadeghi Horst Görtz Institute for IT Security, Ruhr-University Bochum, Bochum, Germany, ahmad.sadeghi@trust.rub.de

Patrick Schaumont ECE Department, Virginia Tech, Blacksburg, VA 24061, USA, schaum@vt.edu

Thomas Schneider Horst Görtz Institute for IT-Security, Ruhr-University Bochum, Bochum, Germany, thomas.schneider@trust.rub.de

Geert-Jan Schrijen Intrinsic-ID, 5656 AE Eindhoven, The Netherlands, geert.jan.schrijen@intrinsic-ID.com

Dave Singelee ESAT-COSIC, Katholieke Universiteit Leuven, Leuven, Belgium, dave.singelee@esat.kuleuven.be

François-Xavier Standaert Crypto Group, Université catholique de Louvain, Louvain-la-Neuve, Belgium, fstandae@uclouvain.be

Berk Sunar Cryptography & Information Security, Worcester Polytechnic Institute, Worcester, MA, USA, sunar@wpi.edu

Mohammad Tehranipoor University of Connecticut, Storrs, CT 06269, USA, tehrani@engr.uconn.edu

Pim Tuyls Intrinsic-ID, 5656 AE Eindhoven, The Netherlands; ESAT/COSIC and IBBT, Catholic University of Leuven, Leuven, Belgium, pim.tuyls@intrinsic-ID.com, pim.tuyls@gmail.com

Markus Ullmann Bonn-Rhein-Sieg University of Applied Sciences, Sankt Augustin, Germany; Bundesamt für Sicherheit in der Informationstechnik, Bonn, Germany, Markus.Ullmann@bsi.bund.de

Ingrid Verbauwhede ESAT-COSIC, Katholieke Universiteit Leuven, Leuven, Belgium, ingrid.verbauwhede@esat.kuleuven.be

Vignesh Vivekraja ECE Department, Virginia Tech, Blacksburg, VA 24061, USA, vigneshv@vt.edu

Matthias Vögeler NXP Semiconductors, Business Line Identification, Hamburg, Germany, Matthias.Voegeler@nxp.com

Ivan Visconti Dipartimento di Informatica ed Applicazioni, University of Salerno, Salerno, Italy, visconti@dia.unisa.it

Christian Wachsmann Horst Görtz Institute for IT-Security (HGI), Ruhr-University Bochum, Bochum, Germany, christian.wachsmann@trust.rub.de

Yu Yu Crypto Group, Université catholique de Louvain, Louvain-la-Neuve, Belgium, yu.yu@uclouvain.be

Moti Yung Department of Computer Science, Columbia University, New York, NY, USA; Google Inc, Mountain View, CA, USA, moti@cs.columbia.edu

Huaiye Zhang Statistics Department, Virginia Tech, Blacksburg, VA 24061, USA, zhanghy@vt.edu

Part I
Physically Unclonable Functions (PUFs)

Physically Unclonable Functions: A Study on the State of the Art and Future Research Directions

Roel Maes and Ingrid Verbauwhede

1 Introduction

The idea of using intrinsic random physical features to identify objects, systems, and people is not new. Fingerprint identification of humans dates at least back to the nineteenth century [21] and led to the field of *biometrics*. In the 1980s and 1990s of the twentieth century, random patterns in paper and optical tokens were used for unique identification of currency notes and strategic arms [2, 8, 53]. A formalization of this concept was introduced in the very beginning of the twenty-first century, first as physical one-way functions [41, 42], physical random functions [13], and finally as *physical(ly) unclonable functions* or PUFs.[1] In the years following this introduction, an increasing number of new types of PUFs were proposed, with a tendency toward more integrated constructions. The practical relevance of PUFs for security applications was recognized from the start, with a special focus on the promising properties of physical unclonability and tamper evidence.

Over the last couple of years, the interest in PUFs has risen substantially, making them a hot topic in the field of hardware security and leading to an expansion of published results. In this work we have made, to the best of our knowledge, an extensive overview of all PUF and PUF-like proposals up to date in an attempt to get a thorough understanding of the state of the art in this topic. Due to the wide variety of different proposals, the different measures used for assessing them, and the different possible application scenarios, making an objective comparison between them is not a trivial task. In order to generalize this and future overview attempts, we identify and concretize a number of properties on which different PUF

R. Maes (✉)
K.U. Leuven, ESAT/COSIC and IBBT, Leuven, Belgium
e-mail: roel.maes@esat.kuleuven.be

This work was supported by the IAP Program P6/26 BCRYPT of the Belgian State and by K.U. Leuven-BOF funding (OT/06/04). The first author's research is funded by IWT-Vlaanderen under grant number 71369.

[1] Note that there is a slight semantical difference between *physical* and *physically* unclonable functions. Further on in this work, we argue why the term *physically unclonable* is more fitting. For the remainder of this text, we will hence speak of PUFs as physically unclonable functions.

A.-R. Sadeghi, D. Naccache (eds.), *Towards Hardware-Intrinsic Security,* Information
Security and Cryptography, DOI 10.1007/978-3-642-14452-3_1,
© Springer-Verlag Berlin Heidelberg 2010

proposals can be evaluated. In the process of listing the different PUFs and their properties, a number of interesting findings and future research and discussion topics will surface.

This chapter is structured as follows: after a necessary introduction in the basic PUF terminology in Sect. 2, an extensive and profound overview of all PUF and PUF-like proposals up to date is presented in Sect. 3. Based on the findings in this overview, we identify a number of fundamental PUF properties in Sect. 4 and assess them for popular PUF proposals. As a result of this comparison, we try to point out the necessary conditions for a construction to be called a PUF. After a brief overview of the basic PUF application scenarios in Sect. 5, we introduce and discuss a number of future research directions in Sect. 6. Finally, we present some concluding remarks in Sect. 8.

2 PUF Terminology and Measures

We introduce a number of commonly used terms and measures used in describing PUFs and their characteristics. We successively describe the challenge–response terminology in Sect. 2.1, the commonly used inter- and intra-distance measures in Sect. 2.2, and point out the problem of environmental effects and possible solutions in Sect. 2.3.

2.1 Challenges and Responses

From its naming it is clear that a PUF performs a functional operation, i.e., when queried with a certain input it produces a measurable output. We immediately stress that in most cases, a PUF is not a true *function* in the mathematical sense, since an input to a PUF may have more than one possible output. It is more appropriate to consider a PUF as a function in an engineering sense, i.e., a procedure performed by or acting upon a particular (physical) system. Typically, an input to a PUF is called a *challenge* and the output a *response*. An applied challenge and its measured response is generally called a *challenge–response pair* or *CRP* and the relation enforced between challenges and responses by one particular PUF is referred to as its *CRP behavior*. In a typical application scenario, a PUF is used in two distinct phases. In the first phase, generally called *enrollment*, a number of CRPs are gathered from a particular PUF and stored in a so-called *CRP database*. In the second phase or *verification*, a challenge from the CRP database is applied to the PUF and the response produced by the PUF is compared with the corresponding response from the database.

For some PUF constructions, the challenge–response functionality is implied by their construction, while for others it is less obvious and particular settings or parameters have to be explicitly indicated to act as the challenge. Also, since for most PUFs a number of post-processing steps are applied, it is not always clear at which point the response is considered. It is preferred to denote both challenges and responses as bit strings; however, this might involve some decoding and quantization, since the physically applied stimuli and measured effects are often analog quantities.

2.2 Inter- and Intra-distance Measures

The fundamental application of PUFs lies in their identification purposes. To that end, the concept of inter- versus intra-(class) distances was inherited from the theory about classification and identification. For a set of instantiations of a particular PUF construction, inter- and intra-distances are calculated as follows:

- For a particular challenge, the *inter-distance* between two different PUF instantiations is the distance between the two responses resulting from applying this challenge once to both PUFs.
- For a particular challenge, the *intra-distance* between two evaluations on one single PUF instantiation is the distance between the two responses resulting from applying this challenge twice to one PUF.

We stress that both inter- and intra-distance are measured on a pair of responses resulting from *the same challenge*. The distance measure which is used can vary depending on the nature of the response. In many cases where the response is a bit string, Hamming distance is used. Often the Hamming distance is expressed as a fraction of the length of the considered strings, and in that case one calls it *relative* or *fractional Hamming distance*.

The value of both inter- and intra-distance can vary depending on the challenge and the PUFs involved. For a particular type of PUF, the inter- and intra-distance characteristics are often summarized by providing histograms showing the occurrence of both distances, observed over a number of different challenges and a number of different pairs of PUFs. In many cases, both histograms can be approximated by a gaussian distribution and are summarized by providing their means, respectively, μ_{inter} and μ_{intra}, and when available their standard deviations, respectively, σ_{inter} and σ_{intra}.

Observe that μ_{intra} expresses the notion of average *noise* on the responses, i.e., it measures the average reproducibility of a measured response with respect to an earlier observation of the same response. It is clear that we would like μ_{intra} as small as possible since this yields very reliable PUF responses. On the other hand, μ_{inter} expresses a notion of *uniqueness*, i.e., it measures the average distinguishability of two systems based on their PUF responses. If the responses are bit strings, the best distinguishability one can achieve is if on average half of the bits differ, i.e., in case μ_{intra} is expressed as relative Hamming distance we would like it to be as close to 50% as possible. The practical use of both notions becomes clear when considering the use of the PUF for identification purposes as explained in Sect. 5.1 and a typical graphical representation is shown in Fig. 7. Note that the goals of minimizing both μ_{intra} and $|50\% - \mu_{inter}|$ can be opposing, and finding an appropriate trade-off is often necessary.

2.3 Environmental Effects

Since producing a PUF response generally involves a physical measurement, there are a number of unwanted physical side effects which could interfere. It was already pointed out in Sect. 2.2 that the same challenge applied to the same PUF does not necessarily produce the same response, giving rise to so-called intra-distance between PUF responses. This might be caused by completely random noise and measurement uncertainties which will inevitably have a random disturbing effect on the measurement. However, certain environmental factors also have a *systematic* effect on the response measurement, e.g., temperature or supply voltage in case of a PUF on an integrated circuit. Average intra-distances will probably increase when measurements are considered over (largely) varying environmental conditions. To enable a fair comparison between different results from literature, it is mentioned when μ_{intra} is obtained from measurements in a fixed or a variable environment.[2]

Because environmental effects are systematic, techniques can be introduced to reduce their influence on the PUF responses. Possible options are as follows:

- If the effects are partially linear and affect the whole device more or less equally, a differential approach can be taken. By considering the relation (difference, ratio, etc.) between two simultaneous measurements instead of one single measurement, one obtains a much more robust measure. This technique was introduced in [11, 13] and is called *compensation*.
- The impact of environmental effects mainly depends on the exact implementation details of the PUF. Certain implementation strategies have a reduced environmental dependency [57]. Another option is to select the environmentally robust responses beforehand and ignoring the unstable ones [52].
- If PUF responses vary heavily over the range of an environmental factor, one can measure this factor with an independent onboard sensor and introduce different operation intervals, narrow enough to minimize the environmental effects within one interval [27, 62].

3 PUF Instantiations

In this section, we provide, to the best of our knowledge, a very thorough overview of all proposed instantiations of PUFs in literature up to now. We also take into account certain constructions which have not been labeled a PUF by their originators,[3] but which we consider to possess certain PUF-like properties. We have

---

[2] Whenever not explicitly mentioned, a fixed environment is assumed.

[3] Possibly because they were proposed before the name *PUF* had been coined, or they were introduced in fields other than cryptographic hardware, where the notion of PUFs has not yet been introduced. When the name of a PUF in the section headings is between quotation marks, it means that we have introduced this name in this work for simplicity and easy reference.

divided this extensive list of PUFs into a number of categories, mainly based on their construction and operation principles. Note that not all proposals are discussed with the same amount of detail, mainly due to a lack of available literature or because some constructions are only mentioned for completeness. Also, within one section, the discussed proposals are sorted in no particular order.

In Sect. 3.1, we describe PUFs or PUF-like proposals whose basic operation is other than electronical. As will become clear, this includes a wide variety of different constructions. Section 3.2 lists a number of constructions consisting of electrical and/or electronic building blocks whose response generation is mainly based on analog measurements. Sections 3.3 and 3.4 describe so-called digital *intrinsic PUFs*, i.e., PUFs which are embedded on an integrated circuit (IC), and of which the basic building blocks are regular digital primitives for the chosen manufacturing technology. This means that intrinsic PUFs are easy to construct, since they do not need any dedicated processing steps during manufacturing and no specialized or external equipment for their operation. For intrinsic PUFs, the measurement setup is often an inherent part of the PUF construction and is integrated on the chip. We discern two types of intrinsic PUFs, i.e., based on delay measurements (Sect. 3.3) and based on the settling state of bistable memory elements (Sect. 3.4). To conclude, we list in Sect. 3.5 a number of conceptual constructions. Some of them are technically not really PUFs, but can be considered as closely related extensions, e.g., POKs, CPUFs, and SIMPL systems. Others are true PUF proposals for which no concrete implementations have been realized, but which possess additional interesting properties distinguishing them from regular PUFs, e.g., quantum readout PUFs and reconfigurable PUFs.

3.1 Non-electronic PUFs

In this section, we give an overview of a number of constructions with PUF-like properties whose construction and/or operation is inherently non-electronic. However, very often electronic and digital techniques will be used at some point anyway to process and store these PUFs' responses in an efficient manner. The common denominator *non-electronic* in this section hence only reflects the nature of the components in the system that contribute to the random structure which makes the PUF unique. It does not say anything about the measurement, processing, and storage techniques which could be using electronics.

3.1.1 "Optical PUFs"

An early version of an unclonable identification system based on random optical reflection patterns, a so-called reflective particle tag, was proposed in [53] well before the introduction of PUFs. They were used for the identification of strategic arms in arms control treaties.

Optical PUFs based on transparent media were proposed in [41, 42] as *physical one-way functions (POWF)*. The core element of their design is an optical token which contains an optical microstructure constructed by mixing microscopic ($500\,\mu m$) refractive glass spheres in a small ($10 \times 10 \times 2.54$ mm) transparent epoxy plate. The token is radiated with a helium–neon laser and the emerging wavefront becomes very irregular due to the multiple scattering of the beam with the refractive particles. The *speckle pattern* that arises is captured by a CCD camera for digital processing. A Gabor hash is applied to the observed speckle pattern as a feature extraction procedure. The result is a string of bits representing the hash value. It is clear and was experimentally verified that even minute changes in the relative orientation of the laser beam and the token result in a completely different speckle pattern and extracted hash. The actual PUF functionality is then completed by a challenge which describes the exact orientation of the laser and the resulting Gabor hash of the arising speckle pattern as the response. The basic implementation and operation of an optical PUF is graphically represented in Fig. 1.

A number of experiments were performed in [41, 42] testing the characteristics of the constructed PUF. Four different tokens were tested using 576 distinct challenges. The inter- and intra-distance measures were evaluated for the obtained Gabor hashes. This resulted in an average inter-distance of $\mu_{\text{intra}} = 49.79\%(\sigma_{\text{intra}} = 3.3\%)$ and an average intra-distance of $\mu_{\text{intra}} = 25.25\%(\sigma_{\text{intra}} = 6.9\%)$. The information-theoretic security aspects of optical PUFs were further studied in [25, 56, 60]. Using the context-tree weighting method (CTW) [61], an average entropy content of 0.3 bit per pixel in the Gabor hash was estimated.

It is clear that the use of an optical PUF as described above is rather laborious due to the large setup involving a laser and a tedious mechanical positioning system. A more integrated design of an optical PUF, largely based on the same concepts, has been proposed in [11] and also in [55].

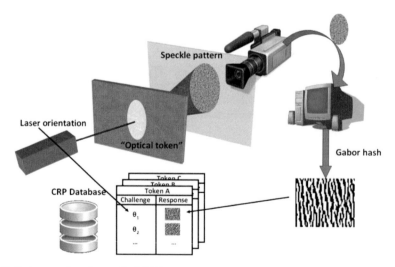

Fig. 1 Basic operation of an optical PUF

3.1.2 "Paper PUFs"

What we call *paper PUFs* are in fact a number of proposals made in literature which basically consist of scanning the unique and random fiber structure of regular or modified paper. As with the optical PUF, also for paper PUFs there were a number of early proposals [2, 8] well before the introduction of the PUF concept, and they were mainly considered as an anti-counterfeiting strategy for currency notes. In [6], the reflection of a focused laser beam by the irregular fiber structure of a paper document is used as fingerprint of that document to prevent forgery. A similar approach is used in [7], but they explicitly introduce ultraviolet fibers in the paper during the manufacturing process which can be measured by a regular desktop scanner. They also introduce a method to strongly link the data on the document with the paper by using a combined digital signature of data and the paper's fingerprint which is printed on the document.

3.1.3 "CD PUFs"

In [18], it was observed that the measured lengths of lands and pits on a regular compact disk contain a random deviation from their intended lengths due to probabilistic variations during the manufacturing process. Moreover, this deviation is even large enough to be observed by monitoring the electrical signal of the photodetector in a regular CD player. This was tested for a large number of CDs and locations on every CD. After an elaborate quantization procedure, an average intra-distance of $\mu_{\text{intra}} = 8\%$ and an average inter-distance of $\mu_{\text{inter}} = 54\%$ on the obtained bit strings is achieved. Using the CTW method, an entropy content of 0.83 bit per extracted bit was estimated. More details on CD fingerprinting are offered in the chapter by Ghaith Hammouri, Aykutlu Dana, and Berk Sunar.

3.1.4 RF-DNA

A construction called radio-frequency- or RF-DNA was proposed in [9]. They construct a small ($25 \times 50 \times 3$ mm) inexpensive token comparable to the one used for the optical PUF, but now they place thin copper wires in a random way in a silicon rubber sealant. Instead of observing the scattering of light as with optical PUFs, they observe the near-field scattering of EM waves by the copper wires at other wavelengths, notably in the 5–6 GHz band. The random scattering effects are measured by a prototype scanner consisting of a matrix of RF antennas. The entropy content of a single token is estimated to be at least 50, 000 bit. A detailed description of RF-DNA is provided in the chapter by Darko Kirovski.

3.1.5 "Magnetic PUFs"

Magnetic PUFs [26] use the inherent uniqueness of the particle patterns in magnetic media, e.g., in magnetic swipe cards. They are used in a commercial application to prevent credit card fraud [36].

3.1.6 Acoustical PUFs

Acoustical delay lines are components used to delay electrical signals. They convert an alternating electrical signal into a mechanical vibration and back. Acoustical PUFs [58] are constructed by observing the characteristic frequency spectrum of an acoustical delay line. A bit string is extracted by performing principle component analysis, and it is estimated that at least 160 bits of entropy can be extracted. The considered construction can constitute to an identification scheme with a false rejection rate of 10^{-4} and a false-acceptance rate at most 10^{-5}.

3.2 Analog Electronic PUFs

In this section, we discuss a number of PUF constructions whose basic operation consists of an analog measurement of an electric or electronic quantity. This in contrast to the constructions in Sect. 3.1, where the measured quantity was inherently non-electronic, and to the proposals in Sects. 3.3 and 3.4, where the measurements are performed digitally, and hence without the need for analog primitives.

3.2.1 "V_T PUFs"

To the best of our knowledge, the first technique to assign a unique identification to every single instance of a regular integrated circuit, without the need for special processing steps or after-fabrication programing, was proposed in [32] and was called ICID. The operation principle is relatively simple. A number of equally designed transistors are laid out in an addressable array. The addressed transistor drives a resistive load and because of the effect of manufacturing variations on the threshold voltages (V_T) of these transistors, the current through this load will be partially random. The voltage over the load is measured and converted to a bit string with an auto-zeroing comparator. The technique was experimentally verified on 55 chips produced in $0.35\,\mu m$ CMOS technology. An average intra-distance under extreme environmental variations of $\mu_{\text{intra}} = 1.3\%$ was observed, while μ_{inter} was very close to 50%.

3.2.2 "Power Distribution PUFs"

In [20], a PUF was proposed based on the resistance variations in the power grid of a chip. Voltage drops and equivalent resistances in the power distribution system are measured using external instruments and it is again observed that these electrical parameters are affected by random manufacturing variability. Experimental results on chips manufactured in 65 nm CMOS technology show $\mu_{\text{inter}} \approx 1.5\,\Omega$ and $\mu_{\text{intra}} \approx 0.04\,\Omega$ for the equivalent resistances.

3.2.3 Coating PUFs

Coating PUFs were introduced in [54] and consider the randomness of capacitance measurements in comb-shaped sensors in the top metal layer of an integrated circuit.

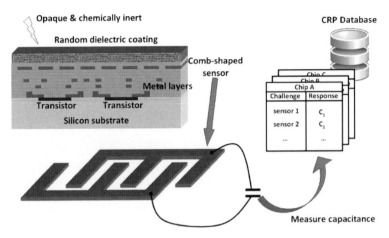

Fig. 2 Basic operation of a coating PUF. The upper left picture shows a schematic cross-section of a CMOS integrated circuit

Instead of relying solely on the random effects of manufacturing variability, random elements are explicitly introduced by means of a passive dielectric coating sprayed directly on top of the sensors. Moreover, since this coating is opaque and chemically inert, it offers strong protection against physical attacks as well. Measurement results on 36 produced chips, each with 31 sensors, show high randomness ($\mu_{intra} \approx 50\%$) and low noise ($\mu_{intra} < 5\%$), after quantization. An experimental security evaluation in [54] reveals that the coating PUF is also tamper evident, i.e., after an attack with a FIB the responses of the PUF are significantly changed. A more theoretical evaluation of coating PUFs was done in [59]. It was estimated that the entropy content of this PUF is approximately 6.6 bit per sensor. The basic implementation and operation of a coating PUF is shown in Fig. 2.

3.2.4 LC PUFs

An LC PUF [17] is constructed as a small ($\approx 1 \text{ mm}^2$) glass plate with a metal plate on each side, forming a capacitor, serially chained with a metal coil on the plate acting as an inductive component. Together they form a passive LC circuit which will absorb an amount of power when placed in a RF field. A frequency sweep reveals the resonance frequencies of the circuit, which depend on the exact values of the capacitive and inductive component. Due to manufacturing variations, this resonance peak will be slightly different for equally constructed circuits. As such, the LC PUF bares a resemblance to the coating PUF of Sect. 3.2.3 in that it measures the value of a capacitance and to the RF-DNA of Sect. 3.1.4 in that it observes the wireless power absorption of a token during a frequency sweep over the RF field. Contrarily to RF-DNA, the LC PUF construction is intrinsically a (passive) electrical circuit and not a random arrangement of copper wire. Experimental data from 500 circuits presented in [17] show a reproducibility of the resonance peak below 1 MHz at a constant temperature and an entropy content between 9 and 11 bits per circuit.

3.3 Delay-Based Intrinsic PUFs

In Sects. 3.1 and 3.2 a number of PUF and PUF-like proposals were discussed. They all basically start from an analog measurement of a random physical parameter, which is later quantized and can be used as an identifier of the whole system. In Sects. 3.3 and 3.4, *intrinsic PUFs* are discussed. Although no formal definition of an intrinsic PUF is provided in literature, we distinguish two prerequisites for a PUF to be called intrinsic:

1. The PUF, including the measurement equipment, should be fully integrated in the embedding device.
2. The complete PUF construction should consist of procedures and primitives which are naturally available for the manufacturing process of the embedding device.

The first condition implies that the device can query and read out its own PUF without the need for external instruments and without the need for the challenge and response to leave the device. Note that some earlier discussed examples already meet this condition, e.g., the coating PUF or the integrated version of the optical PUF. The second condition implies that the complete PUF construction comes at virtually no additional overhead besides the space occupied by the PUF, i.e., no extra manufacturing steps or specialized components are required. This does not hold anymore for the coating PUF and the integrated optical PUF, since they both need highly specialized processing steps. A number of intrinsic PUFs have been proposed so far, all integrated on digital integrated circuits.[4] The big advantage of a PUF integrated on a digital chip is that the PUF responses can be used directly by other applications running on the same device. We distinguish two different classes, i.e., intrinsic PUFs based on digital delay measurements in this section and intrinsic PUFs based on settling memory elements in Sect. 3.4.

3.3.1 Arbiter PUFs

The initial proposal of an arbiter PUF was made in [30, 31]. The basic idea is to introduce a digital race condition on two paths on a chip and to have a so-called *arbiter* circuit decide which of the two paths *won* the race. If the two paths are designed symmetrically, i.e., with the same intended delay, then the outcome of the

---

[4] Note that we do not use the term *silicon PUFs* in this work. It has been used to describe (a class of) PUFs which can be implemented on silicon digital integrated circuits and use the intrinsic manufacturing variability in the production process as a source of randomness. As such, they can be considered a particular case of intrinsic PUFs.

race is not fixed beforehand. During production of the chip, manufacturing variations will have an effect on the physical parameters determining the exact delay of each path and causing a small random offset between the two delays. This leads to a random and possibly device-specific outcome of the arbiter and hence explains the PUF behavior of such a construction. If the offset is too small, the setup-hold time of the arbiter circuit will be violated and its output will not depend on the outcome of the race anymore, but be determined by random noise. This last phenomenon is called *metastability* of the arbiter and introduces noise in the PUF responses.

The initial design of [30, 31] uses so-called *switch blocks* to construct the two symmetrical paths and a latch or flip-flop to implement the arbiter circuit. The switch blocks each have two inputs and two outputs and based on a parameter bit, they are connected *straight* or *switched*. Connecting a number of switch blocks in series creates two parameterizable delay lines feeding into the arbiter. The setting of the switch blocks will be the challenge of the PUF and the output of the arbiter the response. Note that the number of possible challenges is exponential in the number of switch blocks used. The basic arbiter PUF construction is schematically described in Fig. 3. This design was implemented on ASIC, chaining 64 switch blocks. Experimental validation on 37 chips shows $\mu_{inter} = 23\%$ and $\mu_{intra} < 5\%$, even under considerable variations of temperature and supply voltage. Equivalent tests on FPGA show much less unique randomness ($\mu_{inter} = 1.05\%$ and $\mu_{intra} = 0.3\%$), probably due to the discrete routing constraints implied by the FPGA architecture.

Simultaneously with the introduction of delay-based PUFs, it was recognized that digital delay is additive by nature, e.g., in case of the arbiter PUF from [30, 31], the delay of the chain of switch blocks will be the sum of the delays of the separate blocks. This observation leads to so-called *model-building attacks* [11, 13, 30, 31], i.e., one can build a mathematical model of the PUF which, after observing a number of CRP queries, is able to predict the response to an unseen challenge with relatively high accuracy. Such an attack was shown feasible for the basic arbiter PUF design in [14, 30, 31] using simple machine-learning techniques, achieving a prediction error of 3.55% after observing 5,000 CRPs for the ASIC implementation and a prediction error of 0.6% after observing 90,000 CRPs for the FPGA implementation. All subsequent work on arbiter PUFs is basically an attempt to make model-building attacks more difficult, by introducing non-linearities in the delays and by controlling and/or restricting the inputs and outputs to the PUF.

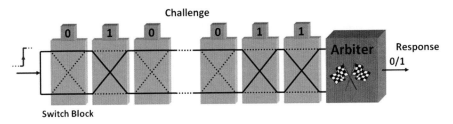

Fig. 3 Basic operation of an arbiter PUF

Feed-forward arbiter PUFs [31] were a first attempt to introduce non-linearities in the delay lines. It is an extension to a regular arbiter PUF, where some challenge bits are not set by the user but are the outcomes of intermediate arbiters evaluating the race at some intermediate point in the delay lines. This was equivalently tested on ASIC leading to $\mu_{inter} = 38\%$ and $\mu_{intra} = 9.8\%$. Note that the responses are much noisier, which is probably caused by the increased metastability since there are multiple arbiters involved. It was shown that the simple model-building attacks which succeeded in predicting the simple arbiter do not work any longer for this non-linear arbiter PUF. However, later results [37, 47] show that with more advanced modeling techniques it is still possible to build an accurate model for the feed-forward arbiter PUF, e.g., [47] achieves a prediction error of less than 5% after observing 49,000 CRPs from a simulated design.

In [38], an elaborate attempt to construct a secure arbiter-based PUF on FPGA was discussed. They use an initial device characterization step to choose the optimal parameters for a particular instantiation and use the reconfiguration possibilities of FPGAs to implement this.[5] To increase randomness and to thwart model-building attacks, they use hard-to-invert input and output networks controlling the inputs and outputs to the PUF, although these are not shown cryptographically secure. By simulation, they show that this construction gives desirable PUF properties and makes model building much harder. However, in [47] and especially in [44] it was again shown that model building of these elaborate structures might be feasible. In [47] they present a model of a slightly simplified structure as the one proposed in [38], which achieves a prediction error of 1.3% after observing 50,000 CRPs from a simulated design.

Finally, a different approach toward model-building attacks for arbiter PUFs was taken in [19, 39, 40]. Instead of preventing the attack, they use the fact that a model of the PUF can be constructed relatively easy to their advantage. They adapt a Hopper–Blum style protocol [24] to incorporate a modelable arbiter PUF.

3.3.2 Ring Oscillator PUFs

Ring oscillator PUFs, as introduced in [11, 13], use a different approach toward measuring small random delay deviations caused by manufacturing variability. The output of a digital delay line is inverted and fed back to its input, creating an asynchronously oscillating loop, also called a *ring oscillator*. It is evident that the frequency of this oscillator is precisely determined by the exact delay of the delay line. Measuring the frequency is hence equivalent to measuring the delay, and due to random manufacturing variations on the delay, the exact frequency will also be partially random and device dependent. Frequency measurements can be done relatively easy using digital components: an edge detector detects rising edges in the

---

[5] Note that there are different meanings given to the term *reconfigurable PUF*. The interpretation used in this work is the one described in Sect. 3.5.3 and is *not* directly related to the use of reconfigurable logic devices like FPGAs as meant in [38].

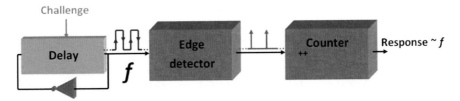

Fig. 4 Basic operation of a ring oscillator PUF

periodical oscillation and a digital counter counts the number of edges over a period of time. The countervalue contains all the details of the desired measure and is considered the PUF response. If the delay line is parameterizable as with the basic arbiter PUF design, the particular delay setting is again considered the challenge. The basic building blocks of the simple ring oscillator construction are shown in Fig. 4.

As explained in Sect. 2.3, some environmental parameters might undesirably affect the PUF responses. In case of delay measurements on integrated circuits, the die temperature and the supply voltage heavily affect the exact delay. For arbiter PUFs, this effect was not so big since they implicitly perform a differential measurement by considering two parallel delay paths simultaneously. For ring oscillator PUFs, these effects are much larger and some sort of compensation is needed. In [11, 13], the proposed compensation technique is to divide the countervalues of two simultaneously measured oscillations, which leads to much more robust responses. This compensation technique is shown in Fig. 5a. They tested a ring oscillator PUF with division compensation on four FPGA devices obtaining $\mu_{inter} \approx 10 \times 10^{-3}$ and $\mu_{intra} \approx 0.1 \times 10^{-3}$ with measurements taken over a $25°C$ temperature interval. It was also shown that supply voltage variations increase μ_{intra} with another 0.003×10^{-3} per mV variation. They use the same delay circuit as in the basic arbiter PUF design from [30, 31] which is hence also susceptible to model-building attacks. Moreover, it has been shown in [34] that in that case, there exists a high correlation, both between responses coming from the same challenge on different FPGAs and responses on the same FPGA coming from different challenges.

In [52], a slightly different approach was taken. The basic frequency measurement by counting rising edges is the same, but now a very simple and fixed delay circuit is used. A number of oscillators with the same intended frequency are

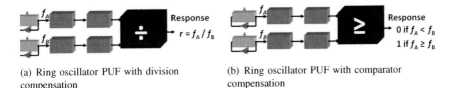

(a) Ring oscillator PUF with division compensation

(b) Ring oscillator PUF with comparator compensation

Fig. 5 Types of ring oscillator PUFs. (**a**) Ring oscillator PUF with division compensation, (**b**) Ring oscillator PUF with comparator compensation

implemented in parallel. The challenge to the PUF selects a pair of oscillators, and the response is produced by comparing the two obtained countervalues. This is a very simple and low-cost form of compensation and is shown in Fig. 5b. Experiments on 15 FPGAs with 1,024 loops per FPGA lead to $\mu_{inter} = 46.15\%$ and $\mu_{intra} = 0.48\%$. It has to be remarked that in order to obtain these results, the authors used a technique called 1-out-of-8 masking, which considers only the most stable response bit from 8 loop pairs. This improves the reproducibility drastically and hence decreases μ_{intra}, but comes at the cost of a relatively large implementation overhead, i.e., 7 out of 8 loop pairs are unused. We note that more details regarding ring oscillator PUFs and more elaborate compensation techniques for them can be found in the chapter by Inyoung Kim et al.

3.4 Memory-Based Intrinsic PUFs

In this section we discuss another type of intrinsic PUFs, based on the settling state of digital memory primitives. A digital memory cell is typically a digital circuit with more than one logically stable state. By residing in one of its stable states it can store information, e.g., one binary digit in case of two possible stable states. However, if the element is brought into an unstable state, it is not clear what will happen. It might start oscillating between unstable states or it might converge back to one of its stable states. In the latter case, it is observed that particular cells heavily prefer certain stable states over others. Moreover, this effect can often not be explained by the logic implementation of the cell, but it turns out that internal physical mismatch, e.g., caused by manufacturing variation, plays a role in this. For this reason, the stable settling state of a destabilized memory cell is a good candidate for a PUF response. We discuss different proposals from literature, based on different kinds of memory cells such as SRAM cells, data latches, and flip-flops.

3.4.1 SRAM PUFs

SRAM PUFs were proposed in [15], and a very similar concept was simultaneously presented in [22]. SRAM or static random-access memory is a type of digital memory consisting of cells each capable of storing one binary digit. An SRAM cell, as shown in Fig. 6a, is logically constructed as two cross-coupled inverters, hence leading to two stable states. In regular CMOS technology, this circuit is implemented with four MOSFETs, and an additional two MOSFETs are used for read/write access as shown in Fig. 6b. For performance reasons, the physical mismatch between the two symmetrical halves of the circuit (each implementing one inverter) is kept as small as possible. It is not clear from the logical description of the cell at what state it will be right after power-up of the memory, i.e., what happens when the supply voltage comes up? It is observed that some cells preferably power-up storing a zero, others preferably power-up storing a one, and some cells have no real preference, but the distribution of these three types of cells over the complete memory is random. As it turns out, the random physical mismatch in the cell, caused

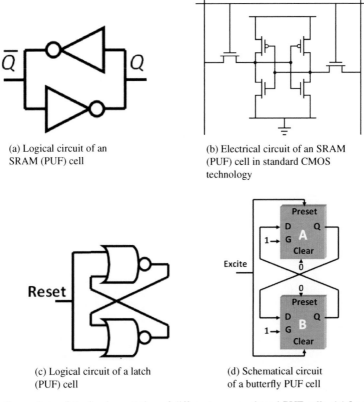

(a) Logical circuit of an
SRAM (PUF) cell

(b) Electrical circuit of an SRAM
(PUF) cell in standard CMOS
technology

(c) Logical circuit of a latch
(PUF) cell

(d) Schematical circuit
of a butterfly PUF cell

Fig. 6 Comparison of the implementation of different memory-based PUF cells. (**a**) Logical circuit of an SRAM (PUF) cell, (**b**) Electrical circuit of an SRAM (PUF) cell in standard CMOS technology, (**c**) Logical circuit of a latch (PUF) cell, (**d**) Schematical circuit of a butterfly PUF cell

by manufacturing variability, determines the power-up behavior. It forces a cell to 0 or 1 during power-up depending on the sign of the mismatch. If the mismatch is very small, the power-up state is determined by stochastical noise in the circuit and will be random without a real preference.

In [15], extensive experiments on SRAM PUFs were done. They collected the power-up state of 8,190 bytes of SRAM from different memory blocks on different FPGAs. The results show an average inter-distance between two different blocks of $\mu_{\text{inter}} = 49.97\%$ and the average intra-distance within multiple measurements of a single block is $\mu_{\text{intra}} = 3.57\%$ for a fixed environment and $\mu_{\text{intra}} < 12\%$ for large temperature deviations. In [16], the authors estimate the entropy content of the SRAM power-up states to be 0.76 bit per SRAM cell. In [22, 23], the SRAM power-up behavior on two different platforms was studied. For 5,120 blocks of 64 SRAM cells measured on eight commercial SRAM chips, they obtained $\mu_{\text{inter}} = 43.16\%$ and $\mu_{\text{intra}} = 3.8\%$ and for 15 blocks of 64 SRAM cells from the embedded memory in three microcontroller chips, they obtained $\mu_{\text{inter}} = 49.34\%$ and $\mu_{\text{intra}} = 6.5\%$.

For more details on SRAM PUFs and a further security analysis we refer to the chapter by Helena Handschuh and Pim Tuyls.

3.4.2 Butterfly PUFs

In [15], SRAM PUFs were tested on FPGAs. However, it turns out that in general this is not possible, since on the most common FPGAs, all SRAM cells are hard reseted to zero directly after power-up and hence all randomness is lost. Another inconvenience of SRAM PUFs is that a device power-up is required to enable the response generation, which might not always be possible. To counter these two drawbacks, butterfly PUFs were introduced in [28]. The behavior of an SRAM cell is mimicked in the FPGA reconfigurable logic by cross-coupling two transparent data latches. The butterfly PUF cell construction is schematically shown in Fig. 6d. Again, such a circuit allows two logically stable states. However, using the clear/preset functionality of the latches, an unstable state can be introduced after which the circuit converges back to one of the two stable states. This is comparable to the convergence for SRAM cells after power-up, but without the need for an actual device power-up. Again, the preferred stabilizing state of such a butterfly PUF cell is determined by the physical mismatch between the latches and the cross-coupling interconnect. It must be noted that due to the discrete routing options of FPGAs, it is not trivial to implement the cell in such a way that the *mismatch by design* is small. This is a necessary condition if one wants the random mismatch caused by manufacturing variability to have any effect. Measurement results from [28] on 64 butterfly PUF cells on 36 FPGAs yield $\mu_{\text{intra}} \approx 50\%$ and $\mu_{\text{intra}} < 5\%$ for large temperature variations.

3.4.3 "Latch PUFs"

What we call a latch PUF is an IC identification technique proposed in [51] which is very similar to SRAM PUFs and butterfly PUFs. Instead of cross-coupling two inverters or two latches, two NOR gates are cross-coupled as shown in Fig. 6c, constituting to a simple NOR latch. By asserting a reset signal, this latch becomes unstable and again converges to a stable state depending on the internal mismatch between the electronic components. Equivalently to SRAM PUFs and butterfly PUFs, this can be used to build a PUF. Experiments on 128 NOR latches implemented on 19 ASICs manufactured in 0.130 μm CMOS technology yield $\mu_{\text{inter}} = 50.55\%$ and $\mu_{\text{intra}} = 3.04\%$.

3.4.4 Flip-flop PUFs

Equivalently to SRAM PUFs, the power-up behavior of regular flip-flops can be studied. This was done in [33] for 4,096 flip-flops from three FPGAs and gives $\mu_{\text{inter}} \approx 11\%$ and $\mu_{\text{intra}} < 1\%$. With very simple post-processing consisting of 1-out-of-9 majority voting, these characteristics improve to $\mu_{\text{inter}} \approx 50\%$ and $\mu_{\text{intra}} < 5\%$.

3.5 PUF Concepts

In the final section of this extensive overview, we discuss a number of proposed concepts which are closely related to PUFs. Some of them are generalizations or even modes of operation of PUFs. Others are actual PUF proposals for which no working implementation has been provided and whose feasibility remains yet unconfirmed.

3.5.1 POKs: Physically Obfuscated Keys

The concept of a physically obfuscated key or POK has been introduced in [11] and has been generalized to physically obfuscated algorithms in [5]. The basic notion of a POK is that a key is permanently stored in a physical way instead of a digital way, which makes it hard for an adversary to learn the key by a probing attack. Additionally, an invasive attack on the device storing the key should destroy the key and make further use impossible, hence providing tamper evidence. It is clear that POKs and PUFs are very similar concepts, and it has already been pointed out in [11] that POKs can be built from (tamper-evident) PUFs and vice versa.

3.5.2 CPUFs: Controlled PUFs

A controlled PUF or CPUF, as introduced in [12], is in fact a mode of operation for a PUF in combination with other (cryptographic) primitives. A PUF is said to be controlled if it can only be accessed via an algorithm which is physically bound to the algorithm in an inseparable way. Attempting to break the link between the PUF and the access algorithm should preferably lead to the destruction of the PUF. There are a number of advantages in turning a PUF into a CPUF:

- A (cryptographic) hash function to generate the challenges of the PUF can prevent chosen challenge attacks, e.g., to make model-building attacks more difficult. However, for arbiter PUFs it has been shown that model-building attacks work equally well for randomly picked challenges.
- An error correction algorithm acting on the PUF measurements makes the final responses much more reliable, reducing the probability of a bit error in the response to virtually zero.
- A (cryptographic) hash function applied on the error-corrected outputs effectively breaks the link between the responses and the physical details of the PUF measurement. This makes model-building attacks much more difficult.
- The hash function generating the PUF challenges can take additional inputs, e.g., allowing to give a PUF multiple personalities. This might be desirable when the PUF is used in privacy-sensitive applications to avoid tracking.

It is clear that turning a PUF into a CPUF greatly increases the security. A number of protocols using CPUFs were already proposed in [12] and more elaborate protocols were discussed in [50]. It must be stressed that the enhanced security of a CPUF strongly depends on the physical linking of the PUF with the access algorithms which can be very arbitrary and might be the weak point of a CPUF.

3.5.3 Reconfigurable PUFs

Reconfigurable PUFs or rPUFs were introduced in [29]. The basic idea behind an rPUF is that it extends the regular CRP behavior of a PUF with an additional operation called *reconfiguration*. This reconfiguration has as effect that the partial or complete CRP behavior of the PUF is randomly and preferably irreversibly changed, hence leading to a new PUF. The authors of [29] propose two possible implementations of rPUFs where the reconfiguration mechanism is an actual physical reconfiguration of the randomness in the PUF. One is an extension of optical PUFs, where a strong laser beam briefly melts the optical medium, causing a rearrangement of the optical scatterers, which leads to a completely new random CRP behavior. The second proposal is based on a new type of non-volatile storage called phase-change memories. Writing to such a memory consists of physically altering the phase of a small cell from crystalline to amorphous or somewhere in between, and it is read out by measuring the resistance of the cell. Since the resistance measurements are more accurate than the writing precision, the exact measured resistances can be used as responses, and rewriting the cells will change them in a random way. Both proposals are rather exotic at this moment and remain largely untested. A third option is actually a logical extension of a regular PUF. By fixing a part of a PUF's challenge with a fuse register, the PUF can be reconfigured by blowing a fuse, which optimally leads to a completely changed CRP behavior for the challenge bits controlled by the user. However, the irreversibility of such a *logical* rPUF might be questionable, since the previous CRP behavior is not actually gone, but just blocked. Possible applications enabled by rPUFs are key zeroization, secure storage in untrusted memory, and prevention of *downgrading*, e.g., of device firmware.

3.5.4 Quantum Readout PUFs

Quantum readout PUFs were proposed in [49] and present a quantum extension to regular PUFs. It is proposed to replace the regular challenges and responses of a PUF with quantum states. Because of the properties of quantum states, an adversary cannot intercept challenges and responses without changing them. This leads to the advantage that the readout mechanism of the PUF does not need to be trusted anymore, which is the case for most regular non-quantum PUFs. Up to now, the feasibility of this proposal has not been practically verified. Moreover, it is unclear if presently existing PUFs can be easily extended to accept and produce quantum states as challenges and responses.

3.5.5 SIMPL Systems and PPUFs

A number of attempts to use PUFs as part of a public-key-like algorithm have been proposed. SIMPL systems were proposed in [45] and are an acronym for

SImulation Possible but Laborious. Two potential implementations of such a system are discussed in [46]. A very similar concept was proposed in [3] as Public PUFs or PPUFs. Both SIMPL systems and PPUFs rely on systems (PUFs) which can be modeled, but for which evaluating the model is laborious and takes a detectable longer amount of time than the evaluation of the PUF itself.

4 PUF Properties

After the extensive overview of the wide variety of different PUF proposals in Sect. 3, it becomes clear that the notion of a physically unclonable function will be hard to capture in one single closed definition. Previous attempts at defining a PUF are often too narrow, excluding certain PUFs, or too broad, including other things than PUFs, and mostly ad hoc, i.e., giving an informal description of the perceived qualities of the proposed construction. Moreover, in many of these attempts, properties are included which are not even validated but just assumed. In this work, we will not yet attempt to come up with a more complete or formal definition of PUFs. Instead, we will first look deeper into proposed PUF properties in Sect. 4.1 and check different PUF proposals against them in Sect. 4.2. Finally, we try to detect a least common subset of necessary properties for a construction to be called a PUF in Sect. 4.3.

4.1 Property Description

Here, we will list the most important properties which we selected from different definition attempts and/or identified in the PUF proposals. Although we do not completely formalize the discussed properties, we give a hint toward a possible formalization and try to make the descriptions as clear as possible to avoid ambiguity in this and future works.

To simplify the property description, we start from a very basic classification for a PUF as a *physical challenge–response procedure*. Note that already this implicitly assigns two properties to PUFs, i.e., an instantiation of a PUF cannot merely be an abstract concept but it is always (embedded in) a physical entity, and a PUF is a procedure (not strictly a function) with some input–output functionality. Since these properties are fundamental and are immediately clear from the construction for every PUF proposal up to now, we will not discuss them further. For brevity, we use the notation $\Pi : \mathcal{X} \rightarrow \mathcal{Y} : \Pi(x) = y$ to denote the challenge–response functionality of a PUF Π.

We begin by listing seven regularly occurring properties identified from multiple attempted PUF definitions and give a concise but accurate description of what we mean by them. We immediately note that these are not completely formal properties, but a hint toward a more formal description is given. In fact, the informal parts of the property descriptions are clearly marked in sans serif font. A more elaborate discussion on each of these properties follows directly below:

1. *Evaluatable:* given Π and x, it is **easy** to evaluate $y = \Pi(x)$.
2. *Unique:* $\Pi(x)$ contains **some** information about the identity of the physical entity embedding Π.
3. *Reproducible:* $y = \Pi(x)$ is reproducible up to a **small error**.
4. *Unclonable:* given Π, it is **hard** to construct a procedure $\Gamma \neq \Pi$ such that $\forall x \in \mathcal{X} : \Gamma(x) \approx \Pi(x)$ up to a **small error**.
5. *Unpredictable:* given only a set $\mathcal{Q} = \{(x_i, y_i = \Pi(x_i))\}$, it is **hard** to predict $y_c \approx \Pi(x_c)$ up to a **small error**, for x_c a random challenge such that $(x_c, \cdot) \notin \mathcal{Q}$.
6. *One-way:* given only y and Π, it is **hard** to find x such that $\Pi(x) = y$.
7. *Tamper evident:* altering the physical entity embedding Π transforms $\Pi \rightarrow \Pi'$ such that with **high probability** $\exists x \in \mathcal{X} : \Pi(x) \neq \Pi'(x)$, not even up to a **small error**.

We now discuss all seven properties in more detail:

1. Whether or not a PUF is evaluatable can be interpreted very broadly. From a theoretical perspective, **easy** can mean that we want the evaluation to be possible within polynomial time and effort. From a practical perspective, it means that we want the evaluation to induce as little overhead as possible, e.g. in the restricted timing, area, power, and energy constraints of an integrated chip. Also note that if a PUF is evaluatable, it is already implied that the PUF is *constructible* to begin with. It is clear that all PUF proposals which provide experimental results are constructible and at least theoretically evaluatable. Whether the overhead of their evaluation is also practically considered feasible depends on the application.
2. Regarding the description of the uniqueness property, there can still be some ambiguity about the meaning of *information* and *identity*. We look at this in an information theoretic sense. If a well-defined set or population of PUF instantiations is considered, the information contained in a PUF response $\Pi(x)$ relates to the partition one can make in the population based on this response. Consecutive responses allow for smaller and smaller partitions of the population until optimally a partition with a single PUF instantiation remains, in which case the considered set of CRPs *uniquely* identifies the PUF in the population. Based on the size of the population and the characteristics of the PUF responses, such a unique identification might or might not be possible. One possible measure of uniqueness which is provided in most experimental results is the inter-distance histogram, summarized by its average value μ_{intra}.
3. The reproducibility property is clear from its description. The responses to different evaluations of the same challenge x on the same PUF Π should be close in the considered distance metric. For experimental results, this is mostly measured by the intra-distance histogram and summarized by its average value μ_{intra}. Reproducibility is the property which distinguishes PUFs from true random number generators (TRNGs).
4. As is clear from its name, unclonability is the core property of a PUF. The provided description is relatively obvious; however, there are many details to be taken into consideration. First, note that the clone Γ is described as a procedure,

but not necessarily a *physical* procedure, since we explicitly distinguish between *physical* and *mathematical* unclonability. If it is hard to come up with a physical entity containing another PUF $\Pi_\Gamma \neq \Pi^6$ such that $\forall x : \Pi_\Gamma(x) \approx \Pi(x)$, we say that Π is *physically unclonable*. Note that the hardness of producing a physical clone even holds for the manufacturer of the original PUF Π and is for that reason also called *manufacturer resistance*. If it is difficult to come up with an (abstract) mathematical procedure f_Γ such that $\forall x : f_\Gamma(x) \approx \Pi(x)$, we say that Π is *mathematically unclonable*. Note that physical and mathematical unclonability are fundamentally different properties since a construction can be easy to clone physically but not mathematically or vice versa. In order to be truly unclonable, Π needs to be both physically and mathematically unclonable. Again, the **hardness** of cloning can be considered from a theoretical and a practical point of view. Practically, cloning can be very hard or infeasible. Demonstrating theoretical unclonability on the other hand is very difficult. The only known systems which can be proven to be theoretically unclonable are based on quantum physics.

5. Unpredictability is in fact a relaxed form of unclonability. If one can correctly predict the outcome of a PUF for a random challenge, only from observing a set of CRPs, it is easy to build a mathematical clone if one has access to the full PUF. Hence, predictability implies mathematical clonability and hence clonability.

6. One-wayness is a classical property coming from cryptography. We include it since the earliest definition of PUFs describes them as a physical variant of one-way functions [41].

7. Over time, a number of notions were proposed in literature regarding tampering and security against tampering. Under tampering, we understand making permanent changes to the integrity of a physical entity. We distinguish between tamper proof systems, i.e., systems for which tampering does not reveal any useful information and tamper-evident systems, i.e., systems for which tampering may be possible but leaves indelible evidence. We call a PUF tamper evident if tampering with the physical entity embedding the PUF with high probability changes the CRP behavior of the PUF.

4.2 Property Check

In this section, we will check a number of PUF proposals against all seven properties identified and discussed in Sect. 4.1. The proposals we consider are basically all proposed digital intrinsic PUFs for which concrete implementation details are available and two well-studied non-intrinsic PUFs. However, we believe that the conclusions of this study in Sect. 4.3 can be generalized to all discussed PUF proposals from

---

[6] By "$\Pi_\Gamma \neq \Pi$" here we mean that Π_Γ and Π are (embedded in) physically distinct entities.

Sect. 3. We begin by summarizing the most important implementation details and experimental results for the discussed PUFs in Table 1.

To draw some sensible conclusions, we have to compare these PUF proposals with some non-PUF reference cases. We check against the following three reference cases which we describe in a challenge–response-like style for easy comparison with PUFs:

- A *true random number generator*. The single challenge is the request for a random number. The response is a random number extracted from a stochastical physical process.
- A very simple *RFID-like identification protocol*. The single challenge is the request for identification. The response is an identifier string which was hard-programed in the device by the manufacturer.
- A *public key signature scheme*. A challenge is a message string. A response is signature on that message generated using a private key which was hard-programed by the device manufacturer.

The result of this study is shown in matrix format in Table 2. Note that we explicitly distinguish between physical and mathematical unclonability since we consider them fundamentally different notions.

4.3 Least Common Subset of PUF Properties

Looking at Table 2, we spot two properties, i.e., evaluatability and uniqueness, which hold for all discussed PUFs, and all reference cases! This means that these are necessary properties for a PUF, but they are certainly not sufficient, since they also allow programed identifiers, public key signatures, and TRNGs. A third necessary property, reproducibility, excludes TRNGs. Finally, the core property of physical unclonability completely distinguishes the PUF proposals from the reference cases based on a hard-programed unique identifier or key. We remark that this observation elegantly justifies the naming of the primitives studied in this work, i.e., *physically unclonable functions*.

Drawing further conclusions from Table 2, we notice that mathematical unclonability is an unachievable property for most of these *naked* PUFs. However, mathematical unclonability can be greatly improved by turning these PUFs into controlled PUFs as described in Sect. 3.5.2, e.g., to prevent exhaustive readout and model-building. One-wayness does not seem to be a good PUF property since no single PUF turns out to be truly one-way. Even for optical PUFs, which were originally introduced as *physical one-way functions* in [41], this property is unclear. Finally, although widely believed to be one of the main advantages of PUF technology, tamper evidence was only experimentally verified for the (non-intrinsic) optical and coating PUFs.

Table 1 Overview of the most important implementation details and experimental results for the PUFs discussed in Sect. 4.2

PUF	Randomness	Challenge	Response	Experiment	μ_{inter} (%)	μ_{intra} (%)	Entropy	Tamper evident?	Model-building?
Optical PUF [41]	Explicitly introduced	Laser orientation	Gabor hash of speckle pattern	576 CRPs on 4 tokens	49.79	25.25	0.3 bit/pixel [25]	Yes [41]	Difficult [56, 60]
Coating PUF [54]	Explicitly introduced	Sensor selection	Quantized capacitance measurement	31 CRPs on 36 ASICs	≈ 50	< 5	6.6 bit/sensor [59]	Yes [54]	Impossible[a]
Basic arbiter PUF [30, 31]	Implicit manuf. var.	Delay line setting	Arbiter decision	10,000 CRPs on 37 ASICs	23	< 5[c]	–	–	$\varepsilon = 3.55\%$ for $q = 5,000$[b] [31]
Feed-forward arbiter PUF [31]	Implicit manuf. var.	Delay line setting	Arbiter decision	10,000 CRPs on 37 ASICs	38	9.8[c]	–	–	$\varepsilon < 5\%$ for $q = 50,000$[b] [47]
Basic ring oscillator PUF [11, 13]	Implicit manuf. var.	Delay line setting	Division compensated countervalues	Many CRPs on 4 FPGAs	≈ 1	≈ 0.01[c]	–	–	Same as basic arbiter PUF
Ring oscillator PUF with comparator [52]	Implicit manuf. var.	Loop pair selection	Countervalues comparison with 1-out-of-8 masking	1,024 loops on 15 FPGAs	46.14	0.48[c]	–	–	Impossible[a]

Table 1 (continued)

PUF	Randomness	Challenge	Response	Experiment	μ_{inter} (%)	μ_{intra} (%)	Entropy	Tamper evident?	Model-building?
SRAM PUF [15]	Implicit manuf. var.	SRAM address	Power-up state of addressed SRAM cell	65440 CRPs on different FPGAs	49.97	< 12	0.76 bit/cell [16][c]	–	Impossible[a]
Latch PUF [51]	Implicit manuf. var.	Latch selection	Settling state of destabilized NOR latch	128 CRPs on 19 ASICs	50.55	3.04	–	–	Impossible[a]
Butterfly PUF [28]	Implicit manuf. var.	Butterfly cell selection	Settling state of destabilized cell	64 CRPs on 36 FPGAs	≈ 50	< 6[c]	–	–	Impossible[a]

[a] We assess model building to be impossible for these PUFs, since the physical random elements contributing to different CRPs are to a very large extent independent of each other.

[b] ε is the prediction error after a model-building attack using q (observed) queries.

[c] For these PUFs, the average observed intra-distance includes at least some minimal environmental variations.

Table 2 Property matrix of different PUF proposals

	Optical PUF	Coating PUF	Basic Arbiter PUF	Feed-forward Arbiter PUF	Basic Ring Oscillator PUF	Comparator Ring Oscillator PUF	SRAM PUF	Latch PUF	Butterfly PUF	TRNG	Simple ID protocol	Public key signature
Evaluatable	√[a]	√[a]	√	√	√	√	√[b]	√	√	√	√/![c]	√/![c]
Unique	√	√	√	√	√	√	√	√	√	√	√	√
Reproducible	√	√	√	√	√	√	√	√	√	×	√	√
Physically unclonable	√	√	√	√	√	√	√	√	√	√	×[d]	×[d]
Mathematically unclonable	√	×[e]	×[f]	×[f]	×[f]	×[e]	×[e]	×[e]	×[e]	√	×[g]	×[g]
Unpredictable	√	√	![h]	![h]	![h]	√	√	√	√	√[i]	×	×
One-way	?	×[j]	×[k]	×[k]	×[k]	×[k]	×[k]	×[k]	×[k]	×	×	√
Tamper evident	√	√	?	?	?	?	?	?	?	?	×[l]	×[l]

√ = proposal meets the property. × = proposal does not meet the property. ! = proposal meets the property under certain conditions. ? = it remains untested whether the proposal meets the property.

[a] Requires extra manufacturing steps and/or external measurement equipment.
[b] Requires device power-up.
[c] Requires explicit hard programing of unique identifier or key.
[d] Physically cloning a hard-programed identifier or key is easy.
[e] For these PUFs, a mathematical clone can be easily created by exhaustively reading out every CRP.
[f] For these PUFs, a mathematical clone can be created by a model-building attack.
[g] An adversary who knows the identifier/private key can easily forge a valid identification/signature.
[h] These PUFs become increasingly easier to predict when an adversary learns more CRPs.
[i] Unpredictability is a key requirement for a good TRNG.
[j] Because these PUFs have so few challenges, a random challenge will with non-negligible probability invert a PUF response.
[k] Because the output of these PUFs is basically one bit, a random challenge will with probability ≈ 50% invert a PUF response.
[l] If there is no additional tamper protection provided, hard-programed identifiers and keys are not tamper evident.

5 PUF Application Scenarios

As a final overview part of this work, we briefly present the three classes of application scenarios which we envision for PUFs, i.e., system identification in Sect. 5.1, secret key generation in Sect. 5.2, and hardware-entangled cryptography in Sect. 5.3.

5.1 System Identification

Because of their physical unclonability property, using PUFs for identification is very interesting for anti-counterfeiting technologies. PUF responses can be used directly for identification very similarly as in a biometrical identification scheme. During an enrollment phase, a number of CRPs from every PUF from the population are stored in a database, together with the identity of the physical system embedding the PUF. During identification, the verifier picks a random CRP from the CRPs stored in the database for the presented system and challenges the PUF with. If the observed response is close enough to the response in the database, the identification is successful, otherwise it fails. In order to prevent replay attacks, each CRP should be used only once for every PUF and has to be deleted from the database after the identification.

The threshold used to decide on a positive identification depends on the separation between the intra-distance and the inter-distance histograms. If both histograms do not overlap, an errorless identification can be made by placing the threshold somewhere in the gap between both histograms. If they do overlap then setting the threshold amounts to making a trade-off between false-acceptance rate (FAR) and false-rejection rate (FRR). The determination of the FAR and FRR based on the overlap of the inter- and intra-distance histograms is shown in Fig. 7. The optimal choice, minimizing the sum of FAR and FRR, is achieved by setting the threshold at the intersection of both histograms, but other trade-offs might be desirable for

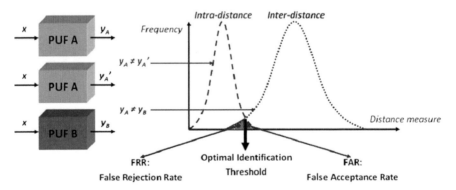

Fig. 7 Details of basic PUF-based system identification. Shown is the inter- and intra-distance distribution and the determination of the FAR and the FRR based on the optimal identification threshold

specific applications. Additionally, it is obvious that a unique identification is only possible with high probability if the response contains enough entropy with regards to the population size.[7]

5.2 Secret Key Generation

Intrinsic PUFs in integrated circuits have interesting properties for use in secret key generation and storage. Since the key is generated from intrinsic randomness introduced by inevitable manufacturing variability, no explicit key-programing step is required, which simplifies key distribution. Moreover, since this randomness is permanently fixed in the (sub-)microscopical physical details of the chip, no conventional non-volatile key memory is required. This also offers additional security against probing attacks and possibly other side-channel attacks, since the key is not permanently stored in digital format, but only appears in volatile memory when required for operation. Finally, possible tamper evidence of the PUF can be used to provide tamper-proof key storage.

For cryptographic algorithms, uniformly random and perfectly reliable keys are required. Since PUF responses are usually noisy and only contain a limited amount of entropy, they cannot be used as keys directly. An intermediate processing step is required to extract a cryptographic key from the responses. This is a problem known in information theory as secret key extraction from close secrets and is generally solved by a two-phase algorithm. During the initial *generation* phase, the PUF is queried and the algorithm produces a secret key together with some additional information often called *helper data*. Both are stored in a secure database by the verifier, but not on the device. In the *reproduction* phase, the verifier presents the helper data to the algorithm which uses it to extract the same key from the PUF as in the generation step. In that way, the device containing the PUF and the verifier have established a shared secret key. It is possible to construct these algorithms such that the key is perfectly secret, even if the helper data is observed, i.e., the helper data can be publicly communicated from the verifier to the device. Practical instances of these algorithms have been proposed, e.g., in [10] and the cost of actual implementations thereof is assessed in [4, 35].

5.3 Hardware-Entangled Cryptography

A recently introduced application scenario transcends the generation of secret keys from PUFs for use in existing cryptographic primitives. Instead, it fully integrates the PUF in the primitive itself, leading to so-called hardware-entangled cryptographic primitives. No key is generated anymore, but the secret element of the primitive is the full unique CRP behavior of the PUF instantiation in the embedding device. The fundamental difference between classical cryptography with a PUF-

---

[7] A response containing n bits of entropy optimally allows for a unique identification in a population with an average size of $2^{\frac{n}{2}}$ because of the birthday paradox.

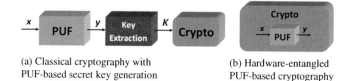

(a) Classical cryptography with   (b) Hardware-entangled
PUF-based secret key generation   PUF-based cryptography

Fig. 8 Schematic comparison of cryptography with PUF-based key generation and hardware-entangled cryptography

based key and hardware-entangled cryptography is conceptually depicted in Fig. 8. The first result [1] based on this principle proposes a PUF-based block cipher and shows that it is possible to prove regular security notions for this construction based on reasonable assumptions for the PUF. In-depth details of this construction can be found in the chapter by Frederik Armknecht et al.

Hardware-entangled cryptographic primitives are basically keyless, i.e., not at any point in the algorithm a secret digital key is stored in memory, neither in non-volatile memory nor in volatile memory. Not only does this offer full security against non-volatile memory attackers, as was already the case for PUF-base secret key generation, but additionally it largely prohibits volatile memory attackers from learning anything useful. In this view, hardware-entangled cryptography is closely related to the field of *provable physical security*, see, e.g., [43]. A discussion on the practical aspects of this field is given in the chapter by François-Xavier Standaert et al.

6 PUF Discussions and Some Open Questions

After the overview and study of PUF instantiations, properties, and application scenarios, respectively, in Sects. 3, 4, and 5, we touch upon some discussion points and open questions. The field of physically unclonable functions has grown a lot over the last couple of years and is still expanding. In this section, we try to point out some interesting future research directions.

6.1 Predictability Versus Implementation Size

From Tables 1 and 2, it is clear that certain PUFs suffer from predictability due to model-building attacks after a relatively small number of CRPs have been observed. This is especially the case for delay-based PUFs such as the arbiter PUF, whereas most memory-based PUFs are reasonably considered to withstand model-building attacks since their responses are based on independent random elements. On the other hand, all memory-based PUFs, and also other PUFs such as the coating PUF and the comparator-based ring oscillator PUF, suffer from another disadvantage. Their implementation size grows exponentially with the desired length of their challenges. In other words, for these PUFs the number of possible CRPs scales linearly with their size, whereas for the modelable arbiter PUF this scales exponentially.

This means that for both types of digital intrinsic PUFs, the number of *unpredictable CRPs* is limited to an amount, which is at best polynomial in the size of the PUF. For arbiter-like PUFs, this limitation is due to model-building attacks, whereas for memory-based PUFs it is simply because of the limited number of available CRPs.

This is a peculiar observation since it is not clear whether this is a sign of an underlying physical bound on the number of unpredictable CRPs, which is obtainable for any intrinsic PUF, or whether this is merely a result of the particular PUF constructions, which have been proposed thus far. Moreover, it limits the possible application scenarios, since an intrinsic PUF with a superlinear or even an exponential amount of unpredictable CRPs could lead to stronger security assumptions. From a physical point of view one could say that since the amount of (thermodynamical) entropy in a physical system is at best polynomial in the size of the system, the number of truly independent random CRPs for a single PUF can never be exponential in the PUF's size. However, for many cryptographical applications we aim for computational rather than perfect measures of security, i.e., even if the true entropy content is limited there still could be a large number of *computationally unpredictable* CRPs. In other words, model building could be possible in theory but infeasible in practice. Further study on this topic, both from a theoretical and from a practical point of view, is definitely recommended.

6.2 Formalization of PUF Properties

The properties which we studied for a number of PUF proposals in Sect. 4 were described informally in Sect. 4.1. In order to make strong claims on the security of PUFs and PUF applications, it is necessary to come up with a formalized version of these property descriptions. This formalization will act as a convenient interface between the people involved in the practical implementation of a physically unclonable function and the people designing PUF-based security primitives and applications. The actual PUF designers should validate their constructions against the proposed properties and further focus on making them as efficient as possible. Application developers can build upon the specified properties without having to worry about the physical details of the underlying PUFs, and they can use the formal nature to prove strong security notions for their proposals. Especially for the further development of hardware-entangled crypto primitives, the need for a formal description of PUF qualities seems inevitable.

We acknowledge that for some of the properties discussed in Sect. 4.1, coming up with a formal definition is far from trivial. Especially the more practical properties, i.e., physical unclonability and tamper evidence, will be hard to fit into a theoretical framework. Moreover, even from a practical point of view it is not yet exactly clear what these properties stand for. With regard to tamper evidence, further experiments on intrinsic PUFs are highly recommended in order to get a better feeling of its feasibility. Physical unclonability, although considered to be the core property of PUFs, is for the moment a rather ad hoc assumption primarily based on the apparent hardness of measuring and controlling random effects during manufacturing processes. However, for a number of intrinsic PUF proposals, it is not clear how difficult this assumption is in reality. Further research into these topics is required.

6.3 Reporting on PUF Implementation Results

In the growing body of literature on the implementation of PUFs, extensively summarized in Sect. 3, a number of different concepts and figures are used to demonstrate the practicality and security of the proposed constructions, some more useful than others. We remark that this poses a possible risk to objectivity. First, without the proper argumentation it becomes rather subjective to assess one's PUF based on one's own proposed measures. Second, a wide variety of measures makes it difficult to objectively compare different PUF proposals. For these two reasons, it is important to agree upon a number of standardized measures which can assess the practical and security-related characteristics of differently constructed PUFs in an objective manner. For some characteristics, this is closely related to a further formalization of different PUF properties as discussed in Sect. 6.2.

We briefly discuss a number of used concepts which we consider to be important for comparison of different PUF proposals.

- *Sample size.* Before even touching upon concrete characteristics, we point out that the sample size used to estimate these characteristics is important, i.e., the number of distinct devices, the number of distinct challenges, the number of distinct measurements of every response, etc. Up to now, most works were conscientious in mentioning the used devices and the size of the sample population that was tested. However, to the best of our knowledge, for none of the PUF proposals a statistical analysis was performed pointing out the confidence level on the estimated characteristics. For further formal analysis of PUFs, this will be of increasing importance.
- *Inter- and intra-distance histograms.* The importance of both inter- and intra-distance as a measure for, respectively, the uniqueness and the noise present in a PUF measurement has been pointed out a number of times earlier in this work. Luckily, these two measures are almost always provided in the body of literature, making at least a partially objective comparison between early proposals possible. However, a number of remarks have to be made. First, these histograms are often assumed to be approximately gaussian and are summarized by their average μ, and sometimes their standard deviation σ. It is important to validate this gaussian approximation with a statistical test. Moreover, it is highly advised to always mention both μ and σ since this at least allows to calculate the optimal FAR and FRR for system identification applications (Sect. 5.1). Second, although inter-distance gives a good *feeling* of the uniqueness of a response, it cannot be used to assess the actual independent entropy present. Some PUFs, e.g., the basic arbiter PUF, which have reasonably large inter-distances suffer from predictability due to the dependence between their responses.
- *Entropy estimations.* To overcome this last problem, a number of works proposing PUFs provide an estimate of the actual entropy present in a large number of PUF responses. Two methods which are used to do this are testing the

compressibility of the response strings, mostly using the context-tree weighting method [61], and running standardized randomness tests such as the Diehard test and the NIST test [48]. We remark that due to the limited length of the available responses, both methods generally offer only a low level of confidence on their outcome. In particular for the randomness tests, which are in fact designed to test the apparent randomness in the output of *pseudo*-random number generators, it is not clear whether the passing of these tests is of any significance for PUFs. Finally, we point out that both methods only estimate the independent entropy within one single PUF, i.e., how much uncertainty does an adversary have about the outcome of an unseen response, even if he has learned all other responses from that PUF. However, for a PUF to be secure, it also has to be unpredictable given responses from other PUFs. This last notion is not assessed by the considered entropy estimates.

- *Environmental influences.* As discussed in Sect. 2.3, PUF responses are subject to unwanted physical influences from their direct environment. For a PUF to be used in a practical application, these effects have to be rigorously quantified in order to prevent unforeseen failures. For intrinsic PUFs on integrated circuits, at least the influence of the die temperature, supply voltage, and device aging on the PUF's responses should be studied.

- *CRP yield and predictability.* In Sect. 6.1 we discussed the remarkable observation that for all proposed intrinsic PUFs up to now, the number of unpredictable CRPs is limited to an amount at best polynomial in the size of the PUF. To assess the usefulness of a particular PUF in some applications, it is important to know how many unpredictable response bits one can optimally obtain from a PUF of a given size. This requires a further study of predictability in PUF responses.

- *Implementation cost and efficiency.* It is evident that, in order to be of any practical use, PUFs and PUF-based applications should be as cost-effective as possible. Measuring the implementation and operation cost in terms of size, speed, and power consumption is an exercise which should be made for any hardware implementation and is not limited to PUFs.

- *Tamper evidence.* A number of remarks concerning the tamper evidence property of PUFs were already discussed in Sect. 4. We point out again that any claims or assumptions regarding the tamper evidence of a particular PUF construction only make sense if they are backed up by an experimental validation. A detailed description of the performed tampering experiments and the resulting effect on the PUF's CRP behavior is invaluable in that case. To the best of our knowledge, such practical results only exist for optical PUFs [41] and coating PUFs [54].

7 Conclusion

In this chapter, we tried to cover the complete field of PUF constructions up to date. From this overview, it becomes clear that a physically unclonable function is not a rigorously defined concept, but a collection of functional constructions which meet a number of naturally identifiable qualities such as uniqueness, physical unclonability, and possibly tamper evidence. A more concrete and comparative study of these properties for different PUFs leads to a common subset of necessary conditions centered around the core property of physical unclonability. This study also reveals some blind spots and open questions which point to a number of interesting future research directions. From a theoretical point of view, a further formalization of the identified properties is necessary to enable the development of strong PUF-based security primitives, notably hardware-entangled cryptography. On a practical level, more concrete and standardized characteristics need to be adapted and verified in order to make objective decisions possible, both in the design of new PUFs and their applications. This will lead to more competitive results on a fair basis, which naturally advances the state-of-the-art research in this field.

References

1. F. Armknecht, R. Maes, A.R. Sadeghi, B. Sunar, P. Tuyls, Memory leakage-resilient encryption based on physically unclonable functions, in *Advances in Cryptology - ASIACRYPT 2009*, ed. by M. Matsui. Proceedings of the15th International Conference on the Theory and Application of Cryptology and Information Security, Tokyo, Japan. Lecture Notes in Computer Science, vol. 5912 (Springer, Berlin, Heidelberg, 2009), pp. 685–702
2. D. Bauder, *An Anti-counterfeiting Concept for Currency Systems*. Technical Report PTK-11990, Sandia National Labs, Albuquerque, NM, 1983
3. N. Beckmann, M. Potkonjak, Hardware-based public-key cryptography with public physically unclonable functions, 2009, pp. 206–220
4. C. Bösch, J. Guajardo, A.R. Sadeghi, J. Shokrollahi, P. Tuyls, in *Efficient Helper Data Key Extractor on FPGA*. CHES, 10–13 August 2008 Washington, DC, USA, 2008, pp. 181–197
5. J. Bringer, H. Chabanne, T. Icart, in *On physical Obfuscation of Cryptographic Algorithms*. INDOCRYPT '09: Proceedings of the 10th International Conference on Cryptology in India, New Delhi, India (Springer, Berlin, Heidelberg, 2009), pp. 88–103
6. J.D.R. Buchanan, R.P. Cowburn, A.V. Jausovec, D. Petit, P. Seem, G. Xiong, D. Atkinson, K. Fenton, D.A. Allwood, M.T. Bryan, Forgery: 'fingerprinting' documents and packaging. Nature **436**(7050), 475 (2005)
7. P. Bulens, F.X. Standaert, J.J. Quisquater, *How to Strongly Link Data and Its Medium: The Paper Case*. IET Information Security (to appear) (2010). http://www.dice.ucl.be/~fstandae/PUBLIS/72.pdf
8. Commission on Engineering and Technical Systems (CETS), *Counterfeit Deterrent Features for the Next-Generation Currency Design, Appendix E* (The National Academic Press, Washington, DC, 1993)
9. G. Dejean, D. Kirovski, in *RF-DNA: Radio-Frequency Certificates of Authenticity*. CHES '07: Proceedings of the 9th International Workshop on Cryptographic Hardware and Embedded Systems, Vienna, Austria, 10–13 September 2007 (Springer, Berlin, Heidelberg 2007), pp. 346–363
10. Y. Dodis, R. Ostrovsky, L. Reyzin, A. Smith, Fuzzy extractors: How to generate strong keys from biometrics and other noisy data. SIAM J. Comput. **38**(1), 97–139 (2008)
11. B. Gassend, *Physical Random Functions*. Master's thesis, MIT, MA, USA, 2003
12. B. Gassend, D. Clarke, M. van Dijk, S. Devadas, in *Controlled Physical Random Functions*. ACSAC '02: Proceedings of the 18th Annual Computer Security Applications Conference (IEEE Computer Society, Washington, DC, 2002), p. 149
13. B. Gassend, D. Clarke, M. van Dijk, S. Devadas, in *Silicon Physical Random Functions*. ACM Conference on Computer and Communications Security (ACM Press, New York, NY 2002), pp. 148–160

14. B. Gassend, D. Lim, D. Clarke, M. van Dijk, S. Devadas, Identification and authentication of integrated circuits: Research articles. Concurr. Comput.: Pract. Exper. **16**(11), 1077–1098 (2004)
15. J. Guajardo, S.S. Kumar, G.J. Schrijen, P. Tuyls, in *FPGA Intrinsic PUFs and Their Use for IP Protection*. Cryptographic Hardware and Embedded Systems Workshop. Lecture Notes in Computer Science, vol. 4727 (Springer, Heidelberg, 2007), pp. 63–80
16. J. Guajardo, S.S. Kumar, G.J. Schrijen, P. Tuyls, in *Physical Unclonable Functions and Public-Key Crypto for FPGA IP Protection*. International Conference on Field Programmable Logic and Applications, 27–30 Aug 2007 (IEEE, Piscataway, NJ, 2007), pp. 189–195
17. J. Guajardo, B. Škorić, P. Tuyls, S.S. Kumar, T. Bel, A.H. Blom, G.J. Schrijen, Anti-counterfeiting, key distribution, and key storage in an ambient world via physical unclonable functions. Inf. Syst. Front. **11**(1), 19–41 (2009)
18. G. Hammouri, A. Dana, B. Sunar, in *CDs Have Fingerprints Too*. CHES '09: Proceedings of the 11th International Workshop on Cryptographic Hardware and Embedded Systems (Springer, Berlin, Heidelberg, 2009), pp. 348–362
19. G. Hammouri, E. Öztürk, B. Birand, B. Sunar, in *Unclonable Lightweight Authentication Scheme*. Proceedings of the 10th International Conference on Information and Communications Security (ICICS 2008) (Springer, Heidelberg, 2008), pp. 33–48
20. R. Helinski, D. Acharyya, J. Plusquellic, in *A Physical Unclonable Function Defined Using Power Distribution System Equivalent Resistance Variations*. DAC '09: Proceedings of the 46th Annual Design Automation Conference (ACM, New York, NY, 2009), pp. 676–681
21. Sir W.J. Herschel, *The Origin of Finger-Printing* (Oxford University Press, London, 1916)
22. D.E. Holcomb, W.P. Burleson, K. Fu, in *Initial SRAM State as a Fingerprint and Source of True Random Numbers for RFID Tags.*. Proceedings of the Conference on RFID Security, Malaga, Spain, 11–13 July 2007
23. D.E. Holcomb, W.P. Burleson, K. Fu, Power-up SRAM state as an identifying fingerprint and source of true random numbers. IEEE Trans. Comput. **58**(9), 1198–1210 (2009)
24. N. Hopper, M. Blum, *A Secure Human-Computer Authentication Scheme*. Technical Report CMU-CS-00-139, Carnegie Mellon University, 2000
25. T. Ignatenko, G.J. Schrijen, B. Škorić, P. Tuyls, F.M.J. Willems, in *Estimating the Secrecy Rate of Physical Unclonable Functions with the Context-Tree Weighting Method*. Proceedings of the IEEE International Symposium on Information Theory, Seattle, WA, USA, 9–14 July 2006, pp. 499–503
26. R.S. Indeck, M.W. Muller, Method and apparatus for fingerprinting magnetic media . U.S. Patent No. 5365586, 1994
27. M.S. Kirkpatrick, E. Bertino, in *Software Techniques to Combat Drift in PUF-Based Authentication Systems*. Workshop on Secure Component and System Identification (SECSI 2010), Cologne, Germany, 2010, p. 9
28. S. Kumar, J. Guajardo, R. Maes, G.J. Schrijen, P. Tuyls, in *Extended Abstract: The Butterfly PUF Protecting IP on Every FPGA*. IEEE International Workshop on Hardware-Oriented Security and Trust, 2008, HOST 2008, Anaheim, CA, USA, 2008, pp. 67–70
29. K. Kursawe, A.R. Sadeghi, D. Schellekens, P. Tuyls, B. Škorić, in *Reconfigurable Physical Unclonable Functions – Enabling Technology for Tamper-Resistant Storage*. 2nd IEEE International Workshop on Hardware-Oriented Security and Trust - HOST 2009, San Francisco, CA, USA (IEEE Computer Society, Los Alamitos, CA, USA, 2009), pp. 22–29
30. J.W. Lee, D. Lim, B. Gassend, G.E. Suh, M. van Dijk, S. Devadas, in *A Technique to Build a Secret Key in Integrated Circuits for Identification and Authentication Application*. Proceedings of the Symposium on VLSI Circuits, 2004, pp. 176–159
31. D. Lim, *Extracting Secret Keys from Integrated Circuits*. Master's thesis, MIT, MA, USA, 2004
32. K. Lofstrom, W.R. Daasch, D. Taylor, in *IC Identification Circuit Using Device Mismatch*. Proceedings of ISSCC 2000, 2000, pp. 372–373
33. R. Maes, P. Tuyls, I. Verbauwhede, in *Intrinsic PUFs from Flip-Flops on Reconfigurable Devices*. 3rd Benelux Workshop on Information and System Security (WISSec 2008), Eindhoven, the Netherlands, 2008

34. R. Maes, P. Tuyls, I. Verbauwhede, in *Statistical Analysis of Silicon PUF Responses for Device Identification*. Workshop on Secure Component and System Identification (SECSI 2008), Berlin, Germany, 2008
35. R. Maes, P. Tuyls, I. Verbauwhede, in *Low-Overhead Implementation of a Soft Decision Helper Data Algorithm for SRAM PUFs*. CHES '09: Proceedings of the 11th International Workshop on Cryptographic Hardware and Embedded Systems (Springer, Berlin, Heidelberg 2009), pp. 332–347.
36. MagneTek(R), MagnePrint(R). http://www.magneprint.com/
37. M. Majzoobi, F. Koushanfar, M. Potkonjak, in *Testing Techniques for Hardware Security*. IEEE International Test Conference (ITC 2008), Santa Clara, CA, USA, 28–30 Oct 2008 pp. 1–10
38. M. Majzoobi, F. Koushanfar, M. Potkonjak, Techniques for design and implementation of secure reconfigurable PUFs. ACM Trans. Reconfigurable Technol. Syst. **2**(1), 1–33 (2009)
39. E. Ozturk, G. Hammouri, B. Sunar, in *Physical Unclonable Function with Tristate Buffers*. IEEE International Symposium on Circuits and Systems (ISCAS 2008), Seattle, WA, USA (IEEE, Washington, DC, 2008), pp. 3194–3197
40. E. Öztürk, G. Hammouri, B. Sunar, in *Towards Robust Low Cost Authentication for Pervasive Devices*. PERCOM '08: Proceedings of the 2008 Sixth Annual IEEE International Conference on Pervasive Computing and Communications (IEEE Computer Society, Washington, DC, 2008), pp. 170–178
41. R.S. Pappu, *Physical One-Way Functions*. Ph.D. thesis, Massachusetts Institute of Technology, 2001
42. R.S. Pappu, B. Recht, J. Taylor, N. Gershenfeld, Physical one-way functions. Science **297**, 2026–2030 (2002)
43. K. Pietrzak, in *Provable Security for Physical Cryptography*. Survey talk at WEWORC'09, Graz, Austria, 7–9 July 2009
44. U. Rührmair, F. Sehnke, J. Sölter, G. Dror, S. Devadas, J. Schmidhuber, *Modeling Attacks on Physical Unclonable Functions*. Cryptology ePrint Archive, Report 2010/251, 2010. http://eprint.iacr.org/
45. U. Rührmair, *Simpl Systems: On a Public Key Variant of Physical Unclonable Functions*. Cryptology ePrint Archive, Report 2009/255, 2009
46. U. Rührmair, Q. Chen, P. Lugli, U. Schlichtmann, G.C. Martin Stutzmann, *Towards Electrical, Integrated Implementations of SIMPL Systems*. Cryptology ePrint Archive, Report 2009/278, 2009
47. U. Rührmair, J. Sölter, F. Sehnke, *On the Foundations of Physical Unclonable Functions*. Cryptology ePrint Archive, Report 2009/277, 2009
48. A. Rukhin, J. Soto, J. Nechvatal, E. Barker, S. Leigh, M. Levenson, D. Banks, A. Heckert, J. Dray, S. Vo, M. Smid, M. Vangel, A. Heckert, J. Dray, L.E.B. Iii, *A Statistical Test Suite for Random and Pseudorandom Number Generators for Cryptographic Applications*. NIST Special Publication 800-22, 2001
49. B. Škorić, *Quantum Readout of Physical Unclonable Functions: Remote Authentication Without Trusted Readers and Authenticated Quantum Key Exchange Without Initial Shared Secrets*. Cryptology ePrint Archive, Report 2009/369, 2009
50. B. Škorić, M.X. Makkes, *Flowchart Description of Security Primitives for Controlled Physical Unclonable Functions*. Cryptology ePrint Archive, Report 2009/328, 2009
51. Y. Su, J. Holleman, B. Otis, in *A 1.6pj/bit 96% Stable Chip-ID Generating Circuit Using Process Variations*. IEEE International Solid-State Circuits Conference, ISSCC 2007. Digest of Technical Papers (IEEE Computer Society, Washington, DC, 2007), pp. 406–611
52. G.E. Suh, S. Devadas, in *Physical Unclonable Functions for Device Authentication and Secret Key Generation*. Design Automation Conference (ACM Press, New York, NY, 2007), pp. 9–14
53. K. Tolk, *Reflective Particle Technology for Identification of Critical Components*. Technical Report SAND-92-1676C, Sandia National Labs, Albuquerque, NM, 1992

54. P. Tuyls, G.J. Schrijen, B. Škorić, J. van Geloven, N. Verhaegh, R. Wolters, in *Read-Proof Hardware from Protective Coatings*. Cryptographic Hardware and Embedded Systems Workshop. Lecture Notes in Computer Science, vol. 4249 (Springer, New York, NY, 2006), pp. 369–383

55. P. Tuyls, B. Škorić, in *Physical Unclonable Functions for Enhanced Security of Tokens and Tags*. ISSE 2006 – Securing Electronic Business Processes, Rome, Italy, 10–12 Oct 2006, pp. 30–37

56. P. Tuyls, B. Škorić, S. Stallinga, A.H.M. Akkermans, W. Ophey, in *Information-Theoretic Security Analysis of Physical Unclonable Functions*. Financial Cryptography and Data Security, Roseau, Dominica, 28 Feb–3 Mar 2005, pp. 141–155

57. V. Vivekraja, L. Nazhandali, in *Circuit-Level Techniques for Reliable Physically Unclonable Functions*. HOST '09: Proceedings of the 2009 IEEE International Workshop on Hardware-Oriented Security and Trust, San Francisco, CA, USA, 27 July 2009, pp. 30–35

58. S. Vrijaldenhoven, *Acoustical Physical Uncloneable Functions*. Master's thesis, Technische Universiteit Eindhoven, the Netherlands, 2005

59. B. Škorić, S. Maubach, T. Kevenaar, P. Tuyls, Information-theoretic analysis of capacitive physical unclonable functions. J. Appl. Phys. **100**(2), 024902 (2006)

60. B. Škorić, P. Tuyls, W. Ophey, in *Robust Key Extraction from Physical Unclonable Functions*. Applied Cryptography and Network Security (ACNS) 2005, New York, NY, USA. Lecture Notes in Computer Science, vol. 3531 (Springer, Berlin, 2005), pp. 407–422

61. F.M.J. Willems, Y.M. Shtarkov, T.J. Tjalkens, The context tree weighting method: Basic properties. IEEE Trans. Inf. Theory **41**, 653–664 (1995)

62. C.E. Yin, G. Qu, in *Temperature-Aware Cooperative Ring Oscillator PUF*. HOST '09: Proceedings of the 2009 IEEE International Workshop on Hardware-Oriented Security and Trust (IEEE Computer Society, Washington, DC, 2009), pp. 36–42

Hardware Intrinsic Security from Physically Unclonable Functions

Helena Handschuh, Geert-Jan Schrijen, and Pim Tuyls

1 Introduction

Counterfeiting of goods in general and of electronic goods in particular is a growing concern with a huge impact on the global economy, the society, and the security of its critical infrastructure. Various examples are known where companies suffer from economic and brand damage due to competition with counterfeit goods. In some cases the use of counterfeit components has even led to tragic accidents in which lives were lost. It has also recently become clear that counterfeit products can penetrate the critical and security infrastructure of our modern societies and hence cause a threat to national security. One of the difficulties to deal with this problem stems from the fact that counterfeit goods can originate from sources that are able to make copies that are very hard to distinguish from their legitimate counterpart. A first well-known aspect of counterfeiting is product cloning. A second much less known but increasingly dangerous aspect consists of overproduction of goods.

A special, but modern, case of counterfeiting is theft of Intellectual Property such as software and designs. The attractive part from the attackers' point of view is that it is relatively easy to steal and has a high value without having to do huge investments in research and development. From a high-level point of view one can state that the attack can be thwarted by using encryption and authentication techniques. Device configuration data or embedded software can, for example, be encrypted such that it will only run on the device possessing the correct cryptographic key. Since encrypted data is still easy to copy, it now becomes essential that the secret key is well protected against copying or cloning.

In order to deal with these two aspects of counterfeiting, a secret unclonable identifier is required together with strong cryptographic protocols. In this chapter we focus on a new way to address these problems: Hardware Intrinsic Security. It is based on the implementation and generation of secret physically unclonable identifiers used in conjunction with cryptographic techniques such as encryption and

H. Handschuh (✉)

Intrinsic-ID, San Jose, CA 95110, USA; ESAT-COSIC, Katholieke Universiteit Leuven, Leuven, Belgium

e-mail: helena.handschuh@intrinsic-ID.com

A.-R. Sadeghi, D. Naccache (eds.), *Towards Hardware-Intrinsic Security,* Information Security and Cryptography, DOI 10.1007/978-3-642-14452-3_2,
© Springer-Verlag Berlin Heidelberg 2010

authentication algorithms which allow to secure the critical information stored in the system. According to common practice in security, the used algorithms are often public but they use a secret key that is stored securely *somewhere* in the system. Using secret physically unclonable identifiers to derive secret keys for the system is our proposed solution to achieve strong anti-counterfeiting and anti-cloning mechanisms in electronic devices.

In every security system, it is essential that the key remains completely secret to keep a high level of protection. The system is broken, i.e. does not guarantee protection anymore when the secret key has leaked. Nowadays, encrypted texts created with state-of-the-art cryptographic algorithms do not leak much information on the secret key. However, since secret keys are stored in everyday objects like smart cards, attackers can easily subject such objects to physical attacks with all kinds of tools in order to get access to the secret keys. Common examples of such tools are very high-resolution microscopes such as optical, atomic force, scanning electron, laser scanning, confocal microscopes, or more destructive tools such as focused ion beams and laser cutters. It has been shown in many occasions that by using these physical means the secret key bits can be visualized and hence the secret key can be retrieved. Although these tools are sophisticated, they are more and more widespread nowadays and affordable for many parties.

Currently an arms race between security IC manufacturers and attackers is taking place to protect the secret keys in improved ways. It turns out, however, that the traditional methods to protect secret keys are approaching their limits and inducing more and more costs and longer time to market. A low cost but strong secret key storage technology is one of the missing links to make affordable but strong security systems. It is a necessary requirement for ICs in smart cards, defense and governmental applications, e-health systems, passports, and so on that protect valuable and sensitive data and that upon failure would cause not only very huge financial losses but also brand and reputation damage and could even expose a nation's critical assets.

Secure key storage is a small but indispensable part of a security system. Since a security system is only as strong as its weakest link, it is important to have a strong key storage mechanism. Moreover, when a secure and *unclonable* key storage mechanism is combined with good cryptography, a strong anti-counterfeiting system can be built. The unclonable key is used as a unique identifier and transfers its unclonability to the product it is embedded in. In order to detect whether a product has been counterfeited, a so-called *authenticity check* is performed. The authenticity check is usually carried out in a protocol between a verifier and the component to be verified. For example, the protocol could be run between a reader and an unclonable smart card or RFID Tag, or between a program running on a processor and the unclonable chip that implements the processor. In the first example, the unclonability of the device guarantees that when the verification succeeds the device is genuine. In the second example, the verifier is embedded in the program. It will authenticate the IC by verification of its secret key in a secure protocol. As a result, the program will not run on a counterfeit IC and protect the Intellectual Property contained in the processor design.

2 Rethinking Secure Key Storage Mechanisms

Current key storage mechanisms produce secret keys that are stored on the device that carries out the security operations. Off-chip storage of a secret key is vulnerable to a competent attacker using a logic analyzer to tap the bus between the external memory and the chip.[1] Therefore the storage mechanisms below have to be considered as embedded on-chip storage systems.

2.1 Limitations of Current Key Storage Mechanisms

A number of approaches exist to permanently store keys in a device. Among these we distinguish between volatile and non-volatile approaches. Non-volatile mechanisms rely on hardwired information or fuse-type technologies or floating gate-type technologies. Volatile approaches based on RAM memory typically use batteries to permanently store information. In this section we provide an overview of the limitations of each type of permanent storage mechanism before highlighting the advantages of our new proposed solution.

- *ROM memory.* ROM (read-only memory) masks are typically generated during manufacturing stages and can thereafter not be erased or modified anymore. This has two implications. First of all, any secret key hidden in ROM is permanently stored there even if the device is powered off and can therefore be extracted with typical failure analysis tools used at manufacturing sites. Second, ROM is about as inflexible as carving the key in stone. Once it has been designed into the IC and taped-out it can never be changed again. In terms of time to market, ROM masks take a number of months to be produced. Since it is impossible to consider that every new device would receive a new key and require a new ROM mask, this implies that ROM stored keys are necessarily master keys and all the more interesting to reverse engineer.
- *Fuse-based storage mechanisms.* Examples of fuse-based storage mechanisms are polyfuses, laser fuses, e-fuses, and anti-fuses. Again, as is the case with ROM memory, the keys stored in these fuses are permanently present in the system even when the device is powered off. Additionally fuses are quite easy to spot in a lay-out because they are quite large; they are all the easier to analyze using typical failure analysis tools from manufacturing sites. Some types of fuses, namely anti-fuses, require an additional charge pump in the system and are thus not as cost-efficient as one might hope for.
- *Floating gate technologies.* These technologies include Flash memory, EEP-ROM, and EPROM cells. The principle is that an electronic charge is trapped on the floating gate between two drains and remains there until a given threshold voltage is applied to remove it. Again, the information is trapped in the device

---

[1] Note that in systems where the external memory is encrypted, there still needs to be an on-chip key to decrypt the data from the memory as it is being read or written.

even when it is powered off and can be read using advanced imaging and failure analysis tools. Floating gate technologies are also vulnerable to fault attacks in which one tries to erase or modify the value trapped on the floating gate while being read or written and infer secret information from the consequences of the modification. Floating gate memory technologies are by no means standard technology components and appear only as process options in the later generations of a new process node. For a customer requiring a new technology node this can cause a substantial delay in the time to market of the product. Floating gate-based technologies also need 6–10 additional mask steps which adds significantly to the product cost. Due to the complicated nature of the processes for these various non-volatile technologies, it is at this point in time believed that it is not economically viable to have all these technologies available in all the process nodes. For example, embedded Flash is only available down to 90 nm technology at this time.

- *Battery-backed RAM.* Battery-backed RAM does not suffer from the security issues most other storage mechanisms have, but has one clear disadvantage compared to all others: It requires an additional component, namely a battery. This induces additional cost and assumes that there is enough room in the system to add a battery. In most embedded ICs, this is not the case. Another drawback of batteries is that they are not always very reliable and the information in the RAM is lost if they fail. This means that such devices can easily become nonfunctional.

As can be seen from the previous discussion, every current key storage mechanism has a number of limitations which cannot be easily overcome, the main one certainly being the permanent presence of the key in the system even when it is powered off.

2.2 A Radical New Approach to Secure Key Storage

Given the drawbacks of the current non-volatile storage mechanisms as described above, there is clearly an exposed gap in hardware security which is playing into the hands of determined attackers. To counter this increasing threat a radically new approach to key storage is needed. Important criteria for this new approach are the following:

1. First of all, the key should not be permanently stored in digital form on the device.
2. Second, it should be extracted from the device only when required. And after having been used, it should be removed from all internal registers, memories, and locations so as to not leave a single trace when the system is powered off again.
3. Third, it should somehow be uniquely linked to a given device such that one cannot reproduce it or manufacture a device with a precise key.

Our new approach that extracts the key from the intrinsic properties of the device overcomes many of the limitations of traditional approaches mentioned above. The implementation of such an approach without the need for technology-dependent components or embedded non-volatile memory has the following advantages:

- *Security*: It offers an unparalleled security level since the key is not even present when the device is switched off. It can be seen as key storage without storing the key.
- *Cost*: It does not require any additional mask steps or additional analog components. Therefore this solution saves cost instead of adding costs as compared to key storage alternatives.
- *Time to Market*: It is ready to use with the newest process nodes without requiring the extensive qualifications required for new process options.
- *Standard Availability*: Clearly, when properties of standard components are used to extract the key, the solution is available in most common process nodes.
- *Flexibility*: It is field upgradeable. Keys based on this principle can be updated in the field even after the device has left the production facility.
- *Reliability*: It offers reliability against a wide range of external influences, such as temperature and voltage variations and humidity. It does not suffer from the presence of an additional component such as a battery and remains stable throughout the device's lifetime without really being there.

3 Hardware Intrinsic Security

3.1 Physically Unclonable Functions

The concept of a physical unclonable function (PUF) forms the basic idea on which the implementation of our new key storage approach is built. PUFs will be used as the hardware from which the key is extracted and can be considered as the intrinsic electronic fingerprint or biometric of a device. We will refer to security mechanisms built on electronic fingerprints as *Hardware Intrinsic Security*. The underlying electronic PUF technology has been extensively investigated in the literature and has been recognized as a new powerful security primitive. The previous chapter in this book by Maes and Verbauwhede provides a complete overview of existing PUF technologies and their essential properties.

An electronic PUF consists of a physical object that is very hard to clone due to its unique micro- or nano-scale properties that originate from the (deep-submicron) manufacturing process variations. An electronic PUF has to meet the following requirements:

1. *Low Cost*: The measurement circuit should be low cost and easy to implement, i.e. with standard components.
2. *Resistance to Physical Attack*: A physical attack meant to find out the behavior of the structure should cause damage to the structure. In particular this implies

that the functional behavior of the PUF should change to such an extent that tampering is detectable. The PUF should not be based on a secret that has to be guarded securely.

3. *Reliable*: The PUF responses should exhibit a low amount of noise in a wide range of circumstances, e.g., when being present in low- and high-temperature environments, environments with electromagnetic radiation, or environments that cause changes in the operating voltage of the device. Finally, still after many years of silicon aging effects, the noise level should be sufficiently low. The next chapter in this book by Schaumont et al. provides a thorough analysis of such aging effects.

3.1.1 Unclonability

PUFs are by definition very hard to clone. This means that it is very difficult, i.e., takes a lot of resources and a lot of time to make either a hardware clone, a mathematical model of the behavior of the structure, or a software program that can compute the response to a challenge in a reasonable amount of time. In order to be able to perform these actions, one would have to know the locations and properties of all the particles in the system with very high accuracy. Since physical systems consist of a very large amount of particles, this becomes a very time-consuming task.

3.1.2 Biometrics

There is a striking analogy between intrinsic PUFs and biometrics, in fact an intrinsic PUF can be seen as the biometric modality, i.e., the intrinsic electronic fingerprint of an IC. Even the ways of working with PUFs and biometrics are very similar. Both require a registration phase: it is necessary to perform some pre-processing before one can work with them. Once the pre-processing has been performed and some reference data based on this has been stored, the biometric/electronic fingerprint can be used for authentication and key storage purposes.

3.2 Examples of PUFs

3.2.1 SRAM PUFs

The best known memory-based intrinsic PUF based on standard available components is the SRAM PUF. Other memory-based intrinsic PUFs are also described in the previous chapter by Maes and Verbauwhede. SRAM or static random access memory is a standard component that is used in most devices (e.g.,ASICs, microprocessors, DSPs, ASSPs) today. It consists of two cross-coupled invertors and two additional transistors for external connection, hence six transistors in total. It is widely used due to its speed for short-term data storage.

When a voltage is applied to a memory cell, it chooses its logical preference state: the logical 1-state or the logical 0-state. Each cell has a unique preference state due to its composition; the composition determines the values of the threshold voltages in the transistors that make up the two cross-coupled invertors. The unique properties of each transistor stem from deep submicron process variations. It is known that the fluctuations in the threshold voltages scale according to the law of Pelgrom: $\Delta(V_T) \sim \frac{1}{\sqrt{LW}}$ where L is the length and W is the width. A complex interaction between all these physical variables determines in the end the logical preference states of the memory cells. The important observation in this example is that the threshold voltages of different transistors may well seem almost identical at the macroscopic level but that it is the difference between two of these threshold voltages that will actually govern the start-up value of each individual cell. Due to tiny local process variations, it is the difference between these differences that leads to a completely random start-up behavior of neighboring SRAM cells on a device as shown in Fig. 1.

The string determined by all the preference start-up values of the memory cells of an SRAM memory array forms a random identifier that identifies the SRAM memory uniquely. This identifier is the PUF response. A schematic representation of the SRAM PUF is shown in Fig. 2. This phenomenon has been verified in many experiments and on many SRAM types. Among the devices we have tested in our own facilities we can list the following: Alliance SRAMs, Cypress SRAMs, IDT SRAM, Faraday Standard Performance SRAM, and Virage Logic SRAMs, both High-Density and High-Speed. All these SRAM memories cover a large range of technology nodes, namely 180, 150, 130, 90, and 65 nm from different foundries, namely UMC and TSMC. A number of experiments were performed for each and every one of them showing that such SRAM memories do indeed start-up in a random fashion and are suitable for PUFs over a large range of environmental conditions.

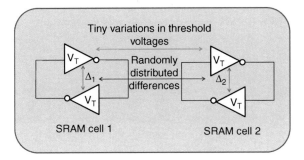

Fig. 1 Schematic representation of the differences in threshold voltages between two neighboring SRAM cells. Even though the threshold voltages are almost identical, their tiny differences are randomly distributed

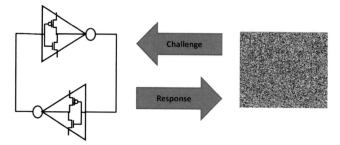

Fig. 2 Schematic representation of an SRAM PUF. The *left side* represents a single SRAM cell, consisting of two cross-coupled inverters. On the *right-hand side*, the SRAM PUF response of a whole SRAM memory is shown, where a *black pixel* can be interpreted as a logical 1 and a *white* as a logical 0

3.2.2 PUFs on FPGAs

Since the SRAM PUF is not available on all mainstream FPGA platforms (because no uninitialized SRAM is available on most types) we present briefly two examples of other types of PUFs that are targeted toward and can be configured on FPGAs: (i) the Butterfly PUF and (ii) the Ring Oscillator-based PUF. These two as well as further examples of memory-based and delay-based electronic PUFs are introduced in the previous chapter of this book by Maes and Verbauwhede.

Butterfly PUF:

The idea behind the Butterfly PUF is similar to the one behind the SRAM PUF. At a high level it consists of two integrated components: (i) an array of Butterfly PUF cells and (ii) a processing component. A single Butterfly PUF cell consists of two cross-coupled latches. Due to this cross-coupling the Butterfly cell has two stable states the logical "0" and the logical "1," just as the SRAM PUF cell. The cell is challenged by bringing this system into an unstable state and letting it converge during a specific time interval to one of the stable states. The preferential stable state is determined by the mismatch defined by the process variations during manufacturing. Its stability with respect to external stresses is guaranteed by tight integration with the processing component. In Fig. 3, a schematic overview of a Butterfly PUF cell is shown.

Ring Oscillator-based PUF:

This PUF consists of an oscillating loop that is constructed by putting a number of delay elements next to each other and feeding the signal back to its starting point. The frequency at which this circuit oscillates is determined by the physical properties of its building blocks and can therefore be used as a basis for a unique identifier. By measuring the unique oscillating frequencies of a number of these loops, a unique identifier is generated that can be translated into a secret key. Since the PUF

Butterfly-PUF-cell

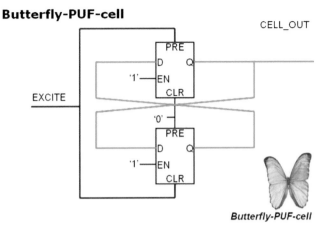

Fig. 3 Schematic representation of a single Butterfly PUF cell on an FPGA

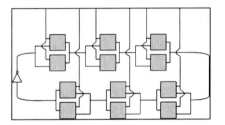

Fig. 4 Schematic representation of one oscillation loop of a ring-oscillator PUF

responses are analog values, the PUF is integrated with a secure and noise-reducing analog to digital conversion algorithm. Apart from security, it guarantees robustness of the PUF responses against external stresses. In Fig. 4 a schematic representation of one oscillation loop of the Ring Oscillator PUF is presented.

3.3 Secure Key Storage Based on PUFs

Our proposed method of deriving the key using a PUF comprises two stages:

- *Noise Cancellation*: Physical measurements are typically noisy. Secret keys used in the context of cryptographic algorithms must always be exactly the same. Otherwise they produce completely corrupt results. Consequently, noise has to be removed from the physical measurements before they can be used to create secret keys.
- *Randomness Extraction*: Even after noise has been removed, a further processing step is required. The security from the cryptographic keys is based on the fact that they are completely random from one device to the next, i.e., very hard to guess. Physical measurements have a high degree of randomness but are usually not uniformly random. By processing the physical data and extracting the randomness

via appropriate compression functions (extractors), a uniformly random key can
be generated.

For a practical implementation of such a key derivation mechanism, in order to
use, for example, an SRAM PUF for Hardware Intrinsic Security, three functional
modules are needed:

- *A PUF Measurement Circuit*: A measurement circuit that is able to read out the
  device-unique characteristics of the PUF and translate this into digital *PUF Data*.
  In case of an SRAM PUF, this is simply a circuit that reads out the start-up values
  of a specific range of SRAM memory that is exclusively reserved for this purpose.
- *A Key Extractor*: This is a module that converts noisy PUF responses into a
  robust secret key. It implements the noise cancellation and randomness extraction algorithms. Besides the PUF responses it needs an *activation code* as input.
  This activation code contains error correction data needed to remove the noise
  from the PUF data and information about the compression function needed to
  extract randomness. The Key Extractor module can be implemented not only as
  an IP block integrated in an IC but also as a software module that runs on an
  (embedded) processor.
- *An Activation Code Constructor*: This module computes the public activation
  code that is needed by the Key Extractor. It takes as input the PUF data and
  optionally a user-selected key that needs to be reconstructed in the future. The
  module can be implemented as an IP block on the same IC as the key extractor is
  located or as part of an external device or service, depending on the application.

Typically the Activation Code Constructor is used only once in a so-called *enrollment* phase. Once the Activation Code is generated, it is stored in a memory that is
accessible by the Key Extractor. Note that this memory may be external to the device
on which the Key Extractor is implemented and does not need to be secure. Each
time the device needs to use the secret key, a new PUF measurement is done and the
Key Extractor is used to reconstruct the key from the measured PUF data and the
stored Activation Code. This is called the *reconstruction* phase. The reconstruction
phase is typically carried out many times during the lifetime of the device (each time
the key is needed).

4 Quality of a PUF

The quality of a PUF is determined by two main parameters which are reliability
and security. Reliability addresses the fact that a PUF has to work under many different external circumstances and has to have a sufficiently long lifetime. On the
other hand, security addresses the level of protection offered against a wide range
of attacks. A very important parameter for security is the amount of randomness or
entropy present in the PUF. A further in-depth discussion of reliability and statistical
modeling of PUFs is performed in the next chapter of this book by Schaumont et al.

4.1 Reliability

Electronic PUFs are based on features of electronic components whose behavior under varying operating conditions is modeled and tested extensively before a PUF is commercially deployed. It must be guaranteed that the cryptographic key or unique identifier derived from the PUF is exactly the same under all circumstances. The following operating conditions can have an influence on the PUF behavior:

- Temperature
- Core voltage
- Electromagnetic radiation.

The influence of these conditions has been investigated by continuously reading out data from PUF implementations while varying the above-mentioned conditions in a climate chamber. The tests we performed included the following: measuring PUF responses under different ambient temperatures and under a gradient of temperatures, typically ranging from $-40°C$ to $+80°C$; measuring the PUF responses at extremely low and extremely high temperatures, sometimes up to $125°C$, and at very high humidity levels; measuring the PUF responses at different core voltage levels; measuring the PUF responses when exposed to different electromagnetic fields. The differences between the measured PUF data and reference measurements taken in a controlled environment were analyzed. It turns out that PUFs are very robust with respect to these variations for a wide range of SRAM types as well as FPGA devices and families of FPGA devices. For some PUF types some data processing is used to make a particular implementation robust against such influences.

Besides dealing with a variety of operating conditions, it is also important to guarantee that a PUF works properly over time. It is known that silicon slowly degrades when in use for a long time. Several mechanisms contribute to this aging effect, the most important ones being

- Electro Migration (EM): the transport of conductor material due to momentum exchange between electrons and the metal lattice.
- Hot Carrier Injection (HCI): carriers generate sufficient kinetic energy to overcome a potential barrier and get injected into the gate oxide, causing interface states and charge traps.
- Time-Dependent Dielectric Breakdown (TDDB): formation of conducting path through the gate oxide.
- Negative Bias Temperature Instability (NBTI): build up of interface charges due to a negative gate-source bias at an elevated temperature.

These mechanisms can influence the behavior of the PUF over time. Depending on the type of PUF, different mechanisms are of importance. For example, the most important aging effects for oscillator-based PUFs are NBTI and HCI. FPGAs incorporating this technology have been submitted to extensive stress tests simulating the aging effect due to both NBTI and HCI. The result is that aging effects have almost no influence on the behavior of an oscillator PUF. As a matter of fact, none of the tests that we performed on the FPGAs both under extreme operating conditions and simulating aging effects ever resulted in the cryptographic keys being wrong. The same key was always reconstructed no matter how the devices were stressed. The most important aging effect for the SRAM PUF is NBTI. In order to investigate its influence, experiments were done where the NBTI effect was accelerated by applying an increased voltage on the SRAM memory and by placing it in an environment with a high ambient temperature for a long time. This way an effective aging of 10–20 years was achieved in only a few months time. The experiments showed that if no countermeasures are taken, the start-up behavior of the SRAM PUF is changing. However, when the right countermeasure (or anti-aging mechanism) is applied, the impact of aging vanishes completely and even the noise on the derived PUF data is reduced.

4.2 Security

Three important security parameters of a PUF are its entropy, its tamper evidence and its unclonability. These properties are discussed below.

4.2.1 Entropy

In order to extract a high-quality secret key from a PUF, a sufficient amount of randomness is needed in the PUF responses. In the literature the amount of entropy present in various PUFs was analyzed. An overview is given in Table 1.

4.2.2 Tamper Evidence

PUFs provide very strong protection against physical attacks and are therefore very well suited to implement *read-proof hardware*. Read-proof hardware is hardware that is very hard to read by an attacker even when a whole arsenal of physical tools is available. Hence, a good key storage mechanism should be implemented by read-proof hardware.

Physical attacks can be *invasive* as well as *non-invasive*. An invasive physical attack is defined as an attack where the attacker physically breaks into a device and thereby modifies its structure. A non-invasive physical attack is one where the attacker performs measurements without modifications to the device's structure.

When a PUF is attacked in a physical manner, its behavior will change. By this we mean that when the same challenge is applied to a PUF, a substantially different response will be generated. A substantially different response is a response whose noise level (w.r.t. to an enrollment measurement) is higher than the noise

Table 1 Entropy of different PUF types

PUF type	Entropy per 1,000 bits
SRAM PUF	950
Delay PUF	130
Butterfly PUF	600

level of responses caused by environmental stresses. The implementation of a detection mechanism of these higher noise levels, allows the device to take appropriate measures when an attack is detected. In case when the PUF responses are used to implement a secure key storage mechanism these higher noise levels lead to a substantially different secret key being generated. Effectively this implies that the secret key in the device is being destroyed and cannot be recovered by an attacker anymore.

In order to assess the security of PUFs against invasive attacks, we submitted our SRAMs to an independent evaluation facility. This lab concluded that the most efficient way to attack these SRAMs consisted of trying to apply voltage contrast attacks. After experimenting for some time and trying different delayering techniques, it turned out that either the chips were functionally destroyed and could not operate anymore or no voltage contrast could be seen on the SRAMs for those which were still functional. This is mainly due to the fact that successful voltage contrast attacks require a very high voltage to be applied to the device and most devices simply do not survive such experiments. As a consequence, SRAM PUFs are shown to be resistant against voltage contrast attacks; the results of voltage contrast attacks on SRAMs will be described in a future detailed publication. We conclude that SRAM PUFs indeed qualify as a read-proof hardware implementation.

4.2.3 Unclonability

The fact that PUFs are unclonable implies that they can be used for anti-counterfeiting purposes and secure key storage.

When PUFs are used for the detection of the authenticity of a product, a physical property of the PUF is measured, translated into a bit string and verified. The physical unclonability of PUFs prevents building of a similar physical structure that upon interrogation produces a similar bitstring that would pass the verification test as the original one.

When the PUF responses are used as a source for secret keys, it is important that the PUF responses are only dealt with within the device to keep them protected from the attackers. In that way, one is protected against attackers that would be able to make a literal clone from a design point of view. Since clones based on an identical design do not translate into literal physical clones, the attacked devices will not have the same secret key or identifier as the original one.

5 Conclusions

In summary, a radically new approach, *hardware intrinsic security*, is available today to prevent cloning of semiconductor products and preserve the revenues of those companies. PUFs are used to generate the intrinsic fingerprint inherent in each device which is combined with a unique activation code to produce the secret key. No key is actually stored in hardware thereby significantly raising the level of security available beyond alternative methods.

References

1. F. Armknecht, R. Maes, A.R. Sadeghi, B. Sunar, P. Tuyls, in *Memory Leakage-Resilient Encryption Based on Physically Unclonable Functions*. Proceedings of ASIACRYPT 2009. Lecture Notes in Computer Science, vol 5912, (Springer, 2009) pp. 685–702
2. C. Bösch, J. Guajardo, A.-R. Sadeghi, J. Shokrollahi, P. Tuyls, in *Efficient Helper Data Key Extractor on FPGAs*. Proceedings of CHES 2008. Lecture Notes in Computer Science, vol. 5154 (Springer, Heidelberg, 2008), pp. 181–197
3. B. Gassend, D.E. Clarke, M. van Dijk, S. Devadas, in *Controlled Physical Random Functions*. Proceedings of the18th Annual Computer Security Applications Conference (ACSAC 2002), Las Vegas, NV, USA, 9–13 Dec 2002 (IEEE Computer Society, Washington, DC, 2002), pp. 149–160
4. M.A. Gora, A. Maiti, P. Schaumont, in *A Flexible Design Flow for Software IP Binding in Commodity FPGA*. IEEE Fourth International Symposium on Industrial Embedded Systems - SIES 2009, Ecole Polytechnique Federale de Lausanne, Switzerland, 8–10 July 2009 (IEEE, Piscataway, NJ, 2009) pp. 211–218
5. J. Guajardo, S.S. Kumar, G.-J. Schrijen, P. Tuyls, FPGA intrinsic PUFs and their use for IP protection, in *Cryptographic Hardware and Embedded Systems - CHES 2007*, ed. by P. Paillier, I. Verbauwhede. Proceedings of the 9th International Workshop on Cryptographic Hardware and Embedded Systems, Vienna, Austria, 10–13 Sept 2007. Lecture Notes in Computer Science, vol. 4727, (Springer, Berlin, Heidelberg, 2007), pp. 63–80
6. J. Guajardo, S.S. Kumar, G.J. Schrijen, P. Tuyls, in *Brand and IP Protection with Physical Unclonable Functions*. IEEE International Symposium on Circuits and Systems - ISCAS 2008, 18–21 May (IEEE, Piscataway, NJ 2008), pp. 3186–3189
7. J. Guajardo, B. Skoric, S.S. Kumar, T. Bel, A.H.M. Blom, G.J. Schrijen, âĂIJAnti-counterfeiting, key distribution and key storage in an ambient world via physical unclonable functions. Inf. Syst. Front. **11**(1), 19–41 (2009).
8. T. Jun, *Circuit Approaches to Physical Cryptography*. Diploma thesis, Technische Universitaet Muenchen, 2009
9. S.S. Kumar, J. Guajardo, R. Maes, G.-J. Schrijen, P. Tuyls, in *The Butterfly PUF: Protecting IP on Every FPGA*, ed. by M. Tehranipoor, J. Plusquellic. Proceedings of the IEEE International Workshop on Hardware-Oriented Security and Trust, HOST 2008, Anaheim, CA, USA, 9 June 2008 (IEEE Computer Society, Washington, DC, 2008), pp. 67–70
10. K. Kursawe, A.R. Sadeghi, D. Schellekens, B. Skoric, P. Tuyls, in *Reconfigurable Physical Unclonable Functions - Enabling Technology for Tamper-Resistant Storage*. 2nd IEEE International Workshop on Hardware-Oriented Security and Trust (HOST 2009), San Francisco CA, USA, 27 July 2009 (IEEE, Piscataway, NJ, 2009), pp. 22–29
11. J.W. Lee, D. Lim, B. Gassend, G.E. Suh, M. van Dijk, S. Devadas, in *A Technique to Build a Secret Key in Integrated Circuits for Identification and Authentication Applications*. Proceedings of the IEEE VLSI Circuits Symposium, Honolulu, HI, June 2004, pp. 176–179
12. D. Lim, *Extracting Secret Keys from Integrated Circuits*. Master thesis, Massachusetts Institute of Technology, May 2004
13. R. Maes, P. Tuyls, I. Verbauwhede, in *Intrinsic PUFs from Flip-Flops on Reconfigurable Devices*. 3rd Benelux Workshop on Information and System Security (WISSec 2008) Eindhoven, The Netherlands, 13–14 Nov 2008 17 pages
14. R. Maes, P. Tuyls, I. Verbauwhede, in *Low-Overhead Implementation of a Soft Decision Helper Data Algorithm for SRAM PUFs*. Proceedings of CHES 2009, Lecture Notes in Computer Science, vol. 5747 (Springer, Berlin, Heidelberg, 2009), pp. 332–347
15. C.W. O'Donnell, G.E. Suh, S. Devadas, *PUF-Based Random Number Generation*, CSAIL CSG Technical Memo 481, Massachusetts Institute of Technology, Mar 2001
16. R.S. Pappu, *Physical One-Way Functions*. Ph.D. thesis, Massachusetts Institute of Technology, Mar 2001

17. B. Skoric, G.-J. Schrijen, W. Ophey, R. Wolters, N. Verhaegh, J. van Geloven, Experimental hardware for coating PUFs and optical PUFs, in *Security with Noisy Data - On Private Biometrics, Secure Key Storage and Anti-Counterfeiting*, ed. byP. Tuyls, B. Skoric, T. Kevenaar (Springer, London, 2007), pp. 255–268
18. B. Skoric, P. Tuyls, W. Ophey, in *Robust Key Extraction from Physical Unclonable Functions*. Applied Cryptography and Network Security (ACNS) 2005, Lecture Notes in Computer Science, vol. 3531 (Springer, Heidelberg, 2005), pp. 407–422
19. G.E. Su, S. Devadas, in *Physical Unclonable Functions for Device Authentication and Secret Key Generation*. Proceedings of the 44th Design Automation Conference, DAC 2007, San Diego, CA, USA, 4–8 June, (IEEE, Piscataway, NJ, 2007), pp. 9–14
20. P. Tuyls, G.J. Schrijen, F. Willems, T. Ignatenko, Secure key storage with PUFs, in *Security with Noisy Data - On Private Biometrics, Secure Key Storage and Anti-Counterfeiting* ed. by P. Tuyls, B. Skoric, T. Kevenaar. (Springer, London, 2007), pp. 269–292
21. P. Tuyls, B. Skoric, T. Kevenaar, *Security with Noisy Data. Private Biometrics, Secure Key Storage and Anti-Counterfeiting* (Springer, London, 2007)
22. P. Tuyls, G.-J. Schrijen, B. Skoric, J. van Geloven, N. Verhaegh, R. Wolters, in *Read-Proof Hardware from Protective Coatings*. Proceedings of the Eighth International Workshop on Cryptographic Hardware and Embedded Systems (CHES '06), Yokohama, Japan, 10–13 Oct, pp. 369–383
23. P. Tuyls, B. Skoric, S. Stallinga, A.H.M. Akkermans, W. Ophey, Information-theoretic security analysis of physical unclonable functions. in *Financial Cryptography and Data Security*, ed. by A.S. Patrick, M. Yung. 9th International Conference, FC 2005, Roseau, The Commonwealth of Dominica, 28 Febr –3 Mar 2005. Revised Papers, Lecture Notes in Computer Science, vol. 3570 (Springer, Berlin, Heidelberg, 2005), pp. 141–155

From Statistics to Circuits: Foundations for Future Physical Unclonable Functions

Inyoung Kim, Abhranil Maiti, Leyla Nazhandali, Patrick Schaumont, Vignesh Vivekraja, and Huaiye Zhang

1 Introduction

Identity is an essential ingredient in secure protocols. Indeed, if we can no longer distinguish Alice from Bob, there is no point in doing a key exchange or in verifying their signatures. A human Alice and a human Bob identify one another based on looks, voice, or gestures. In today's networked world, Alice and Bob are computer programs. Their identity relies on the computer hardware they execute on and this requires the use of a trusted element in hardware, such as a Trusted Platform Module (TPM) [1]. In the future, Alice and Bob will include tiny embedded computers that can sit anywhere – in a wireless key, in a cell phone, in a radio-frequency identifier (RFID) [2]. Such tiny electronics cannot afford a full-fledged TPM for identification and authentication. There is thus a great need to develop cost-efficient, reliable, stable, and trustworthy circuit identifiers that can fit in a single chip combined with the rest of the system.

Traditional approaches to hardware identity, such as non-volatile memories, increase system cost, may be tamperable, and at times are not trustworthy. We are therefore investigating electronic fingerprints that are based on the existing, small and random manufacturing variations of electronic chips.

Physical unclonable functions (PUFs) are a known solution to create an on-chip fingerprint. However, the issues of scalability, cost, reliability, and the threats of reverse engineering have not been fully investigated. We advocate a cross-disciplinary approach to combine recent advances in the field of statistics with those in circuits design. Our main concerns in PUF design are to come up with designs that (a) are cheap to build and integrate, (b) show a high reliability toward environmental changes and aging, and (c) are able to prevent some common attacks. In this chapter, we will report several steps toward achieving these objectives.

I. Kim (✉)

Statistics Department, Virginia Tech, Blacksburg, VA 24061, USA

e-mail: inyoungk@vt.edu

A.-R. Sadeghi, D. Naccache (eds.), *Towards Hardware-Intrinsic Security,* Information
Security and Cryptography, DOI 10.1007/978-3-642-14452-3_3,
© Springer-Verlag Berlin Heidelberg 2010

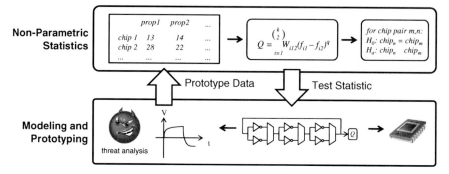

Fig. 1 Combining statistics, architecture, and circuit in the PUF design process

Figure 1 illustrates the two layers in our cross-disciplinary approach: statistics and modeling and prototyping. Novel data-processing ideas are created in the statistics layer and passed down to the architecture layer. The architecture layer, in turn, implements prototypes and specific optimizations and returns prototype test data to the statistics layer.

- *Statistics*: Based on data obtained from prototype architectures, we are developing a novel *test statistic* that can be used as a unique on-chip fingerprint. A test statistic (TS) is an expression that transforms the measurement data into a single number. Besides a TS, we are also working on adequate hypothesis testing techniques to distinguish chips. Based on non-parametric statistics, the measurement data can be directly used, and assumptions on the underlying statistical distributions are avoided [3].
- *Modeling and Prototyping*: We are also working on prototype architectures in CMOS technology. The target architectures include field-programable gate arrays and standard cells. The designs are driven by the TS requirements, but they extend it with architecture-level optimizations and circuit-level optimizations. The modeling layer supports threat analysis by investigation of the response of the resulting PUF designs to specific active and passive attacks.

This chapter is structured as follows. In the next section, we describe a generic model for a PUF architecture. We identify the major phases of PUF operation (sample measurement, identity mapping, and quantization), and we demonstrate how each phase can be handled using different techniques. Next, we discuss related work, with specific attention to the research that has been done to improve reliability, cost, and threat sensitivity. We then cover our research efforts in PUF design. This includes a discussion on circuit-level optimization techniques based on subthreshold voltage operation (Sect. 3), a discussion on architecture-level optimization techniques to efficiently implement redundancy (Sect. 4), and a discussion on statistical techniques for identity mapping an testing (Sect. 5). We provide outlook and conclusions in Sect. 6.

2 Components and Quality Factors of a PUF Design

This section describes a generic template for CMOS-based PUF designs. We provide numerical expressions for the various PUF quality factors. We also discuss sources of wanted and unwanted variability in CMOS technologies. For a review of existing PUF technologies, we refer to the chapter by Maes and Verbauwhede.

2.1 Components of a PUF

Figure 2a illustrates the three components of a PUF. They include sample measurement, identity mapping, and quantization. Figure 2b shows the example of a Ring Oscillator PUF (RO PUF), which takes a 2-bit challenge C and which produces a single-bit response R.

- A *Sample Measurement* converts a digital challenge C into a vector of physical measurements that reflects the identity of the device. In the example of the RO PUF in Fig. 2b, the challenge selects two out of four oscillators. To complete the measurement, the frequency of both selected ROs is measured. The frequencies are determined by the round-trip delay of each RO, which in turn depends on the process manufacturing variations of the digital components for the RO. Hence the pattern of frequencies is unique for each chip.
- In *Identity Mapping*, a vector of measurements is converted into a single, real valued, decision variable. In the example of the RO PUF, the test statistic

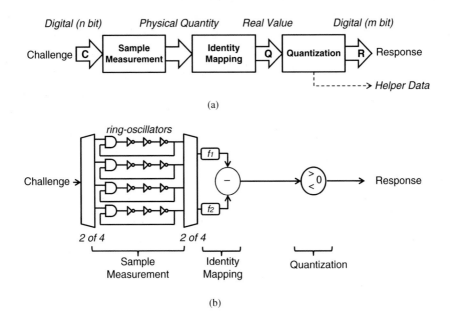

Fig. 2 (a) Generic PUF structure (b) A sample implementation: the ring oscillator PUF

(TS) is the frequency difference of the two selected RO. Because physical measurements can be noisy, the TS will be influenced by noise as well.

- The *Quantization* step maps the real-valued TS into a digital response R. In the example of Fig. 2b, the quantization function is a simple check on the sign of the TS value: negative yields a 0 response and, positive or zero yields a 1 response. Several researchers have generalized the quantization step as a fuzzy-extraction process that generates, besides the response R, additional helper-data bits [4, 5]. The role of these helper-data bits is to correct for the effects of noise in the TS. Indeed, using channel-coding techniques, the helper-data bits can be used to correct for bit errors in future noisy responses [6, 7]. This guarantees that a given challenge will always map to the same response.

Figure 2a can be used as a template for different types of PUF technologies. In recent years, proposals have been made that exploit the startup state of SRAM and flip-flops [8–10], on-chip logic delay [11–13], and the resistance of the on-chip power grid [14, 15]. Hence, there is general agreement that modern silicon technology contains ample amounts of process variation to implement Identity Mapping.

The typical use of a PUF is as follows. The C/R pairs available from a PUF are used for authentication of the PUF. A verifier that wants to use a PUF obtains a table with known C/R pairs. This table is provided by a trusted source that can *enroll* PUF devices before they are deployed. To authenticate the PUF, the verifier then selects a challenge from the table, sends it to the PUF, and compares the response with the enrolled response in the table. In order to prevent playback attacks, each C/R pair in the table may be used only once. Hence, the trustworthy lifetime of a PUF is determined by the number of C/R pairs in the table. In addition, in order to prevent aliasing between PUFs, each C/R pair must be unique over the PUF population. Therefore, the total number of unique C/R pairs determines the number of PUF circuits that can be fielded.

2.2 PUF Quality Factors

In their chapter, Handschuh and Tuyls already defined reliability and security as two PUF quality factors. We can add a third metric, namely design cost. A good PUF design will seek a trade-off between these three factors.

- *Cost*: A low-cost on-chip fingerprint means that the smallest possible circuit area is used to identify a given population of chips or that a given PUF design can distinguish the largest possible population of chips.
- *Reliability*: A stable on-chip fingerprint requires that a given PUF characteristic is insensitive to environmental variations (temperature, voltage, noise) and to temporal variations (aging).
- *Security*: A high-quality PUF should provide a high entropy (a large amount of secret bits), should be resistant against tampering, and should provide unclonabil-

ity. Majzoobi describes unclonability as being resistant to reverse engineering, as well as being resistant to emulation [16].

We need to be able to quantify these factors. This is not an easy task. The most straightforward factor is design cost, which is directly proportional to the silicon area of a PUF design. A smaller design is cheaper to manufacture.

Maes and Verbauwhede described two metrics in their chapter that can be used to estimate the PUF reliability and entropy. They introduce *intra-distance* as the Hamming distance between two evaluations in a single PUF under the same challenge. They also define *inter-distance* as the Hamming distance between the responses of two different PUFs under the same challenge.

We can use these metrics as an estimate of *reliability* and *security*. Assuming we have a set of PUF, then the reliability can be defined as the average number of bits from a response that will stay unchanged under the same challenge. Increased reliability simplifies the quantization step in Fig. 2. Numerically, reliability is estimated as follows.

$$S|_{C_1} = 100\% - \max_{i,j} \frac{\{HD(R_i, R_j)\}}{m} \times 100\% \tag{1}$$

with HD equal to the Hamming distance between any two responses R_i and R_j from the same PUF to the same challenge C_1, and m the number of response bits. The optimal reliability is 100%. Measuring reliability implies that one is able to control the environmental factors that can affect a PUF's responses.

A similar metric is created from the inter-distance metric, which is called *uniqueness* by Maes and Verbauwhede in their chapter. Uniqueness is an estimate for the entropy available from a PUF. Over a similar population of PUF, uniqueness can be calculated as follows:

$$U|_{C_1} = \frac{2}{k(k-1)} \sum_{i=1}^{i=k-1} \sum_{j=i+1}^{j=k} \frac{HD(R_i, R_j)}{m} \times 100\% \tag{2}$$

with HD equal to the Hamming distance between any two responses R_i and R_j from *different* PUFs to the same challenge C_1, k the number of PUFs in the population under test, and m the number of response bits. The optimal uniqueness is 50%.

We refer to the earlier chapters by Maes and Verbauwhede and by Handschuh and Tuyls, for further discussion on PUF quality factors.

2.3 Sources of CMOS Variability and Compensation of Unwanted Variability

In order to design new CMOS PUF architectures or to improve upon existing ones, we need to understand the sources of variability in digital circuits. We distinguish several different sources in Fig. 3.

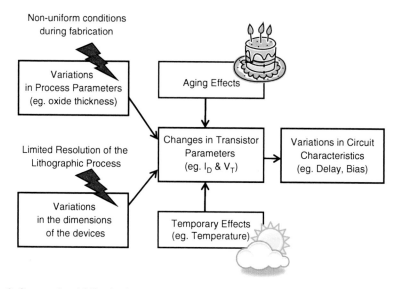

Fig. 3 Causes of variability in circuit characteristics

- *Process Manufacturing Variations*: The random and permanent deviation from the designed, nominal value of a circuit structure, caused by random effects during manufacturing [25]. Process manufacturing variations (PMV) can be separated into two categories. The first category covers variations in process parameters, such as impurity concentration densities, oxide thicknesses, diffusion depths. These result from non-uniform conditions during the deposition and/or the diffusion of the dopants. The second category covers variations in the dimensions of the devices. These result from limited resolution of the photo-lithographic process which in turn causes width and length variations in transistors.

- *Environmental Variations*: These are temporary variations caused by changes in the environmental parameters, including temperature, operating voltage, and external noise coupling. Because the PUF environment cannot always be controlled, the effect of these variations should be minimized. A increase in temperature, and a decrease in power supply, will slow down a circuit. Moreover, the performance loss is not linear, and it may affect different parts of a circuit differently. This will affect the reliability of the PUF.

- *Aging*: Ultra-slow, but eventually permanent, variations that generally deteriorate circuit performance. Aging results in slower operation of circuits, irregular timing characteristics, increase in power consumption, and sometimes even in functional failures [26–28]. Circuit aging is accelerated by use of increased voltages. Aging is largely dependent on how often the circuit is used. Therefore, blocks of a design that are used more often suffer a larger deviation of characteristics with time.

As shown in Fig. 3, each of these variabilities eventually has a similar impact on the observable circuit characteristics. In order to create a stable PUF behavior, we

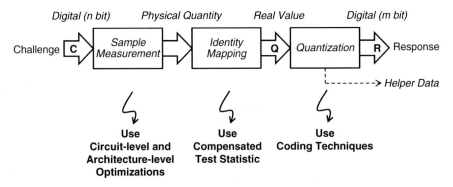

Fig. 4 Compensation of unwanted variability in a PUF

need to detect the process manufacturing variations while tolerating or compensating as much as possible for environmental variations and aging. Figure 4 demonstrates three different strategies to remove the effect of these unwanted variations. Each of these three techniques is implemented on different parts of a PUF design. We briefly describe the general concepts in this section, and we will address their detailed implementation in later sections.

The first approach is to use circuit-level and architecture-level optimizations to provide sample measurement with improved reliability. This will not entirely remove the effect of environmental variations, but it will reduce their effect on the response bits R. The second approach is to improve the identity mapping step, by selecting a test statistic that takes environmental parameters into account. Such a test statistic will estimate the process manufacturing parameters, while ignoring others. The third approach is to use coding techniques that generate helper data in addition to the response. The helper data can be used to reconstruct the correct response at a later time [6, 7]. Each of these three approaches thus work at a different level of abstraction, and therefore they can be combined.

3 Circuit-Level Optimization of PUF

In a CMOS-based PUF, a design built as a CMOS circuit is used to produce random responses, which constitute the basis of the PUF. It is, therefore, very important that the circuit is optimized to improve the quality of the PUF. Conventional circuit design strives to achieve quality factors such as low power and high speed for circuits. Another goal of conventional circuit design is to build the circuit such that a large population of chips have a similar set of characteristics so that they can be conveniently used in their targeted application. In PUF circuit design, although power consumption and overall speed of the PUF have some importance, they are second-hand citizens to quality factors discussed in Sect. 2.2. Therefore, circuit-level decisions have to be made in a completely different fashion.

These decisions include circuit operating voltage, body bias voltage, technology node (transistor length), gate size (transistor width), gate family (high-speed low-

V_T versus low-speed high-V_T gates), and layout decisions (placement of the gates). These decisions vary in terms of their effectiveness in optimizing the PUF quality. In this chapter, we study the effect of the first two in this list, namely, circuit operating (supply) voltage and body bias voltage.

3.1 Methodology

The graphs and numbers presented in this chapter were collected using SPICE simulations. The simulations were performed on a 90 nm technology node, using transistors models and process variation models from UMC [29]. The test circuit is an RO PUF (see Sect. 2.1) with 32 ring oscillators of 11 NAND stages each. The characteristics of 20 different PUF ICs were obtained by applying Monte Carlo simulation on the SPICE model, while enabling both the intra-die and inter-die process variation flags. We ensured that our results are as close as possible to the actual implementation by using the simulation libraries from a commercial foundry and by running the simulation at the highest possible accuracy setting.

3.2 Background: Operating Voltage and Body Bias

It is well known that the power consumption and the frequency of a CMOS circuit are critically controlled by the supply voltage and to some extent by the body bias. The purpose of this section is to explain the impact of these two parameters, not on power and performance, but on the sensitivity of the circuit to process variation, which is key to quality of a PUF. In this regard, we use a key metric: *Coefficient Variation (CV)*. CV is defined as the ratio of standard deviation of a population to its average. For example, the CV of a group of 'n' ring oscillators is the ratio of the standard deviation of its characteristic frequency 'f' to the average characteristic frequency of all ring oscillators, as shown in formula-(3). In statistics, CV is used to compare two different populations with two different averages, in order to see which one has a population whose members are more spread apart. Consequently, in our example, if we build our PUF with two different circuit configurations and the first configuration has a higher value of CV, it indicates that the frequency of the ring oscillators in different chips for this configuration is more spread apart. In other words, the design is more susceptible and less tolerant toward process variation. The motivation behind studying CV is that we believe higher variability can result in higher uniqueness. We investigate this claim in the rest of this section.

$$\text{CV} = \frac{\sigma(f_1, f_2, \ldots f_n)\,(n)}{\Sigma(f_1, f_2, \ldots f_n)}. \tag{3}$$

3.2.1 Operating Voltage

Traditionally, the reduction of supply voltage, also known as voltage scaling, has been successfully employed to reduce the power consumption of a circuit. How-

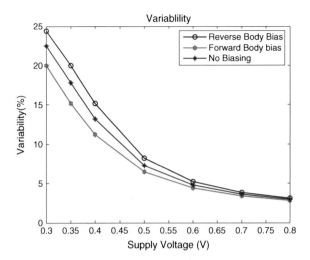

Fig. 5 Scaling of coefficient variation with supply voltage and body bias

ever, lowering the supply voltage increases the sensitivity of the circuit to process variation. This has been shown in Fig. 5 (the middle line with body bias of 0 V), which shows the CV of a ring oscillator with respect to operating voltage. The graph was obtained through Monte Carlo SPICE simulation of a ring oscillator using the setup explained previously in this chapter.

In recent years, it has been shown that the supply voltage of a CMOS circuit can be scaled even further, to voltages below the threshold voltage, which is called subthreshold operation. Various subthreshold circuits have been successfully designed, fabricated, and tested to prove the effectiveness and viability of subthreshold operation [30–32]. However, the sensitivity of the circuit to process variation increases drastically in this region. As can be seen in Fig. 5, the CV increases at a slow but steady pace as we reduce the voltage from nominal voltage to the threshold voltage, around 500 mV. However, around this point the circuit starts to show significant increase in susceptiblity toward process variation. The reason behind this is that below the threshold voltage, transistors are not switching as usual and rely heavily on leakage current for charging and discharging the load capacitance. Leakage current is more affected by process variation, which results in overall higher CV in subthreshold region. This effect is considered a drawback of subthreshold operation in general designs. However, we believe this effect can be employed to our advantage when designing a PUF circuit.

3.2.2 Body Bias Voltage

Figure 6 identifies the source, drain, gate, and bulk contacts of the PMOS and NMOS transistors in a CMOS circuit. Reverse body biasing (RBB) is the process of raising the voltage of the PMOS N-wells with respect to supply voltage or lowering the voltage of the substrate relative to ground. In a forward body bias (FBB) config-

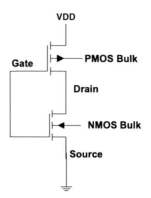

Fig. 6 A CMOS inverter with substrate nodes identified

uration, the PMOS is biased with a voltage lower than supply voltage or the voltage of the NMOS substrate is made negative relative to ground. Traditionally, RBB [33] is employed to reduce the leakage current of the circuit, thereby reducing its leakage power. But this configuration makes the design more susceptible to inherent process variations and decreases its performance. Forward body biasing on the other hand has been used for increasing the frequency of operation and making the design more tolerant toward process variation. Figure 5 shows the effect of body bias on a ring oscillator's CV.

3.3 Effect of Operating Voltage and Body Bias on PUF

Figure 7 presents the scaling of uniqueness with a varying supply voltage for a circuit with three different levels of body bias: zero, forward, and backward. It can be seen that reverse body biasing results in higher uniqueness in a PUF design. However, the effect of reducing operating voltage is much more pronounced and return of using reverse body bias is almost insignificant in subthreshold voltages. It can be concluded from this graph that operating the PUF in subthreshold voltages, which is a relatively cheap technique and requires a very small amount of hardware overhead, is a very effective approach in improving uniqueness of a PUF. This can be explained by the fact that population of frequencies of ROs in subthreshold have a much higher CV compared to a population in nominal voltage. In fact, in the subthreshold region, the circuit's sensitivity toward process variation is so high that the value of uniqueness tends to reach the theoretical maximum of 50%.

Of course, increasing a circuit's sensitivity towards process variation may also increase its sensitivity toward other sources such as variations of temperature and operating voltage. This may reduce the stability of the PUF design. Figure 8 shows the stability of the design as the temperature is varied between -15 and $+85°C$. Initially, decreasing the operating voltage of the circuit decreases the Stability under temperature variations. This is caused by RO frequencies "crossing" each other, which affects the C/R characteristic of the PUF [34]. However, below 500 mV the stability recovers. This is explained as follows. By reducing the voltage, the cir-

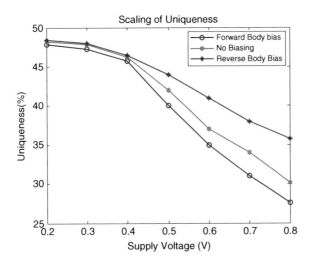

Fig. 7 Scaling of uniqueness with supply voltage for three different body bias levels

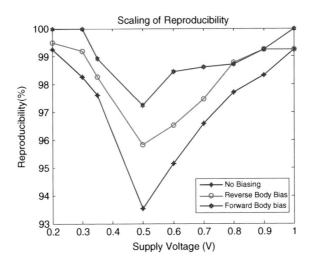

Fig. 8 A CMOS inverter with substrate nodes identified

cuit is exponentially more sensitive toward process variations (Fig. 7) exceeding its sensitivity toward temperature. Therefore, the "crossing" of RO frequencies can be suppressed, and the effects of temperature can be avoided.

4 Architecture-Level Optimization of PUF

Architecture-level optimizations on a PUF aim to optimize PUF quality factors, such as uniqueness and reliability, while at the same time minimizing the circuit cost. An architecture-level optimization distinguishes itself from a circuit-level optimization in that it considers the architecture of the entire PUF. Therefore, it is sensitive to the spatial organization of the PUF. Architecture-level optimizations can be used to address two quality issues in PUF design: compensation of environmental effects and compensation of correlated process manufacturing variations (PMV). We describe each of these two aspects.

4.1 Compensation of Environmental Effects

The impact of environmental effects can be addressed at architectural level through redundancy. First, we should observe that the effect of a temperature change or a voltage change on a PUF is only problematic when the quantization of the response bits has a low signal-to-noise ratio. In this case, the relative magnitude of the effect of PMV is similar or smaller than the effect of environmental variations. As a result, changes in environmental conditions have a dominant effect on the PUF output.

Lim describes a redundant ring oscillator PUF as follows [13]. Each ring-oscillator is replicated four or eight times, thereby creating several redundant oscillators. Each set of redundant oscillators is one group. Next, instead of comparing individual ring oscillator frequencies, one will compare groups of ring oscillators. For each comparison, one ring oscillator is chosen from each group such that their frequency difference is maximal. This will ensure that the contribution of PMV is maximized, and hence will create a stable response.

The disadvantage of redundancy is increased architecture cost. We developed a simple, economical implementation of the redundancy technique for ring oscillators [35]. Figure 9a illustrates a configurable ring oscillator. In this design, each stage of the ring oscillator is configurable and can select one of two inverters. The design in the figure has three stages and therefore can implement eight different ring oscillators. Of particular interest for -ield-programable gate array designs is that this structure can be efficiently implemented. Figure 9a illustrates the mapping of this configurable ring oscillator in a single configurable logic block (CLB) of a Xilinx FPGA. The configurable ring oscillator occupies the same configurable area as a non-configurable ring oscillator, which means that this redundancy technique comes for free in an FPGA.

The configurable ring oscillator is used as follows. When, in a given PUF design, two oscillators A and B need to be compared, we will select A's and B's configuration such that their difference is maximal. Figure 9b illustrates the effect of this simple strategy on the PUF reliability as a function of circuit voltage and temperature, obtained over a population of five different XC3S500E FPGA's. By adaptively choosing a configuration that maximizes the frequency different, the reliability remains almost perfect. In contrast, when having a non-adaptive strategy that fixes a single, fixed configuration, the reliability is as below 70% for voltage variations and below 90% for temperature variations. Note that the overall reliability of the fixed configuration can be worse than the reliability of a single case because different environmental conditions can affect different PUF response bits.

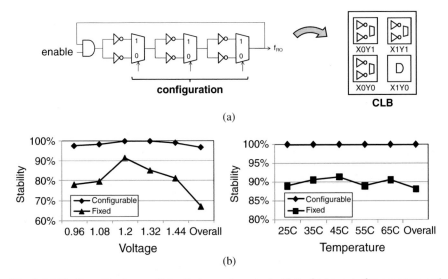

Fig. 9 (**a**) Configurable ring oscillator that maps in a single CLB; (**b**) impact of temperature and voltage on the reliability of the resulting PUF. These results were obtained over a population of 5 Spartan XC3S500E FPGA's

Another recently proposed scheme to compensate of environmental effects is the use of so-called cooperative ring oscillators [34]. In this case, a ring oscillator group is adaptively constructed, depending on the response of ring oscillators. This approach also provides significant area savings. However, since the group-forming data depends on the environmental parameter (such as temperature), it may reveal details about the internal PUF structure and therefore this data must be adequately protected.

4.2 Compensation of Correlated Process Variations

Architectural techniques are also useful to address correlated process variations. In this case, we will make use of our knowledge on the spatial organization of the PUF.

First, we briefly clarify *correlated* PMV. Figure 10a shows 256 ring oscillators arranged as a 16-by-16 matrix in an FPGA fabric. These 256 oscillators form a ring oscillator-based PUF, and the comparison of their frequencies will lead to the response bits R. If one would observe the oscillator frequencies of a chip with *ideal* PMV, the oscillator frequencies would look like Fig. 10b. In this case, the challenge/response scheme can select any two oscillators for comparison, and the comparison outcome would depend on the PMV only. In a real FPGA, however, one can observe a spatial dependency of the average ring oscillator frequency, as illustrated in Fig. 10c. This is caused by various effects, including the detailed structure of the power grid, the irregularities of the reconfigurable fabric due to hard macro's, and the systematic intra-die variations in the FPGA chip.

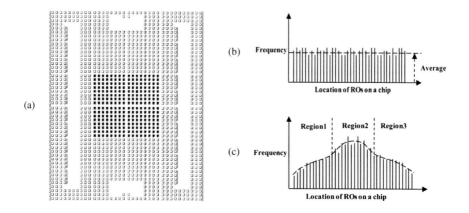

Fig. 10 (**a**) 256 ring oscillators arranged as a 16-by-16 matrix in a Spartan 3ES500 FPGA (**b**) Oscillator frequencies of an FPGA with ideal PMV (**c**) Spaital dependency of the average ring oscillator frequency of a real FPGA

Correlated PMV degrade the entropy that can be extracted from the FPGA. Considering Fig. 10c, it is clear that the comparison of ring oscillators, located far apart from each other, is likely to be biased due to systematic variations. If we need to minimize the effect of correlated variations, we should only compare frequencies of ring oscillators that are close together. Note that this simple strategy will reduce the amount of entropy as well [35].

An alternate strategy to cancel correlated variations is to make use of a common centroid technique [36]. In this case, redundant ring oscillators are used to establish a common average, and then the differential on this average is analyzed. For example, for a group of four ring oscillators with frequency A, B, C, D, the differential $(A + D) - (C + B)$ will be evaluated.

To summarize, architecture-level optimization is complementary to circuit-level optimization, and it can be used to remove environmental effects as well as correlated process variations. In the next section, we focus on the design and implementation of the identity mapping step using a test statistic and non-parametric statistics.

5 Identity Mapping and Testing

Recall that the design of a PUF scheme includes three steps: sample measurement, identity mapping, and quantization. The previous sections described strategies for sample measurement using circuit-level and architecture-level techniques. In this section, we describe a new approach to identity mapping, based on a new test statistic (TS). We will demonstrate that this TS improves over known schemes in terms of extracted entropy. Further, we will also show that the nonlinear nature of the TS provides protection against reverse engineering.

Complementary to the TS, we also propose a non-parametric hypothesis test to evaluate it, and we demonstrate experimental results obtained from five different

FPGA. We wrap up the chapter with an overview of pending challenges in Identity Mapping and Testing.

5.1 Statistical Preliminaries

We first define some notations and introduce a statistical model to explain our approach in detail. Let f_{ijl} be the frequency value for the lth measurement of the jth RO in chip i, where $i = 1, \ldots, n$, $j = 1, \ldots, m$, and $l = 1, \ldots, r$. Using these frequency data, we consider the following statistical model which is expressed as a function of unknown parameters and an additive error term,

$$f_{ijl} = f_{ij} + \varepsilon_{ijl},$$

where f_{ij} is the fixed unknown mean frequency for the ith chip and the jth RO and ε_{ijl} is a random measurement error following unknown distribution.

5.1.1 Bootstrapping the RO frequency

Since f_{ij} is unknown parameter, we need to estimate f_{ij} without assuming any distribution of ε_{ijl}. For this kind of situation, the bootstrapping approach [37] can be applicable. The bootstrapping approach is a resampling-based approach; we start from one sample which represents the unknown population from which it was drawn and then create many re-samples by repeatedly sampling with replacement. Each re-sample is the same size m as the original random sample. Using each re-sample, we estimate f_{ij} which minimizes the least square estimation criterion

$$\mathrm{argmin}_{f_{ij}} \sum_{ijl} \varepsilon_{ijl}^2 = \mathrm{argmin}_{f_{ij}} \sum_{ijl} (f_{ijl} - f_{ij})^2.$$

The bootstrap distribution of a f_{ij} collects its values from many re-samples. The bootstrap distribution gives information about the sampling distribution. The procedure of bootstrapping approach is summarized in Fig. 11.

Using the bootstrap distribution, we can estimate f_{ij} and also obtain a confidence interval of f_{ij} which indicates the precision with which the parameter is estimated. Because bootstrap approach can be used without assuming distribution of ε_{ijl}, it is called a *non-parametric approach*. This approach is especially useful in situation where less is known with small sample size.

5.1.2 Hypothesis Testing

Statistical test plays a key role in experimental research and aids in problem-solving and decision-making processes. Statistical test enables us to draw inferences about a phenomenon or process of interest. These inferences are obtained by using TS to draw conclusions about postulated models of the underlying data generat-

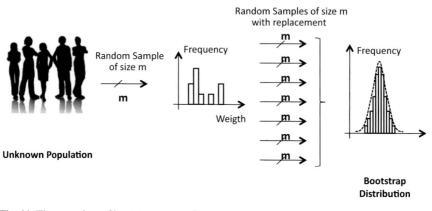

Fig. 11 The procedure of bootstrap approach

ing mechanism. To do this, we need to specify statistical hypotheses which are statements about theoretical models or about probability or sampling distributions. There are two hypotheses that must be specified in any statistical testing procedure: the *null* hypothesis and *alternative* hypothesis. In the context of a statistically designed experiment, the null hypothesis, denoted H_0, defines hypothesis of no change or experimental effect and alternative hypothesis, denoted H_a, indicates change or experimental effect. One hypothesis of interest in our study is whether chips are different from each other. Thus one set of hypotheses of interest in comparison of all chips is H_0: *the distribution of all chips are the same* vs H_a: *the distribution of all chips are different*. This set of hypotheses is equivalent to that H_0: *the distribution of any two chips are the same* vs H_a: *the distribution of some two chips are different*. To decide whether or not we reject H_0, we need to make a decision rule which is constructed by TS. Hence, for our testing, we develop a new TS and propose to use its distribution as a unique on-chip fingerprint in this chapter.

For testing our hypothesis, H_0: *the distribution of any two chips are the same* vs H_a: *the distribution of some two chips are different*, we use two statistical methods. One is bootstrap-based confidence interval and the other is Kolmogorov–Smirnov non-parametric test [38].

The first method which we implemented is the bootstrap-based confidence interval. For each chip, we calculate Q values (The Q value is the result of our proposed identity mapping step and will be explained in Sect. 5.2). Next, we obtain the difference between two Q values of two chips which we want to test whether they are different. Using bootstrapping approach, we obtain the distribution of the difference between two chips and calculate a bootstrap-based confidence interval which is 95% percentile-based confidence interval [2.5%, 97.5%] for the mean of difference between two Q values. We then obtain these confidence intervals for all possible pairs of chips. If the confidence interval does not include 0, we make a decision of rejecting H_0 which means that there are statistical evidence that any two chips are different.

The second method which we used is Kolmogorov–Smirnov test. It is used whether two underlying distributions differ. Since this test does not specify what that common distribution is, it is a non-parametric test. It is a form of minimum distance estimation used as a non-parametric test of equality of two distributions. The Kolmogorov–Smirnov statistic quantifies the distance between two distribution of Q values obtained from two chips. If the distance is larger than a critical values of the Kolmogorov distribution, we make a decision of rejecting H_0 which supports that there are statistical evidence that two chips are different.

5.2 A New Test Statistic: Q

5.2.1 Motivation

The need for a new TS can be motivated by means of an example. Let us say that we are evaluating a ring oscillator PUF design. For a given challenge, we select four ring oscillators. The response R needs to be derived from the resulting frequencies. Assume that two different chips return the following four frequencies:

$$\text{chip}_1 = (5, 2, 1, 3),$$
$$\text{chip}_2 = (9, 3, 2, 5).$$

These chips are clearly different. We will show, however, that these chips are indistinguishable using the conventional approach. Indeed, the conventional approach compares the magnitude of the ring oscillator frequencies with one another, and thus builds the response R based on the *rank* of each ring oscillator frequency. The rank is the relative order of a number in a list. The frequency ranks of both chips are given as follows:

$$\text{chip}_1 = (4, 2, 1, 3),$$
$$\text{chip}_2 = (4, 2, 1, 3).$$

In this case, the rank list of both chips is the same. The conventional identity mapping approach will therefore treat these two chips as identical, while they are clearly not the same! We will therefore develop a TS which can look across the frequency rank, and which directly considers the frequency values in terms of their distance. Each distance is then evaluated using a nonlinear power function, and the distribution of the resulting distance is used as the chip response R. Before giving a formal derivation, we illustrate this approach with an example.

First, we derive all frequency distances in each list. These are defined based on the creation of frequency subsets. The two chips have the following subsets:

$$\text{chip}_1 = \{(5, 2), (5, 1), (5, 3), (2, 1), (2, 3), (1, 3),$$
$$(5, 2, 1), (5, 2, 3), (5, 1, 3), (2, 1, 3), (5, 2, 1, 3)\},$$

$$\text{chip}_2 = \{(9, 3), (9, 2), (9, 5), (3, 2), (3, 5), (2, 5),$$
$$(9, 3, 2), (9, 3, 5), (9, 2, 5), (3, 2, 5), (9, 3, 2, 5)\}.$$

Next, we evaluate the set of distances between the frequency subsets. For example, an Euclidean metric for tuples and triples of frequencies would be as follows:

$$d(f_1, f_2) = (f_1 - f_2)^2,$$
$$d(f_1, f_2, f_3) = (f_1 - f_2)^2 + (f_2 - f_3)^2 + (f_1 - f_3)^2.$$

This way, the set of distances for each chip leads to a distribution defined as the Q value:

$$Q_{\text{chip}_1} = (9, 16, 4, 1, 1, 4, 26, 14, 24, 6, 36),$$
$$Q_{\text{chip}_2} = (36, 49, 16, 1, 4, 9, 86, 56, 74, 14, 119).$$

A kernel density plot of these two distributions is shown in Fig. 12. This plot shows that one distribution is sharper than the other, which suggests that the two distributions are different. Note that the Q value is not the actual bitpattern that defines the response R. The bitpattern is only obtained after the quantization step. In the following sections, we provide a formal derivation of the Q-value definition and explain the ideas that define it.

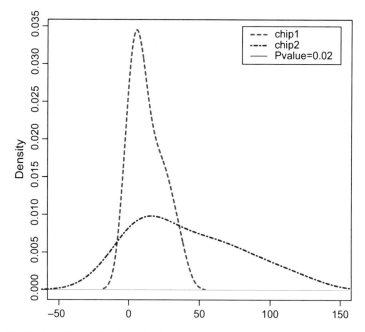

Fig. 12 Distribution of two chips in motivating example

5.2.2 Proposed Q Test Statistic

We define the sample spaces $S_2, \ldots, S_k$ which are the sets of possible outcomes of a particular experiment and random variables $Q_2, \ldots, Q_k$ which are functions mapping from the sample space to a real number (R), where $k \leq m$.

Let S_2 be the sample space which contains all possible frequency pairs. The S_3 is the sample space for all possible frequency triples. Then finally we have $S_k = \{(f_{i1}, f_{i2}, \ldots, f_{ik})\}$ for all possible k frequency values. Denoting $\| \cdot \|$ as the distance, let Q_k be a random variable which is a nonlinear function mapping from S_k to R,

$$Q_k : S_k \to R \text{ such that } Q_k(f_{i1}, f_{i2}, f_{i3}, \ldots, f_{ik}) = \sum_{t=1}^{\binom{k}{2}} w_{i12} \cdot \| f_{i1} - f_{i2} \|^q.$$

The weight factor w_{i12} can include additional information on the design, such as the Euclidian distance between the two ROs under consideration. q may be any real number, e.g., $q = 1/2, 2$ for Euclidian distance. The choice of k depends on m, the number of RO's in the chip. While the ideal value of k is m, the construction of Q_k can be computationally expensive when m is large. In that case, we can reduce k while ensuring that the entropy of the resulting Q_k still remains large. Once q and k are chosen, our Q value is defined as $Q = (Q_2, \ldots, Q_k)$. Q can be quantized as needed to create a suitable response space. For example, we can use (1% percentile, 3% percentile, 5% percentile,…,99% percentile) of Q as the identifier of each chip, or we can use five summary values (minimum, 25% percentile, 50% percentile, 75% percentile, maximum) as the identifier. This Q value is a natural extension of the traditional method based on ordering RO frequencies. Hence, we propose this Q value for TS.

The distribution of Q values can be obtained using bootstrapping approach. We create many re-samples by repeatedly sampling with replacement. Using each re-sample, we calculate Q values. The bootstrap distribution of a Q collects its values from many re-samples. The distribution of TS can be used as a unique on-chip fingerprint.

5.2.3 Entropy of the Q Test Statistic

We now show that our approach based on the Q value has a larger entropy compared to the traditional approach based on rank. The most obvious effect of increased entropy is that the wordlength of the response R can grow. Thus, for a given PUF architecture, the probability of aliasing reduces, which means that the C/R space is more efficiently used. The entropy of the proposed approach, $H(Q)$, is obtained from $H(Q_2), \ldots, H(Q_k)$ as follows. Since all possible subsets are $_mC_2 = \binom{m}{2}$ in Q_2, the probability of choosing one set with equal probability is $1/_mC_k$. Hence, the entropy is

$$H(Q_2) = \sum_{e=1}^{_mC_2} \left\{ -\frac{1}{_nC_2} \log \left(\frac{1}{_mC_2} \right) \right\} = \log {_mC_2}.$$

Similarly, we can calculate $H(Q_3) = \log {}_m C_3, \ldots, H(Q_k) = \log {}_m C_k$. Then the entropy $H(Q)$ is

$$
\begin{aligned}
H(Q) &= \sum_{e=1}^{{}_m C_2 \, {}_m C_3 \cdots {}_m C_k} \left\{ -\frac{1}{{}_m C_2 \, {}_m C_3 \cdots {}_m C_k} \right\} \log \left(\frac{1}{{}_m C_2 \, {}_m C_3 \cdots {}_m C_k} \right) \\
&= \log {}_m C_2 + \cdots + \log {}_m C_k = H(Q_2) + \cdots + H(Q_k) \\
&= \log_2 \left({}_m C_2 \cdot_m C_3 \cdots_m C_k \right)
\end{aligned}
$$

On the other hand, the entropy of a traditional RO-based PUF is $\log {}_m C_2$. The entropy for an arbited-based PUF is given by its challenge space 2^m. Hence, the probability choosing one of them with equal probability is $1/2^m$. The entropy $H(T_d)$ of an arbiter-based approach is therefore

$$
H(T_d) = \sum_{e=1}^{2^m} \left(-\frac{1}{2^m} \right) \log_2 \left(\frac{1}{2^m} \right) = \log_2 2^m = m.
$$

The comparison between traditional approach and our proposed approach is summarized in Table 1. This result clearly shows that our proposed approach yields a larger entropy for the PUF.

5.3 Experimental Results

We have implemented five instances of an FPGA-based PUFs which contain 128 ROs. We obtained 25 frequency measurements for each RO. For given $q = 1$ and $k = 2$, we obtained Q values and used them as a TS value. We then test the hypothesis whether the distributions of Q values are the same among chips. Using Kolmogorov–Smirnove test, all chips are distinguished well. The KS result is summarized in the Table 2 which suggests that there is strong statistical evidence that all chips are quite different because all p values for all possible comparison between any two chips are very small. A p value is the probability of obtaining a value for a test statistic that is as extreme as or more extreme than the observed value, assuming the null hypothesis is true.

In order to confirm this result, we compared bootstrap-based 95% confidence intervals of the mean of the difference between Q values of two chips, which is

Table 1 Summary of entropy comparison

Traditional approach	Entropy	Comparison	Our proposed approach
Ring oscillation	$\log_2({}_m C_2)$	$<$	$\log_2({}_m C_2 \, {}_m C_3 \cdots_m C_k)$
Delay-based approach	$\log_2(2^m)$	$<$	$\log_2({}_m C_2 \, {}_m C_3 \cdots_m C_k)$
Example, $m = 4$,	$\log_2(2^4)$	$<$	$\log_2\left(2^4 \cdot \frac{3}{2}\right) = \log_2({}_4 C_2 \cdot_4 C_3)$

Table 2 The result of Kolmogorov–Smirnov test

Chip #	Chip #	p Value of KS test	Are two chips significantly different?
1	2	1.11e-07	Yes
	3	3.75e-09	Yes
	4	1.45e-06	Yes
	5	4.35e-34	Yes
2	3	2.30e-19	Yes
	4	3.97e-13	Yes
	5	1.24e-38	Yes
3	4	3.55e-02	Yes
	5	1.95e-16	Yes
4	5	9.94e-15	Yes

p Value is the probability of obtaining a value for a test statistic that is as extreme as or more extreme than the observed value, assuming the null hypothesis is true.

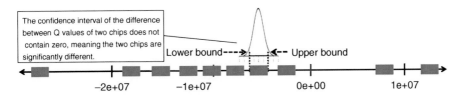

Fig. 13 95% bootstrap-based confidence interval for the mean of the difference between Q values of two chips

shown in Fig. 13. In this figure, each band represents the confidence interval. All intervals are far from 0, which strongly suggests that all chips are significantly different from each other.

5.4 Compensation of Environmental Effects

The proposed approach has three advantages compared to traditional methods. First, it is a nonlinear mapping, and it is therefore more effective against reverse engineering. Second, it is able to increase the entropy of the PUF which makes the PUF lifetime longer. Third, it is very flexible to be extended easily to control several source of variations such as temperature, voltage, and aging. For example, let us assume that f_{ijl} is measured at temperature T_{ijl}, voltage V_{ijl}, and aging A_{ijl}. Then we can model as

$$f_{ijl} = f_{ij} + \beta_1 T_{ijl} + \beta_2 V_{ijl} + \beta_3 A_{ijl} + \varepsilon_{ijl}, \ l = 1, \ldots, r.$$

Using bootstrapping approach we can estimate $f_{ij}, \beta_1, \beta_2, \beta_3$. We then adjust frequency, $f_{ijl}^* = f_{ijl} - \hat{\beta}_1 - \hat{\beta}_2 V_{ij} - \hat{\beta}_3 A_{ij}$. Using this adjusted frequency value we can construct Q value with weight. This way, we can control the effect of several sources of variation. The resulting identifier will have a higher accuracy, and the testing can be done with higher confidence.

5.5 *Open Challenges*

Our present efforts address the following open challenges.

1. For a desired level of entropy, we need to select the optimal number of RO that minimizes the overall cost.
2. Q can be constructed in many ways, based on selection of measurements, the choice of q and k, and the conversion into an identifier. We need to find what exact format is the best feasible one as well as the optimal one.
3. We need to develop a testing technique to evaluate the distribution of the difference of Q values among chips. Since Q is a multidimensional quantity, not a scalar number, we also need to find the optimal testing approach. Although our experiment result shows that the distribution of the difference of Q values distinguishes all chips very well using both Kolmogorov–Smirnove statistic and bootstrap-based confidence interval, we need to investigate what testing approaches will be optimal.
4. To implement the identity mapping step, the Q test statistic formula needs to be implemented in hardware. We will develop efficient architectures to do this, based on efficient signal processing architectures. We will also evaluate the possibility of using on-chip embedded software post-processing.

6 Conclusions

The design and implementation of reliable and efficient PUF covers many different aspects, including circuit-level optimization, architecture-level optimization, and statistical analysis. Through our research, we find that a cross-disciplinary approach is important to cover this very large design space. For example, by employing sub-threshold circuits, we can increase the sensitivity of the design to process manufacturing variations. By using clever redundancy at architecture level, we can then compensate any non-desirable sensitivities to environment variables (such as to temperature and operating voltage). Finally, using an appropriate test statistic, we can harvest the entropy in a statistically optimal way. Clearly, this type of design relies on a range of skills rather than a point specialty. We are currently developing prototypes of the ideas described in this chapter, using FPGA as well as ASIC technology.

Acknowledgments This work was supported in part by the Institute for Critical Technology and Applied Science (ICTAS) and the National Science Foundation with grant no. CNS-0964680.

References

1. Trusted Computing Group, TCG Trusted Network Connect - Federated TNC, 2009. http://www.trustedcomputinggroup.org/resources/federated_tnc_version_10_revision_26
2. D.D. Hwang, P. Schaumont, K. Tiri, I. Verbauwhede, Securing embedded systems. IEEE Security and Privacy, **4**(2), 40–49 (2006)

3. P.H. Kvam, B. Vidakovic, *Nonparametric Statistics with Applications to Science and Engineering* (Wiley-Interscience, Hoboken, NJ, 2007)

4. Y. Dodis, L. Reyzin, A. Smith, in *Fuzzy Extractors: How to Generate Strong Keys from Biometrics and Other Noisy Data*. Proceedings of EUROCRYPT'04 on Advances in Cryptology, Lecture Notes in Computer Science, vol. 3027 (Springer, Berlin, Heidelberg, 2004), pp. 523–540

5. P. Tuyls, B. Skoric, T. Kevenaar, *Security with Noisy Data: Private Biometrics, Secure Key Storage and Anti-Counterfeiting* (Springer-Verlag New York, Inc., Secaucus, NJ, 2007)

6. R. Maes, P. Tuyls, I. Verbauwhede, in *Low-Overhead Implementation of a Soft Decision Helper Data Algorithm for SRAM PUFs*. Cryptographic Hardware and Embedded Systems - CHES 2009, Lausanne, Switzerland, 6–9 Sept 2009 (Springer Verlag, Berlin, Heidelberg, New York)

7. C. Bösch, J. Guajardo, A.-R. Sadeghi, J. Shokrollahi, P. Tuyls, in *Efficient Helper Data Key Extractor on FPGAs*. Cryptographic Hardware and Embedded Systems - CHES 2008, Washington, DC, USA, 10–13 Aug 2008 (Springer Verlag, Berlin, Heidelberg, New York) pp. 181–197

8. Y. Su, J. Holleman, B. Otis, in *A 1.6pj/bit 96% Stable Chip-ID Generating Circuit Using Process Variations*. Solid-State Circuits Conference, 2007. ISSCC 2007. Digest of Technical Papers. IEEE International, Feb 2007, pp. 406–611

9. S. Kumar, J. Guajardo, R. Maes, G.-J. Schrijen, P. Tuyls, in *Extended Abstract: The Butterfly PUF Protecting IP on Every FPGA*. IEEE International Workshop on Hardware-Oriented Security and Trust, 2008. HOST 2008, Anaheim, CA, USA, 9 June, 2008, pp. 67–70

10. D.E. Holcomb, W.P. Burleson, K. Fu, Power-up SRAM state as an identifying fingerprint and source of true random numbers. IEEE Trans. Comput. **58**(9), 1198–1210 (Sept 2009)

11. G.E. Suh S. Devadas, in *Physical Unclonable Functions for Device Authentication and Secret Key Generation*. DAC '07: Proceedings of the 44th Annual Design Automation Conference (ACM, New York, NY, 2007), pp. 9–14

12. E. Ozturk, G. Hammouri, B. Sunar, in *Physical Unclonable Function with Tristate Buffers*. IEEE International Symposium on Circuits and Systems, 2008 (ISCAS 2008), Seattle, WA, 18–21 May 2008 (IEEE, Piscataway, NJ, 2008), pp. 3194–3197

13. D. Lim, J. Lee, B. Gassend, G. Suh, M. van Dijk, S. Devadas, Extracting secret keys from integrated circuits. IEEE Trans. VLSI Syst. **13**(10), 1200–1205 (Oct 2005)

14. R. Helinski, J. Plusquellic, Measuring power distribution system resistance variations. IEEE Trans. Semicond. Manuf. **21**(3), 444–453 (Aug 2008)

15. R. Helsinki, D. Acharyya, J. Plusquellic, in *A Physical Unclonable Function Defined Using Power Distribution System Equivalent Resistance Variations*. Proceedings of the 46th Design Automation Conference (DAC '09), San Francisco, CA, USA (ACM, New York, NY, 2009), pp. 676–681

16. M. Majzoobi, F. Koushanfar, M. Potkonjak, Techniques for design and implementation of secure reconfigurable PUFs. ACM Trans. Reconfigurable Technol. Syst. **2**(1), 1–33 (2009)

17. R. Pappu, B. Recht, J. Taylor, N. Gershenfeld, Physical one-way functions. Science **297**(5589), 2026–2030 (2002)

18. J.D.R. Buchanan, R.P. Cowburn, A.V. Jausovec, D. Petit, P. Seem, G. Xiong, D. Atkinson, K. Fenton, D.A. Allwood, M.T. Bryan, Forgery: 'fingerprinting' documents and packaging. Nature **436**(7050), 475–475 (2005)

19. G. Hammouri, A. Dana, B. Sunar, in *CDs Have Fingerprints Too*. Cryptographic Hardware and Embedded Systems - CHES 2009 (Springer, Heidelberg, 2009), pp. 348–362

20. G. DeJean, D. Krovski, in *RF-DNA: Radio-Frequency Certificates of Authenticity*. Cryptographic Hardware and Embedded Systems - CHES 2007 (Springer, Heidelberg, 2007), pp. 346–363

21. F. Kousanfar, A. Candore, O. Kocabas, in *Robust Stable Radiometric Fingerprinting for Wireless Devices*. IEEE International Workshop on Hardware Oriented Security and Trust 2009 (HOST 2009), San Francisco, CA, USA, July 2009, pp. 43–49

22. S. Jana, S.P. Nandha, M. Clark, S.K. Kasera, N. Patwari, S. Krishnamurty, in *On the Effectiveness of Secret Key Extraction Using Wireless Signal Strength in Real Environments*.. Proceedings of the ACM Sigmobile International Conference on Mobile Computing and Networking (MOBICOM), Beijing, 20–25 September 2009

23. B. Skoric, S. Maubach, T. Kevenaar, P. Tuyls, Information-theoretic analysis of capacitive physical unclonable functions. J. Appl. Phys. **100**(2), 024902 (2006).

24. B. Skoric, G.-J. Schrijen, W. Ophey, R. Wolters, N. Verhaegh, J. van Geloven, Experimental hardware for coating PUFs and optical PUFs. in *Security with Noise Data*, ed. by P. Tuyls, B. Skoric, T. Kevenaar (Springer, New York, NY, 2008)

25. P. Gupta, A.B. Kahng, in *Manufacturing-Aware Physical Design*. ICCAD '03: Proceedings of the 2003 IEEE/ACM International Conference on Computer-Aided Design (IEEE Computer Society, Washington, DC, 2003), p. 681

26. N. Shah, R. Samanta, M. Zhang, J. Hu, D. Walker, in *Built-In Proactive Tuning System for Circuit Aging Resilience*. IEEE International Symposium on Defect and Fault-Tolerance in VLSI Systems, Cambridge, MA, USA, 1–3 October 2008, pp. 96–104

27. P. Lee, M. Kuo, P. Ko, C. Hu, *BERT - Circuit Aging Simulator (CAS)*. Technical Report UCB/ERL M90/2, EECS Department, University of California, Berkeley, 1990

28. W. Wang, V. Reddy, B. Yang, V. Balakrishnan, S. Krishnan, Y. Cao, in *Statistical Prediction of Circuit Aging Under Process Variations*. Custom Integrated Circuits Conference, 2008. CICC 2008. (IEEE, Piscataway, NJ, Sept 2008), pp. 13–16

29. UMC Foundry, http://www.umc.com. Accessed 11/2009

30. A. Wang, A. Chandrakasan, A 180-mv subthreshold FFT processor using a minimum energy design methodology. IEEE J. Solid-State Circuits **40**(1), 310–319 (Jan 2005)

31. V. Sze, R. Blazquez, M. Bhardwaj, A. Chandrakasan, in *An Energy Efficient Sub-Threshold Baseband Processor Architecture for Pulsed Ultra-Wideband Communications*. Acoustics, Speech and Signal Processing, 2006. ICASSP 2006 Proceedings. 2006 IEEE International Conference on, vol. 3, Toulouse, 2006

32. C.H.I. Kim, H. Soeleman, K. Roy, Ultra-low-power DLMS adaptive filter for hearing aid applications. IEEE Trans.VLSI Syst. **11**(6), 1058–1067 (Dec 2003)

33. J. Tschanz, J. Kao, S. Narendra, R. Nair, D. Antoniadis, A. Chandrakasan, V. De, in *Adaptive Body Bias for Reducing Impacts of Die-to-Die and Within-Die Parameter Variations on Microprocessor Frequency and Leakage*. Solid-State Circuits Conference, 2002. Digest of Technical Papers. ISSCC. 2002 IEEE International, vol. 1, San Francisco, CA, USA, 2002, pp. 422–478

34. C.E. Yin, G. Qu, in *Temperature-Aware Cooperative Ring Oscillator PUF*. IEEE International Workshop on Hardware-Oriented Security and Trust, 2009. HOST '09, San Francisco, CA, USA, July 2009, pp. 36–42

35. A. Maiti, P. Schaumont, in *Improving the Quality of a Physical Unclonable Function Using Configurable Ring Oscillators*. 19th International Conference on Field Programmable Logic and Applications (FPL 2009), 2009

36. H. Yu, P.H.W. Leong, M. Glesner, H. Hinkelmann, L. Moller, P. Zipf, in *Towards a Unique FPGA-Based Identification Circuit Using Process Variations*. Proceedings of the 19th International Conference on Field Programmable Logic and Applications 2009 (FPL09), September 2009

37. B. Efron, R.J. Tibshirani, *An Introduction to the Bootstrap* (Chapman & Hall, London, England,1993)

38. I.M. Chakravarti, R.G. Laha, J. Roy. *Handbook of Methods of Applied Statistics*, vol. I (Wiley, New York, NY, 1967), pp. 392–394

Strong PUFs: Models, Constructions, and Security Proofs

Ulrich Rührmair, Heike Busch, and Stefan Katzenbeisser

1 Introduction

Electronic devices have pervaded our everyday life to a previously unseen extent, and will likely continue to do so in the future. But their ubiquity also makes them a potential target for adversaries and brings about privacy and information security issues.

The tools that classical cryptography offers in order to fight these issues all rest on the concept of a secret binary key. They assume that devices can contain a piece of information that is, and remains, unknown to the adversary. Unfortunately, this assumption can be difficult to uphold in practice: Physical attacks such as invasive, semi-invasive, or side-channel attacks, as well as software attacks like API attacks and viruses, can lead to key exposure. The fact that the employed devices should be inexpensive, mobile, and cross-linked aggravates the problem.

The described situation was one of several motivations that inspired researchers to develop the concept of a *physical unclonable function* (PUF). A PUF is a physical system S that can be challenged with so-called stimuli or challenges C_i and which reacts with corresponding responses R_{C_i}. The responses shall depend on manufacturing variations or structural disorder in S that is beyond the control of the original manufacturer and which cannot be cloned or reproduced exactly. The tuples (C_i, R_{C_i}) are often called *challenge–response pairs* (CRPs) of the PUF. More details on the foundations of PUFs can be found in the chapter by Maes and Verbauwhede.

Two important subtypes of PUFs are so-called *Strong PUFs* and *Weak PUFs* [8]; the latter have also been called Physically Obfuscated Keys (POKs) in [6]. Strong PUFs must possess a very large number of possible challenges. A complete determination/measurement of all challenge–response pairs within a limited time frame (such as several days or even weeks) must be impossible. Furthermore, it must be difficult for an adversary to numerically predict or guess the response R_C of a Strong PUF to a randomly selected challenge C. This should hold even if many other challenge–response pairs are known to him. Thus, a Strong PUF's

U. Rührmair (✉)
Technische Universität München, München, Germany
e-mail: ruehrmai@in.tum.de

A.-R. Sadeghi, D. Naccache (eds.), *Towards Hardware-Intrinsic Security,* Information
Security and Cryptography, DOI 10.1007/978-3-642-14452-3_4,
© Springer-Verlag Berlin Heidelberg 2010

challenge–response behavior must be complex and difficult to imitate and "learn." A well-known example of a Strong PUF is the Optical PUF of [16, 17], which also historically is the first PUF that has been suggested.

Typical applications of Strong PUFs are key establishment and identification protocols [17]. The latter usually work in the following manner: A central authority (CA) holds a secret list of many CRPs of a PUF S. The PUF is assumed to be embedded in a hardware system or contained on a security token. In order to identify the PUF, the CA sends k randomly chosen challenges $C_1, \ldots, C_k$ from the CRP list. If the hardware/token can return the correct, corresponding PUF responses $R_{C_1}, \ldots, R_{C_k}$, then the identification is successful. Note that such an approach avoids the storage of secret binary keys in the PUF-embedding hardware. It also avoids the use of standard symmetric or asymmetric cryptosystems, whose security depends on a small set of well-known, but unproven assumptions. It also obviates the potentially costly implementation of standard cryptosystems in mobile devices.

Weak PUFs, on the other hand, have few challenges – in the extreme case just one, fixed challenge. Their response(s) are used to derive a classical binary secret key, which is subsequently processed by the embedding system in a standard fashion, i.e., as a secret input for classical cryptosystems. This makes Weak PUFs similar to a non-volatile key storage. Their advantage is that they may be harder to read out invasively than common non-volatile memory such as EEPROM. Since they depend on inherent manufacturing variations, they can individualize hardware without costly, dedicated individualization steps in the production. Typical examples of Weak PUFs are the SRAM PUF [8], Butterfly PUF [11], and Coating PUF [22].

Since Weak PUFs are nothing else than a special form of secret key storage, they can be used for essentially all cryptographic schemes and applications. Please note, however, that this also makes them susceptible to side-channel attacks like power consumption or emanation analysis, in just the same manner as classical schemes. Protocols based on Weak PUFs usually show the same dependency on computational assumptions as standard cryptoschemes built on secret binary keys. Furthermore, since zero errors in the derivation of the secret key from the PUF are tolerable, error correction plays a critical role for Weak PUFs.

In this chapter, we focus on Strong PUFs and investigate their formal foundations and their application for identification purposes. We start by an overview of currently existing Strong PUF implementations in Sect. 2. Then, we analyze currently existing definitions of (Strong) PUFs in Sect. 3 and devise new adversarial models and definitions in Sect. 4. We introduce PUF-based identification schemes in Sect. 5. Subsequently, we perform a formal security proof for identification based on Strong PUFs in one of our models. We conclude the chapter in Sect. 6.

2 Implementations of Strong Physical Unclonable Functions

We start by briefly surveying the current candidates for Strong Physical Unclonable Functions; more details can be found in the chapter by Maes and Verbauwhede in this book. In 2001, Pappu [16] suggested an optical system as the historically first

PUF. It consists of a laser beam, which is directed at a transparent scattering token comprising of many randomly distributed scatterers. The laser light is scattered multiple times in the token and interferes constructively and destructively with itself. This leads to an interference pattern of bright and dark spots on a subsequently placed CCD. This pattern sensitively depends not only on the location of the scatterers in the token but also on the angle and point of incidence of the laser light (and on other parameters of the setup).

The angle and point of incidence of the laser beam are usually regarded as the challenge of this PUF, while the interference pattern (or a suitably chosen image transformation of it) is interpreted as its response. This optical Strong PUF offers high internal complexity and security. On the downside, it cannot be integrated easily into an electronic microsystem, and requires an external, precise readout apparatus.

Relatively soon afterward, integrated, electrical candidates for Strong PUFs have been suggested. One important example is the so-called Arbiter PUF [5, 14], which exploits the natural variations in the runtime delays of integrated circuits. The Arbiter PUF consists of a sequence of k stages (e.g., multiplexers), which are conditioned by a corresponding sequence of k external bits $(b_1, \ldots, b_k)$. An incoming electrical signal is split into two signals, which race against each other in parallel through the sequence of stages. Their exact paths are thereby determined by the values b_i. At the end of the structure, an "arbiter element" (consisting of a latch) determines whether the top or bottom path arrived first, and correspondingly outputs a zero or a one. The Arbiter PUF thus maps a k-bit input challenge $C_i = (b_1, \ldots, b_k)$ to a 1-bit response R_{C_i}.

However, it has been noted early by its inventors that the Arbiter PUF can be attacked successfully by standard machine learning (ML) methods, such as Support Vector Machines or Perceptrons [5, 13]. An attacker collects a number of CRPs and uses them to train the ML algorithm. If trained successfully, the algorithm will subsequently be able to predict the correct responses to other challenges with high probability.

In order to improve their resilience to ML attacks, several variants of the basic Arbiter PUF have been proposed. Examples include XOR Arbiter PUFs [4], Feed-Forward Arbiter PUFs [5, 12], and Lightweight Secure PUFs [15]. All of them are based on runtime delays, but employ the basic Arbiter PUF as a building block in more complex architectures. Nevertheless, it has been shown recently that even these improved variants can be attacked by more sophisticated ML techniques [20, 21], at least, for instances, of medium lengths.

Another potential Strong PUF candidate that must be considered is the Power Grid PUF of [9]. It exploits the resistance variations in the power grid of integrated circuits. A Power Grid PUF can in principle be used both as Weak PUF and as Strong PUF. Due to its simple linear model, however, the Power Grid PUF is presumably susceptible to ML attacks just like other linear PUF structures. This makes it more useful as Weak PUF, as already noted in [9].

Two approaches that follow new routes to machine learning resilient Strong PUFs have been suggested just recently. In [2, 3], analog circuits, in particular so-called

Cellular Nonlinear Networks (CNNs), have been introduced as Strong PUFs. CNNs are two-dimensional, cellular, analog computing arrays, in which every cell is coupled in an analog fashion to its direct neighbors. Commercially available, programable CNNs contain millions of transistors, while operating in a stable fashion. CNN-PUFs promise to allow stable PUFs with a very large number of interacting components, whose output strongly depends on a very large number of random components. Furthermore, their internal models are driven by complex differential equations, which complicates re-modeling and machine learning attacks.

A second recent approach to machine learning resistant Strong PUFs [10, 18, 19] is to employ as many densely packed, independent random subunits as possible, which are read out individually and independently of each other at slow readout rates. It was shown in [10, 18] that large, monolithic, memory-like crossbar structures based on random diodes can practically implement this approach. They reach optimal information densities of up to 10^{10} bits per cm^2 and can be designed such that the slow readout rate is not enforced by an artificially slow access module or the like, but by the inductive and resistive capacitances of the structure itself [18]. Faster readout leads to overloading and immediate destruction of the wiring, rendering the remaining structure unusable.

The resulting Crossbar PUFs are provably immune against machine learning and any other computational attacks. Their security merely depends on the access time of the adversary and on the ratio of the already readout bits vs. the number of overall bits stored in the structure. Modeling attacks subsequent to readout are fruitless, since all components are independent of each other. Whether the limited readout speed is a severe disadvantage depends on the intended application of this Strong PUF. Rührmair et al. [19] suggested the term SHIC PUFs (pronounce as *chique PUFs*) for this new category of Strong PUFs, where SHIC stands for Super High Information Content.

3 Physical Unclonable Functions: Toward a Formal Definition

In the following we take a closer look at formal definitions proposed for PUFs, which is a necessary prerequisite to being able to formally reason about the security of PUF-based protocols. Our discussion follows [21].

3.1 Physical One-Way Functions

We start our overview with the historically first definition, which is the definition of physical one-way functions [16]. The following Notation 1 and Definition 1 are taken directly from [16].

Notation 1 (Notation for Physical One-Way Functions) *Let Σ be a physical system in an unknown state $X \in \{0, 1\}^l$. X could also be some property of the physical system. l is a polynomial function of some physical resource such as volume, energy, space, matter.*

Let $z \in \{0, 1\}^k$ be a specific state of a physical probe P such that k is a polynomial function of some physical resource. Henceforth, a probe P in state z will be denoted by P_z.
Let $y = f(X, P_z) \in \{0, 1\}^n$ be the output of the interaction between system Σ containing unknown state X and probe P_z.

Definition 1 (Physical One-Way Functions) $f : \{0, 1\}^l \times \{0, 1\}^k \rightarrow \{0, 1\}^n$ is a PHYSICAL ONE-WAY FUNCTION if

- $\exists$ a deterministic physical interaction between P and Σ which outputs y in $O(1)$, i.e., constant time.
- Inverting f using either computational or physical means requires $\Omega(\exp(l))$ queries to the system Σ.
  This may be restated in the following way: The probability that any probabilistic polynomial time algorithm or physical procedure A' acting on $y = f(X, P_r)$, where $y \in \{0, 1\}^n$ is drawn from a uniform distribution, is able to output X or P_r is negligible. Mathematically,

$$\Pr[A'(f(X, P_r)) \text{ outputs } X \text{ or } P_r] < \frac{1}{p(l)}$$

  where $p(\ )$ is any positive polynomial. The probability is taken over several realizations of r.

We also stipulate that for any physical one-way function f

- Simulating y, given X and P, requires either $O(\text{poly}(l))$ or $O(\exp(l))$ in time/space resources depending on whether f is a WEAK or STRONG physical one-way function.
- Materially constructing a distinct physical system Σ' such that its unknown state $X' = X$ is hard.

Definition 1 is reminiscent of the well-known definitions of mathematical one-way functions. It transfers the concepts known from that area (such as polynomial time and negligible probability) to the physical, finite context of PUFs. We would like to stress that the definition certainly owns the great merit of being the first formalization attempt in the field. It is associated with the highly respected, seminal work of [16] that established the whole field. But nevertheless, it touches upon a few noteworthy issues.

First, let us address some formal aspects. Definition 1 employs the concept of polynomial resources and of negligible probability. However, these concepts cannot directly be applied to finite functions $f : \{0, 1\}^l \rightarrow \{0, 1\}^k$. In particular, no such function can meet the "hard to invert" condition of Definition 1, since there is always an algorithm that contains f hard coded as a lookup table in its code. Since l and k are constant, this table has constant size; browsing the table for inverting f thus requires constant time as well.

These formal problems could be resolved by a suitable asymptotic treatment. Such a treatment might work along similar lines as collections of one-way functions

[7] and could consider infinite families $(f_i)_{i \in I}$ of physical one-way functions. Perhaps such a treatment was already attempted in Notation 1, when it is stated that l and k are a polynomial function of some physical resource. However, in Definition 1, l and k are treated as constants, and also the other relevant parameters of Definition 1, such as the polynomial runtime of A', are not functions of an external parameter.

Still, even if we assume that Definition 1 was meant to be asymptotic in some physical resource, and if l and k were intended to be functions of this resource, another interesting issue arises. The parameter n, which describes the length of $f(X, P_r)$, is a constant both in Definition 1 and Notation 1. At the same time, the runtime of A' is required to be polynomial in $|f(X, P_r)| = n = \text{const}$. This is not meaningful, since A' should be given longer computation time for a growing size of the considered PUF instances. If Definition 1 was intended asymptotically, it might be better to formulate the non-invertability condition in the following manner:

$$\Pr[A'(f(X, P_r), 1^{k+l}) \text{ outputs } X \text{ or } P_r] < \frac{1}{p(l)}. \tag{1}$$

There are some interesting conceptual aspects of Definition 1 as well. For example, it can be observed that Definition 1 excludes functions with small ranges, e.g., ones with a binary output {0, 1}. Such functions are not hard to invert in the sense of Definition 1. The reason is that for each of the two possible output values, *some* preimage can always be found efficiently – simply by testing several randomly chosen challenges for their response until there is a match. Most electrical candidates for PUFs (e.g. variations of the Arbiter PUF or the Ring Oscillator PUF) have only a single bit or a fixed number of bits as output. They are hence excluded by Definition 1. Note that it cannot be regarded as a flaw of Definition 1 that it excludes electrical PUFs, as they were only introduced after the definition was written. Nevertheless, our observation points at two facts: (i) The concept of Physical One-Way Functions cannot serve as a comprehensive PUF definition today. (ii) The non-invertibility condition of Definition 1 might not be the essential feature that makes PUF applications work. To our knowledge, the only PUF application where the non-invertibility of f plays a role is the bit commitment protocol that was described in [16].

3.2 Physical Unclonable Functions

Another characterization of PUFs was given in [8]. It is not marked as a formal definition in the original text of [8], whence we term it a description here. It distinguishes between Strong PUFs and Weak PUFs and is as follows.

Description 1 (Physical Unclonable Functions) *Physical unclonable functions consist of inherently unclonable physical systems. They inherit their unclonability from the fact that they consist of many random components that are present in the manufacturing process and cannot be controlled. When a stimulus is applied to the system, it reacts with a response. Such a pair of a stimulus C and a response R is*

called a challenge–response pair (CRP). In particular, a PUF is considered as a function that maps challenges to responses.
The following assumptions are made on the PUF:

1. *It is assumed that a response R_i (to a challenge C_i) gives only a negligible amount of information on another response R_j (to a different challenge C_j) with $i \neq j$.*
2. *Without having the corresponding PUF at hand, it is impossible to come up with the response R_i corresponding to a challenge C_i, except with negligible probability.*
3. *Finally, it is assumed that PUFs are tamper evident. This implies that when an attacker tries to investigate the PUF to obtain detailed information of its structure, the PUF is destroyed. In other words, the PUF's challenge–response behavior is changed substantially.*

We distinguish between two different situations. First, we assume that there is a large number of challenge–response pairs (C_i, R_i), $i = 1, \ldots, N$, available for the PUF; i.e., a strong PUF has so many CRPs such that an attack (performed during a limited amount of time) based on exhaustively measuring the CRPs only has a negligible probability of success and, in particular, $1/N \approx 2^{-k}$ for large $k \approx 100$. We refer to this case as Strong PUFs. *If the number of different CRPs N is rather small, we refer to it as a* Weak PUF. *Due to noise, PUFs are observed over a noisy measurement channel, i.e., when a PUF is challenged with C_i a response R'_i which is a noisy version of R_i is obtained.*

Description 1 stipulates that Strong PUFs shall have an exponential number N of CRPs, with $1/N \approx 2^{-k}$ for some k with $k \approx 100$. In addition, item 1 of Description 1 demands that all CRPs of the PUF shall only reveal a negligible amount of information about each other. It is not fully clear how and under which conditions these two requirements can be met simultaneously. For example, it is argued in detail in [21] that the information content of any physical system is bounded polynomially in its size. If this is true, then the two above requirements mutually exclude each other. Again, this definition excludes PUFs whose output consists only of a single bit (such as the aforementioned Arbiter PUF and the Ring Oscillator PUF), as the probability to guess the PUF output correctly is at least $1/2$. This is better than negligible. Therefore, all these PUFs are excluded as Strong PUFs by Description 1.

The concept of Weak PUFs in the sense of Description 1 is logically consistent. But it is a relatively restrictive notion. From all currently known PUFs, only coating PUFs and SRAM-based PUFs are Weak PUFs. The reason is that (i) they only have very few possible challenges and (ii) their responses to different challenges are fully independent of each other, since they read out single, non-interacting subunits of the PUF (isolated SRAM cells in the case of SRAM PUFs and spatially isolated sensor arrays in the case of coating PUFs). Therefore the mutual information that different responses give about each other is essentially zero. For all other known PUFs (in particular the Arbiter PUF including all of its variants, Ring Oscillator PUFs, and Pappu's Optical PUFs), most responses to different challenges contain a non-negligible mutual amount of information about each other. This contradicts item 1 of Definition 1.

3.3 Physical Random Functions

Another PUF definition, taken from [5], is as follows:

Definition 2 (Physical Random Functions) A PHYSICAL RANDOM FUNCTION (PUF) is a function that maps challenges to responses, that is embodied by a physical device, and that verifies the following properties:

1. Easy to evaluate: The physical device is easily capable of evaluating the function in a short amount of time.
2. Hard to predict: From a polynomial number of plausible physical measurements (in particular, determination of chosen challenge–response pairs), an attacker who no longer has the device, and who can only use a polynomial amount of resources (time, matter, etc.) can only extract a negligible amount of information about the response to a randomly chosen challenge.

The terms short and polynomial are relative to the size of the device.[1]

Definition 2 is very compact and intuitively appealing. It also stipulates some sort of asymptotic treatment, with the parameter being the size of the system. A few interesting conceptual aspects can be observed. The underlying security model allows an adversary to measure polynomially many challenge–response pairs (CRPs). This has the consequence that several PUFs cannot meet the definition, since they only possess polynomially many challenges at all. An adversary can create a full lookup table without breaking the polynomial CRP bound and can use this table to imitate/predict the PUF. This applies, for example, to the Ring Oscillator PUF [4], which has only a quadratic number of challenges. It also holds for the Optical PUF of [16, 17]: Its number of CRPs is directly proportional to the dimensions of the scattering token, multiplied by the number of distinct laser angles realizable by the measurement set-up. This means that this Optical PUF only has polynomially many challenges. Similar considerations also apply to the Crossbar PUF [18, 19], which only has quadratically many challenges, too. All these PUFs are excluded by Definition 2, while especially the Optical PUF and the Crossbar PUF seem secure in practice.

4 Alternative Attack Models

Our discussion in the last section showed that the familiar notion of polynomial resources and the usual asymptotic treatment of mathematical cryptography cannot be transferred to PUFs easily. We therefore work out an alternative treatment in this chapter, which is based on concrete time bounds. We start with semi-formal models in this section and provide a more formal version later.

---

[1] In the original text this sentence is placed after the definition.

4.1 Semi-formal Models for Strong PUFs

We start by some fundamentals and some notation for PUFs.

Specification 1 (SEMI-FORMAL SPECIFICATION OF PUFs) *A PUF is a physical system S that maps stimuli or challenges C_i to responses R_{C_i}. The set of all possible challenges of S is denoted as $\mathbf{C}_S$, and the set of all possible responses as $\mathbf{R}_S$. Without loss of generality, $\mathbf{C}_S$ and $\mathbf{R}_S$ are assumed to be finite subsets of $\{0, 1\}^*$. By its challenge–response behavior, S implements a function F_S with*

$$F_S : \quad \mathbf{C}_S \rightarrow \mathbf{R}_S, \quad C \mapsto R_C.$$

We further assume that in each PUF, the responses are notably influenced by fabrication variations beyond the control of the manufacturer (a fact that distinguishes PUFs from purely digital systems based on secret binary keys).

We suggest that apart from these basic requirements, no further security features should be required from a (plain) PUF. In our opinion, such additional security features should strictly be associated with special subnotions derived from PUFs, such as Strong PUFs and Weak PUFs. We will now present a first, semi-formal security model for Strong PUFs (see also [21]):

Specification 2 (SEMI-FORMAL SPECIFICATION OF STRONG PUFs) *Let S be a PUF according to Specification 1. S is called a STRONG $(t_L, t_A, t_P, q, \varepsilon)$ PUF if no cryptographic adversary Eve limited by the current state of technology will succeed in the following experiment with a probability of at least ε.*

SecExp (S, t_L, t_A, t_P, q):

> PHASE *1:* LEARNING. *Eve is given a time period t_L for learning the PUF S. Within that period, she is given physical access to S at adaptively chosen points in time and for time periods of adaptive length. The sum of all time periods for which she had access to S must not excel t_A. Further, Eve can adaptively query an oracle $\mathcal{O}_{F_S}$ for the function F_S at most q times. After the end of the learning phase, Eve cannot access S or $\mathcal{O}_{F_S}$ any more.*
> PHASE *2:* PREDICTION. *A challenge C_0 is chosen uniformly at random from the set $\mathbf{C}_S$ and is given to Eve. Within time t_P, she must produce an output V_{Eve}.*

Thereby the experiment is called successful if $V_{\text{Eve}} = R_{C_0}$. The probability ε is taken over the uniformly random choice of the challenge C_0, and the random choices or procedures that Eve might employ during Phases 1 and 2.

The specification models real application scenarios relatively closely. Typically, Eve will have a relatively long "learning period" t_L in practice. During this phase, she may gather information about the PUF in several ways: (i) She can obtain standard CRPs, for example, through protocol eavesdropping, via direct physical access to the PUF, or remotely (e.g., by a virus in the embedding system). These possibilities are comprehensively included via the adaptive oracle access and the physical access period t_A that we grant Eve. (ii) Eve may attempt arbitrary measurements (beyond mere CRP determination) on the PUF, including measurement of internal system parameters and invasive probing. This possibility is included in our model through the physical access time t_A. Note that Eve will often not be able to execute her physical access at adaptively chosen points in time or for periods of adaptive time length. But specifying our model in this way includes worst-case scenarios and puts us on the safe side.

In practice, t_L is typically relatively long and is only limited by the lifetime of the device embedding the PUF and/or the relevance of the data that was encrypted with a key derived from the PUF. Contrary to that, the physical access time t_A is usually short and costly. This motivates a distinction between these two parameters in our model.

The further distinction between t_L and t_P, on the other hand, is not relevant for all applications of Strong PUFs, but plays a role in many of them. In order to obtain definitions with maximal generality, we opted to include it in our model. To illustrate this point, let us consider two typical applications of Strong PUFs, namely key establishment and identification. In key establishment, the main security requirement is that Eve will not be able to predict the responses R_{C_i} that were used to derive a key between the cryptographic players. Usually no distinction between t_L and t_P is necessary here – we are only interested in the sum $t_L + t_P$ of the two values, and hope for the sake of security that $t_L + t_P$ is impractically large. In PUF-based identification, however, an adversarial attack strategy that leads to very long prediction times t_P is worthless. It can be countered in practice through measuring the response time of the identifying party. In other words, Eve's attacks on PUF-based identification protocols are only successful if they deliver the R_{C_i} fast.

Specification 2 provides a workable model for Strong PUFs. The definition is non-asymptotic, whence it allows statements about concrete PUF instances. For example, as machine learning results show [20, 21], we can make the following statements[2]:

- A 64-bit Arbiter PUF is no (0.6 s, 0 s, 0.001 s, 18050, 0.99)-Strong PUF.
- A 64-bit Arbiter PUF that produces CRPs at a 1 MHz frequency is no (0.6 s, 0.01805 s, 0.001 s, 0, 0.99)-Strong PUF.

The formalism can also be used to make positive statements, not only negations:

- Assuming that its readout speed cannot be accelerated,[3] a Crossbar PUF of size $10^5 \times 10^5$ and readout speed of 100 bits/s is a $(t_L, 10^7 \text{ s}, t_P, 0, 0.6)$-Strong PUF for arbitrary values of t_L and t_P.

---

[2] The statements follow from machine learning experiments-based simulation data, which were reported in Table 1 of [20]. They show that a 64-bit Arbiter PUF can be broken (in simulations) with the respective parameters in terms of learning times, access times, prediction times, CRPs, and prediction rates.

[3] It is argued in [18] in all detail that such an acceleration is indeed practically impossible if the crossbar's design is chosen appropriately.

- Assuming that its readout speed cannot be accelerated, a Crossbar PUF of size $10^5 \times 10^5$ and readout speed of 100 bits/s is a $(t_L, 0, t_P, 10^9, 0.6)$-Strong PUF for arbitrary values of t_L and t_P.

Specification 2 also has its limitations. Most importantly: How do we model Eve? Since we allow Eve arbitrary physical actions, a standard Turing machine is insufficient. This lack of a formal model leads to two problems. First, in a strict sense, we do not know over which set we quantify when we state in Specification 2 that "...no cryptographic adversary Eve limited by the current state of technology will succeed in the following experiment ... .". This logical problem is awkward. But it could perhaps be acceptable under the following provisions: (i) Specification 2 is not understood as a formal definition, but as a semi-formal specification. (ii) The main purpose of Specification 2 is to put down a precise, but not overly technical description of the essence of Strong PUFs, which can be used as a common basis by all communities involved in PUF research. The second problem that results from the lacking model for Eve is perhaps more severe, at least for theoreticians. Without a formal model for Eve, we cannot perform reductionist security proofs.

In order to resolve this dilemma, we could introduce a new computational model, which captures arbitrary physical actions (some sort of "Physical Turing Machine"). But this seems very intricate. Alternatively, we may restrict the attack model; this route is taken in the rest of the chapter.

4.2 The Digital Attack Model

In the *digital attack model*, we follow the basic adversarial model that was put down in Specification 2, with one exception: We do not grant Eve direct physical access to the PUF S, and do not allow arbitrary physical measurements on S. Instead, we restrict her to the measurement of CRPs of the PUF. This restriction is not as unrealistic as it may seem: The natural tamper sensitivity of many PUFs enforces this setting by itself. If a PUF is embedded in a device, separating it from the device to make arbitrary measurements will often be impossible.

The advantage of the digital model is that Eve can be formalized by a standard probabilistic Turing machine with an oracle that provides her with CRPs of the PUF S, or, more precisely, with an oracle for the function F_S. This will allow us to carry over reductionist techniques from the security proofs of mathematical cryptography to PUFs.

Let us now define what it means to break the security properties of a Strong PUF in the digital model.

Definition 3 (BREAKING STRONG PUFs IN THE DIGITAL ATTACK MODEL) Let S be a PUF. Let an adversary $\mathcal{A}$ be given by a tuple $(\mathcal{L}, \mathcal{M})$, where $\mathcal{L}$ is a probabilistic oracle Turing machine, and $\mathcal{M}$ is a probabilistic Turing machine. We say that $\mathcal{A}$ $(t_L, t_P, q, \varepsilon)$-BREAKS S AS A STRONG PUF IN THE DIGITAL ATTACK MODEL if $\mathcal{A}$ succeeds in the following security experiment with a probability of at least ε:

$SecExp\,(S, t_L, t_P, q)$:

> PHASE 1: LEARNING. $\mathcal{L}$ is provided with an oracle $\mathcal{O}_{F_S}$ for the function F_S and is started with an empty input. It may make at most q adaptive queries to $\mathcal{O}_{F_S}$. After at most t_L Turing steps, it outputs a string z and halts.
>
> PHASE 2: PREDICTION. A challenge C_0 is chosen uniformly at random from the set $\mathbf{C}_S$, and $\mathcal{M}$ is started with input (z, C_0). Within t_P Turing steps, $\mathcal{M}$ must outputs a string V_{ADV} and halts.

Thereby the experiment is called successful if $V_{\text{ADV}} = R_{C_0}$. The probability ε is taken over the uniformly random choice of the challenge C_0, and the random coins that $\mathcal{L}$ and $\mathcal{M}$ might employ during Phases 1 and 2.

5 Identification Schemes Based on Strong PUFs

We will now work toward a security proof of PUF-based identification schemes in the digital model. We start by defining the concept of a PUF-based identification scheme.

5.1 PUF-Based Identification Schemes

Roughly speaking, a PUF-based identification scheme is a protocol where one party $\mathcal{P}$, known as prover, tries to convince another party $\mathcal{V}$, known as verifier, of its identity. The prover possesses a PUF that he can query at will. The protocol should both assert the identity of the prover and the physical availability of the PUF. More precisely, a PUF-based identification scheme is defined as a tuple $(\mathcal{K}, \mathcal{P}, \mathcal{V})$:

Definition 4 (PUF-BASED IDENTIFICATION SCHEME) Let S be a PUF. An identification scheme based on S is a tuple $(\mathcal{K}, \mathcal{P}, \mathcal{V})$, where $\mathcal{K}$ is a probabilistic oracle Turing machine, $\mathcal{P}$ is a probabilistic interactive oracle Turing machine, and $\mathcal{V}$ is a probabilistic interactive Turing machine, which together fulfill the following properties:

> INITIALIZATION. On input 1^k, and provided with an oracle $\mathcal{O}_{F_S}$ for the function F_S, the instantiation algorithm $\mathcal{K}$ returns a string x_{in}.
>
> INTERACTIVE IDENTIFICATION. In the identification process, $\mathcal{P}$ and $\mathcal{V}$ execute a joint interactive computation, where 1^k is their joint input, $\mathcal{P}$ is provided with an oracle $\mathcal{O}$, and $\mathcal{V}$ gets a private input x. At the end of the computation, $\mathcal{V}$ outputs a bit $d \in \{0, 1\}$.
>
> COMPLETENESS CONDITION. We require that if in the identification process, $\mathcal{P}$ is run with oracle $\mathcal{O}_{F_S}$, and $\mathcal{V}$ is run with x_{in} as private input, the output of $\mathcal{V}$ at the end of the interactive computation is "1" with probability 1.

VERIFIER $(\mathcal{V})$ PROVER $(\mathcal{P})$

Private input: Oralce: $\mathcal{O}_{F_S}$
$(C_i, Rc_i), i = 1, \dots, k$

$\xrightarrow{\hspace{1cm} C_1, \dots, C_k \hspace{1cm}}$ Set $V_i \longleftarrow \mathcal{O}_{F_S}(C_i)$

Accept, if $Rc_i = V_i$ *for all* i $\xleftarrow{\hspace{1cm} V_1, \dots, V_k \hspace{1cm}}$
Reject, otherwise.

Fig. 1 Canonical identification scheme based on a PUF S

Let us consider the following "canonical" PUF-based identification scheme, illustrated in Fig. 1. In a setup phase, the verifier $\mathcal{V}$ chooses a set of k random challenges (where k denotes a security parameter) $C_1, \dots, C_k$ and measures the PUF response for each challenge. $\mathcal{V}$ stores the set of all chosen challenge–response pairs as private data. Subsequently, the device is given to the prover $\mathcal{P}$. The interactive identification protocol starts when $\mathcal{P}$ presents its device to a reader: the verifier $\mathcal{V}$ sends the challenges $C_1, \dots, C_k$ to the prover, who answers with responses $V_1, \dots, V_k$ which are derived from PUF measurements with challenges $C_1, \dots, C_k$. $\mathcal{V}$ accepts if all responses match the pre-recorded responses $R_{C_1}, \dots, R_{C_k}$ during initialization.

More formally, we define the canonical identification scheme based on a PUF S as the tuple $(\mathcal{K}, \mathcal{P}, \mathcal{V})$, where the algorithms $(\mathcal{K}, \mathcal{P}, \mathcal{V})$ implement the above protocol. In particular,

- $\mathcal{K}$ takes as input 1^k and as oracle $\mathcal{O}_{F_S}$. It chooses $C_1, \dots, C_k$ uniformly at random from the set $\mathbf{C}_S$ and produces as output $x_{\text{in}} = (C_1, R_{C_1}, \dots, C_k, R_{C_k})$.
- $\mathcal{V}$ gets the public input 1^k and the private input $x_{\text{in}} = (C_1, R_{C_1}, \dots, C_k, R_{C_k})$. It sends $C_1, \dots, C_k$ to $\mathcal{P}$. Subsequently, it receives values $V_1, \dots, V_k$ from $\mathcal{P}$, and outputs "1" if and only if

$$V_i = R_{C_i} \quad \text{for all } i = 1, \dots, k.$$

- $\mathcal{P}$ gets as public input 1^k and as oracle $\mathcal{O}_{F_S}$. Upon receiving values $C_1, \dots, C_k$, it queries $\mathcal{O}_{F_S}$ for the values $V_1 = F_S(C_1), \dots, V_k = F_S(C_k)$ and sends the oracle responses to $\mathcal{V}$.

5.2 Security of PUF-Based Identification in the Digital Attack Model

We now state what it means to break a PUF-based identification scheme in the digital attack model. We closely follow the IMP-CA security notion of traditional identification schemes [4]. Thereby, the adversary's goal is to impersonate the prover, that is to make the verifier accept, despite he has no access to the PUF. More precisely, the definition consists of two phases: a *learning* and an *impersonation* phase. In the learning phase, the adversary has access to an oracle $\mathcal{O}_{F_S}$ in order to collect PUF measurements up to a certain maximum number. Furthermore, the adversary is allowed to play a cheating verifier which can interact with an honest prover for an arbitrary number of independent protocol runs. In the impersonation phase, the adversary tries to impersonate the prover such that the verifier accepts the false proof.

Definition 5 Let S be a PUF, and let $\text{ID}_S = (\mathcal{K}, \mathcal{P}, \mathcal{V})$ be an identification scheme based on S. Let an adversary $\mathcal{A}$ be a tuple $(\mathcal{V}^*, \mathcal{P}^*)$, where $\mathcal{V}^*$ is a probabilistic oracle Turing machine, and $\mathcal{P}^*$ is a probabilistic Turing machine. We say that $\mathcal{A}$ $(t_L, t_P, q, r, \varepsilon)$-BREAKS ID_S FOR THE SECURITY PARAMETER k if $\mathcal{A}$ succeeds in the following security experiment with a probability of at least ε:

$SecExp\,(S, t_L, t_P, q, r, k)$:

> PHASE 1: INITIALIZATION. $\mathcal{K}$ is run on input 1^k and produces an output x_{in}.
> PHASE 2: LEARNING. $\mathcal{V}^*$ is provided with an oracle $\mathcal{O}_{F_S}$ for the function F_S and is started with input 1^k. It may make at most q adaptive queries to $\mathcal{O}_{F_S}$. Furthermore, it may interact at most r times with the honest prover $\mathcal{P}$, instantiated with a fresh random tape, whereby $\mathcal{P}$ gets $\mathcal{O}_{F_S}$ as oracle and 1^k as input at each of these interactions. After at most t_L Turing steps, $\mathcal{V}^*$ must output a string z.
> PHASE 3: IMPERSONATION. $\mathcal{P}^*$ is provided with the private input z. $\mathcal{V}$ is provided with the private input x_{in}. Both get as joint input 1^k and execute a joint computation. Within t_P Turing steps, $\mathcal{V}$ outputs a bit $d \in \{0, 1\}$.

We say that the experiment was successful if $\mathcal{V}$ outputs "1" at the end of Phase 3. The probability ε is taken over the random coins that $\mathcal{K}, \mathcal{V}, \mathcal{P}, \mathcal{V}^*, \mathcal{P}^*$ employ during Phases 1–3.

We will now perform a reductionist security proof for the canonical PUF-based identification scheme. The following statement informally says that if S is a Strong PUF, then the canonical identification scheme based on S is secure.

Theorem 1 *Let S be a PUF. Then there is a generic black-box reduction that constructs from any adversary $\mathcal{A} = (\mathcal{V}^*, \mathcal{P}^*)$, which $(t_L, t_P, q, r, \varepsilon)$-breaks the canonical identification scheme based on S for the security parameter k, another adversary $\mathcal{A}' = (\mathcal{L}, \mathcal{M})$, which $(t_L + c \cdot \lceil k/\varepsilon \rceil, \lceil k/\varepsilon \rceil (t_P + c \cdot k), q + kr + \lceil k/\varepsilon \rceil, 0.6\sqrt[k]{\varepsilon}/k)$ breaks S as a Strong PUF. Thereby c is a small constant independent of k.*

Proof In the following, we show how to build an adversary $\mathcal{A}' = (\mathcal{L}, \mathcal{M})$ that predicts the response to a given challenge by running black-box simulations of $\mathcal{A} = (\mathcal{V}^*, \mathcal{P}^*)$.

More precisely, $\mathcal{L}$ runs a black-box simulation of $\mathcal{V}^*$. Whenever $\mathcal{V}^*$ makes a query to $\mathcal{O}_{F_S}$, $\mathcal{L}$ simulates this query by using his oracle $\mathcal{O}_{F_S}$. $\mathcal{L}$ keeps track of all oracle queries and their responses in a list crp to avoid duplicate oracle queries. Whenever $\mathcal{V}^*$ engages in a protocol run with the prover, $\mathcal{L}$ simulates this interaction as follows: upon receipt of a list of k challenges $(C_1, \ldots, C_k)$, $\mathcal{L}$ creates a corresponding list of PUF responses $(R_{C_1}, \ldots, R_{C_k})$, either by looking up the result in crp or by querying $\mathcal{O}_{F_S}$. Once $\mathcal{V}^*$ stops with output z, $\mathcal{L}$ proceeds to draw $\ell = \lceil k/\varepsilon \rceil$ further (previously unused) challenges randomly from the set of challenges and obtains their answers by querying $\mathcal{O}_{F_S}$; all challenges and responses collected in this last step are collected in a list $CR = (\hat{C}_1, \hat{R}_1, \ldots, \hat{C}_\ell, \hat{R}_\ell)$. Subsequently $\mathcal{L}$ halts and outputs (z, CR).

On receiving a challenge C_1, $\mathcal{M}$ performs the following operations:

1. $\mathcal{M}$ selects uniformly at random a position k_0 with $1 \leq k_0 \leq k$ and constructs a list of k challenges $(\bar{C}_1, \ldots, \bar{C}_k)$ as follows: he sets $\bar{C}_{k_0} = C_1$ and $\bar{C}_i = \hat{C}_i$ for $1 \leq i \leq k_0 - 1$; furthermore he sets all $\bar{C}_i$ with $k_0 + 1 \leq i \leq k$ to random challenges from $\mathbf{C}_S$.
2. $\mathcal{M}$ runs $\mathcal{P}^*$ on $(\bar{C}_1, \ldots, \bar{C}_k)$ and input z to obtain $(\bar{R}_1, \ldots, \bar{R}_k)$.
3. If $\bar{R}_i = \hat{R}_i$ for $1 \leq i \leq k_0 - 1$, algorithm $\mathcal{M}$ outputs $\bar{R}_{k_0}$ and stops. Otherwise, $\mathcal{M}$ deletes the first (used) $k_0 - 1$ entries of the list CR and re-starts the operation at step 1. After $m = \lceil k/\varepsilon \rceil$ unsuccessful runs, $\mathcal{M}$ stops and fails.

We denote by A_i the probability that $\mathcal{P}^*$ (when run in the game of Definition 5) outputs the correct response for the ith challenge it is given. We thus have $\mathrm{Prob}[\mathcal{P}^* \text{ succeeds}] = \mathrm{Prob}[\bigcap_{i=1,\ldots,k} A_i] > \varepsilon$. We can write $\mathrm{Prob}[\bigcap_{i=1,\ldots,k} A_i]$ as

$$\mathrm{Prob}[A_1]\,\mathrm{Prob}[A_2 \mid A_1]\,\mathrm{Prob}[A_3 \mid A_2 \cap A_1] \ldots \mathrm{Prob}[A_k \mid A_{k-1} \cap \ldots \cap A_1].$$

Since $\mathrm{Prob}\left[\bigcap_{i=1,\ldots,k} A_i\right] > \varepsilon$, we know that one of the factors in the above product must be larger than $\sqrt[k]{\varepsilon}$. Thus, there exists a position $1 \leq \bar{k} \leq k$ in which $\mathcal{P}^*$ succeeds with a higher probability, under the condition that the algorithm has predicted all prior challenges correctly. The reduction attempts to exploit this fact. It guesses this position $\bar{k}$. Then, it outputs the response of $\mathcal{P}^*$ for the $\bar{k}$th challenge, but only in case $\mathcal{P}^*$ has predicted all previous challenges correctly. Otherwise, this (sub-)run of $\mathcal{M}$ fails, and a new run is started, up to $m = \lceil k/\varepsilon \rceil$ overall iterations.

The probability that $\mathcal{M}$ succeeds to guess this position $\bar{k}$ in one iteration is $1/k$; the probability that $\mathcal{M}$ outputs a correct guess in this round is $\mathrm{Prob}[A_{\bar{k}} \mid A_{\bar{k}-1} \cap \ldots \cap A_1] \geq \sqrt[k]{\varepsilon}$, since the reduction is constructed in a way that it outputs a guess only if $\mathcal{P}^*$ predicts all challenges $\bar{C}_1, \ldots, \bar{C}_{\bar{k}}$ correctly. Due to the independence of successive runs of $\mathcal{P}^*$, we can estimate the overall success probability of $\mathcal{M}$ as

$$\begin{aligned}
\mathrm{Prob}[\mathcal{M} \text{ succeeds}] &\geq 1/k(1 - (1 - \varepsilon)^{k/\varepsilon})\sqrt[k]{\varepsilon} \\
&\geq 1/k(1 - (1/e)^k)\sqrt[k]{\varepsilon} \\
&\geq 0.6\sqrt[k]{\varepsilon}/k.
\end{aligned}$$

The bounds on the run times of $\mathcal{V}^*$ and $\mathcal{P}^*$ can easily be obtained by observing that the simulation requires an overhead that scales linearly in the security parameter k. The precise scaling constant c is dependent on the machine model and is independent of k. Furthermore, $\mathcal{V}^*$ makes at most $q + kr + \lceil k/\varepsilon \rceil$ oracle queries. This proves the theorem.

6 Conclusions

We investigated the formal foundations and applications of Strong physical unclonable functions. The problem of defining PUFs and Strong PUFs in a formally sound way turns out to be complicated. One reason for the occurring obstacles is that PUFs are a hybrid between physics and computer science. Some of their properties, such as their unclonability or the dependence of their output on uncontrollable manufacturing variations, can hardly be expressed in a formalism based on standard Turing machines. On the other hand, some other central features of Strong PUFs – such as their unpredictability, even if many CRPs are known – are closely related to computational complexity. Expressing them formally requires some sort of computational model. Finally, a purely information-theoretic approach to PUF security is not going to work for all Strong PUFs: Several electrical Strong PUF candidates contain only a relatively small amount of relevant structural information.

We made the following contributions. We started by analyzing existing definitions of physical one-way functions, physical random functions, and physical unclonable functions and noted some interesting aspects in these definitions. We subsequently proposed new semi-formal specifications for Strong PUFs. They have some limitations from a strictly formal point of view and do not enable reductionist proofs. But they are intuitive and not overly technical and also specify an adversarial model and its security relevant parameters relatively exactly. The specifications also have the asset of being non-asymptotic, meaning that they can be applied directly to concrete PUF instances.

Next, we introduced a restricted adversarial model, the *digital attack model*, and gave a security definition for breaking Strong PUFs, which eventually enabled reductionist proofs. In principle, it is based on the adversarial scenario of the above informal specifications. But it restricts the adversary's measurements on the PUF to mere CRP determination. This constraint allowed to model the adversary by oracle Turing machines and made classical reductionist techniques applicable. Finally, we showed that the security of the "canonical" PUF identification scheme can be provably based on the security of the underlying Strong PUF without any complexity theoretic assumptions.

References

1. M. Bellare, A. Palacio, in *GQ and Schnorr Identification Schemes: Proofs of Security Against Impersonation Under Active and Concurrent Attacks*. Advances in Cryptology (CRYPTO 2002), Proceedings. Lecture Notes in Computer Science, vol. 2442 (Springer, London, 2002), pp. 162–177

2. Q. Chen, G. Csaba, X. Ju, S.B. Natarajan, P. Lugli, M. Stutzmann, U. Schlichtmann, U. Rührmair, *Analog Circuits for Physical Cryptography*. 12th International Symposium on Integrated Circuits (ISIC), Singapore, 14–16 Dec, 2009
3. G. Csaba, X. Ju, Z. Ma, Q. Chen, W. Porod, J. Schmidhuber, U. Schlichtmann, P. Lugli, U. Rührmair, in *Application of Mismatched Cellular Nonlinear Networks for Physical Cryptography*. 12th IEEE CNNA - International Workshop on Cellular Nanoscale Networks and their Applications, Berkeley, CA, USA, 3–5 Feb 2010
4. S. Devadas G.E. Suh, in *Physical Unclonable Functions for Device Authentication and Secret Key Generation*. Proceedings of the Design Automation Conference (DAC 2007), San Diego, CA, USA, 4–8 June 2007, pp. 9–14
5. B. Gassend, D. Lim, D. Clarke, M. van Dijk, S. Devadas, Identification and authentication of integrated circuits. Concurr. Comput. **16**(11), 1077–1098 (2004)
6. B.L.P. Gassend, *Physical Random Functions*. Master thesis, Massachusetts Institute of Technology, Feb 2003
7. O. Goldreich, *Foundations of Cryptography: Volume 1, Basic Tools* (Cambridge University Press, New York, NY, 2001)
8. J. Guajardo, S.S. Kumar, G.J. Schrijen, P. Tuyls, in *FPGA Intrinsic PUFs and their Use for IP Protection*. Cryptographic Hardware and Embedded Systems - CHES 2007, 9th International Workshop, Proceedings. Lecture Notes in Computer Science, vol. 4727 (Springer, Heidelberg, 2007), pp. 63–80
9. R. Helinski, D. Acharyya, J. Plusquellic, in *A Physical Unclonable Function Defined Using Power Distribution System Equivalent Resistance Variations*. Proceedings of the 46th Design Automation Conference (DAC 2009) (ACM, New York, NY, 2009), pp. 676–681
10. C. Jaeger, M. Algasinger, U. Rührmair, G. Csaba, M. Stutzmann, Random pn-junctions for physical cryptography. Appl. Phys. Lett. **96**, 172103 (2010)
11. S.S. Kumar, J. Guajardo, R. Maes, G.J. Schrijen, P. Tuyls, in *The Butterfly PUF: Protecting IP on Every FPGA*. International Symposium on Hardware-Oriented Security and Trust (HOST 2008), Anaheim, CA, USA, 2008, pp. 67–70
12. J.W. Lee, D. Lim, B. Gassend, G.E. Suh, M. van Dijk, S. Devadas, in *A Technique to Build a Secret Key in Integrated Circuits for Identification and Authentication Applications*. Proceedings of the IEEE VLSI Circuits Symposium, Honolulu, HI (IEEE, Piscataway, NJ, 2004), pp. 176–179
13. D. Lim, *Extracting Secret Keys from Integrated Circuits*. Master's thesis, Massachusetts Institute of Technology, 2004
14. D. Lim, J.W. Lee, B. Gassend, G.E. Suh, M. van Dijk, S. Devadas, Extracting secret keys from integrated circuits. IEEE Trans.VLSI. Syst. **13**(10), 1200–1205 (Oct 2005)
15. M. Majzoobi, F. Koushanfar, M. Potkonjak, in *Lightweight Secure PUFs*. International Conference on Computer-Aided Design (ICCAD'08) (IEEE Computer Society Press, Washington, DC, 2008), pp. 670–673
16. R.S. Pappu, *Physical One-Way Functions*. Ph.D. thesis, Massachusetts Institute of Technology, March 2001
17. R.S. Pappu, B. Recht, J. Taylor, N. Gershenfeld, Physical one-way functions. Science **297**(5589), 2026–2030 (Sept 2002)
18. U. Rührmair, C. Jaeger, Ma. Bator, M. Stutzmann, P. Lugli, G. Csaba, Applications of high-capacity crossbar memories in cryptography. IEEE Trans. Nanotechnol. 99, 1(2010)
19. U. Rührmair, C. Jaeger, C. Hilgers, M. Algasinger, G. Csaba, M. Stutzmann, in *Security Applications of Diodes with Unique Current-Voltage Characteristics*. 14th International Conference on Financial Cryptography and Data Security (FC 2010), Tenerife, Spain, 25–28 Jan 2010
20. U. Rührmair, F. Sehnke, J. Sölter, G. Dror, S. Devadas, J. Schmidhuber, *Modeling Attacks on Physical Unclonable Functions*. Technical Report 251, IACR Cryptology E-print Archive, 2010

21. U. Rührmair, J. Sölter, F. Sehnke, *On the Foundations of Physical Unclonable Functions.* Technical Report 227, IACR Cryptology E-print Archive, 2009
22. P. Tuyls, G.J. Schrijen, B. Škorić, J. van Geloven, N. Verhaegh, R. Wolters, in *Read-Proof Hardware from Protective Coatings.* Proceedings of Cryptographic Hardware and Embedded Systems (CHES 2006). Lecture Notes in Computer Science, vol. 4249 (Springer, Heidelberg 2006), pp. 369–383

Part II
Hardware-Based Cryptography

Leakage Resilient Cryptography in Practice

François-Xavier Standaert, Olivier Pereira, Yu Yu,
Jean-Jacques Quisquater, Moti Yung, and Elisabeth Oswald

1 Introduction

Theoretical treatments of physical attacks have recently attracted the attention of the cryptographic community, as witnessed by various publications, e.g., [1, 17, 22, 24, 29, 31, 33, 34, 42]. These works consider adversaries enhanced with abilities such as inserting faults during a computation or monitoring side-channel leakages. They generally aim to move from the ad hoc analysis of the attacks toward stronger and more systematic security arguments or proofs. Quite naturally, these more general approaches also have limitations that are mainly caused by the versatility of physical phenomenons. Namely, since it is impossible to prove the security against an all powerful physical adversary, one has to find ways to meaningfully restrict them. This is in fact similar to the situation in classical cryptography, where we need to rely on computational assumptions. That is, when moving to a physical setting, we need to determine what are the physical limits of the adversary. Therefore, the question arises of how relevant these physical models are and to which extent they capture the engineering experience. In order to tackle this question, it is useful to first introduce some usual terminology, e.g., from the side-channel lounge [14]:

1. *Invasive vs. non-invasive attacks.* Invasive attacks require depackaging the chip to get direct access to its inside components. A typical example of this is the connection of a wire on a data bus to see the data transfers. A non-invasive attack only exploits externally available information (the emission of which is, however, often unintentional) such as running time and power consumption.

   One can go further along this axis by distinguishing local and distant attacks: a local attack requires close but external (i.e., non-invasive) proximity to the device under concern, for example, by a direct connection to its power supply. As opposed, a distant attack can operate at a larger distance, for example, by measuring an electromagnetic field several meters (or hundreds of meters) away.

F.-X. Standaert (✉)
Crypto Group, Université catholique de Louvain, Louvain-la-Neuve, Belgium
e-mail: fstandae@uclouvain.be

Olivier Pereira is a Research Associate of the Belgian Fund for Scientific Research (FNRS - F.R.S.).

A.-R. Sadeghi, D. Naccache (eds.), *Towards Hardware-Intrinsic Security,* Information Security and Cryptography, DOI 10.1007/978-3-642-14452-3_5,
© Springer-Verlag Berlin Heidelberg 2010

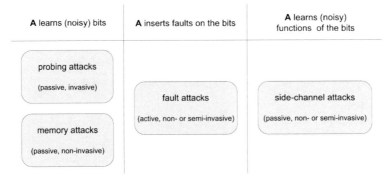

Fig. 1 Informal classification of physical attacks

2. *Active vs. passive attacks.* Active attacks try to tamper with the devices proper functioning. For example, fault induction attacks will try to introduce errors in the computation. By contrast, passive attacks will simply observe the devices behavior during its processing, without disturbing it.

As an illustration, Fig. 1 classifies different physical attacks according to these two axes. With this respect, it is important to remark that seemingly similar abilities can have very different costs in practice. For example, probing attacks such as described by Anderson and Kuhn [3] and the recent memory attacks based on data remanence issues [20] both allow the adversary to learn the value of certain bits in a cryptographic device. But the first one can only target depackaged chips and may require very expensive tools to probe circuits, e.g., when realized in advanced (65 nm or smaller) technologies. By contrast, memory remanence-based attacks can take advantage of cheap "cold boot" techniques to read memory. In addition, the cost of an attack is not the only factor to consider when discussing its applicability. The likelihood to find its scenario in real-life conditions is obviously as important. As a result, side-channel attacks are usually considered as the most dangerous type of physical attacks. They are at the same time low cost and realistic, e.g., when applied against small embedded devices, as in the Keeloq case study [13].

In this report, we consequently investigate the relation between theoretical models and practical engineering works in the area of side-channel attacks. In particular, we consider the pseudorandom generators (PRG) proposed at ASIACCS 2008 [33] and Eurocrypt 2009 [34], respectively. These constructions are based on the same core ideas. First, they assume a bounded leakage for each iteration of the PRG. Second, they rely on frequent key updates in order to avoid the application of standard DPA attacks. In fact, these ideas were not new. Directly after the publication of the first power analysis attack [26], Paul Kocher listed possible countermeasure in which key updates combined with bounded leakage were explicitly mentioned [27, 28]. Hence, the novelty in the previous PRGs is not really in the design principles but rather in the advanced techniques for their analysis, leading to a better confidence in their security levels.

Both papers have pros and cons. Summarizing, the ASIACCS PRG was the first one to provide a systematic analysis of a block cipher-based construction in a physically observable setting. It initiated a study of forward secure cryptographic primitives with their relation to side-channel issues. The underlying model in this work [42] is a specialization of Micali and Reyzin [31] and is motivated by the need to evaluate side-channel attacks on a fair basis. As will be shown in Sect. 5.4 of this report, this connection to the practice of side-channel attacks is required anyway, anytime a leakage bound needs to be assumed (and hence, quantified). In other words, the framework presented at Eurocrypt 2009 [42] is also useful to the work of Pietrzak [34] and more generally to any construction based on a λ-bit leakage: it can be used to estimate practical values for λ. On the negative side, the analysis in [33] considers black box security and physical security separately. It also relies on the existence of ideal ciphers.

The main advantage of [34] is to analyze the security of a PRG in a combined fashion, mixing black box and physical security issues, and in the standard model. It also introduces useful tools for the systematic investigation of physically observable devices (e.g., the quantification of the leakages with the HILL pseudoentropy). On the negative side, the PRG of Eurocrypt 2009 lacks a secure initialization process (as discussed in Sect. 6.1). It is also designed in order to face unrealistic (i.e., too powerful) leakages, e.g., the so-called future computation attacks that we describe in Sect. 5.1. As a result, it exploits a (less efficient) "alternating structure" for which it is not clear if it is really required to prevent actual side-channel attacks or if it is only caused by proof technicalities. Eventually, its security proofs rely on the assumption that "only computation leaks" (or relaxed but related assumptions) which may not always be respected.

Following these observations, the goal of this report is threefold.

First, we aim to connect Micali and Reyzin's definition of leakage function to the practice of side-channel attacks. Doing so, we review the intuitions behind some important results, e.g., the non-equivalence between the unpredictability and indistinguishability of PRGs implemented in leaking devices. Second, we aim to investigate the underlying assumptions and the concrete security provided by two PRGs, with and without alternating structure, in a systematic manner. Doing so, we provide a proof of security for an efficient construction similar to the one of Bellare and Yee [5], using the random oracle methodology. We also introduce definitions allowing us to formalize the practical security of an implementation, inspired by the q-limited adversaries used in Vaudenay's decorrelation theory [48]. Third, we analyze the initialization of a leakage resilient PRG with a public seed. Doing so, we put forward the incompatibility of a secure initialization with a fully adaptive selection of the leakage functions. We also emphasize that standard constructions of pseudorandom functions (PRF) [18] can be shown leakage resilient, in the random oracle model. We conclude this chapter with a negative result. We show with a simple example that the leakage resilience of Bellare and Yee's PRG cannot be directly (meaning, without additional computational and physical hypotheses) proven in the standard model. We leave as an open problem to determine the minimum black box assumptions and restrictions of the leakage function that would be required for this purpose.

Note that this work mainly focuses on the formal techniques used to analyze two PRGs, namely [33] and [34]. Obviously, they are not the only attempts to study side-channel attacks from a more theoretical point of view. Several other references could be acknowledged, e.g., [8, 41], typically. However, we believe these two PRGs are emblematic of one approach for proving the security against side-channel attacks that we denote as the "global approach" in Sect. 4.

2 Background

2.1 Notations

We take advantage of notations from [39, 42]. In particular, let $x \xleftarrow{R} \mathcal{X}$ be a uniformly distributed plaintext picked up from a set $\mathcal{X}$ and $k \xleftarrow{R} \mathcal{K}$ be a uniformly distributed key picked up from a set $\mathcal{K}$. For simplicity, we take $\mathcal{X} = \mathcal{K} = \{0, 1\}^n$. Let also $\mathsf{E}_k(x)$ be the encryption of a plaintext x under a key k with a n-bit block cipher. In classical cryptanalysis, an adversary is able to query the block cipher in order to obtain pairs of the form $[x_i, \mathsf{E}_k(x_i)]$. In side-channel attacks, he is additionally provided with the output of a leakage function of which exemplary outputs are illustrated in Fig. 2. Let $\mathbf{x}_q = [x_1, x_2, \ldots, x_q]$ be a vector containing a sequence of q input plaintexts to a target implementation. In our notations, the measurements resulting from the observation of the encryption of these q plaintexts are stored in a leakage vector that we denote as $\mathbf{l}_q = [l_1, l_2, \ldots, l_q]$. Each element l_i of the leakage vector is referred to as a leakage trace and is in the set of leakages $\mathcal{L}$. Typically, $\mathcal{L} = \mathbb{R}^{N_l}$, where N_l is the number of samples per trace (possibly coming from multiple channels [44]). For example, Fig. 2 represents four leakage traces, corresponding to four input plaintexts encrypted under the same key. Eventually, we denote the tth leakage sample of a trace as $l_i(t)$.

2.2 Definition of a Leakage Function

Following [31], a leakage function is an abstraction that models all the specificities of a side channel (e.g., the power consumption or the EM radiation of a chip), up to the measurement setup used to monitor the physical observables.

Using the previous notations, it means that each leakage sample $l_i(t)$ in a leakage trace l_i is the output of a leakage function L_t. In our block cipher example, this leakage function takes at least a plaintext x_i and a secret key k as inputs. But in theory, the leakages take many more parameters into account. For this reason, Micali and Reyzin consider three input arguments, namely the target device's internal configuration C, a measurement parameter M, and a random string R. Note that in order to be reflective of actual physical leakages, C has to contain all the configurations of the device before the computation corresponding to its current inputs, i.e., before

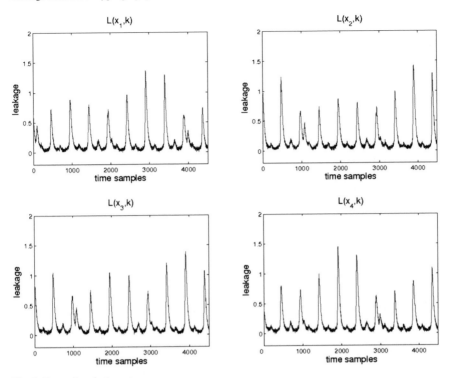

Fig. 2 Exemplary leakage traces

the tth sample of input x_i has been produced in our block cipher example. This incorporates the fact that the leakages can in principle be dependent on anything that has been computed prior to this time sample t, e.g., on the transitions between a former state and a current state in standard CMOS devices. Since Micali and Reyzin define a leakage function as a polynomial time function of its input arguments, this "history" is necessary to include forward secure primitives for which previous states are not polynomial time computable from current states. It yields

$$l_i(t) = \mathsf{L}_t(\underbrace{(x_i, k, \ldots)}_{C}, M, R). \tag{1}$$

For convenience and in order to avoid unnecessarily heavy notations, we will omit the parameters of Eq. (1) that are not directly useful for our discussions in the following of this chapter (e.g., M, typically). We will also sometimes replace the generic state C by the parts of the state for which the leakage dependencies are exploited in a side-channel attack (e.g., inputs and keys).

We note that polynomial time functions actually correspond to more powerful leakages than what is usually observed in practice. As will be discussed in Sect. 5.1, leakage functions generally have a limited complexity of which the exact specification is an interesting direction to obtain more efficient side-channel resistant constructions. Also, a consequence of the previous generic definition is that we potentially have a different leakage function for every implementation.[1] It implies that any security analysis that one may perform in the context of side-channel attacks has to be positioned between two extremes:

1. On the one hand, we can analyze cryptographic primitives with respect to a generic adversary who is enhanced with an arbitrary leakage function. But then, generic and positive statements are hard to obtain and prove. They are also difficult to interpret for a specific device.
2. On the other hand, we can completely fix one instance of leakage function (i.e., perform an experimental attack against a given target device). But then, the conclusions obtained are only valid within this particular setting.

Quite naturally, an interesting approach would be to investigate the existence of intermediate contexts, i.e., to restrict the leakage function in such a way that conclusions can be drawn as long as the leakages fulfill a set of practically meaningful conditions. It is typically the approach followed, e.g., by [22, 33, 34].

Note that a concurrent solution for the analysis of physical security issues in cryptography is to rely on the existence of some minimum primitives from which it is possible to build secure devices. This corresponds, e.g., to the "minimal one way function" of Micali and Reyzin [31] or the "tamper proof" pieces of hardware in [17, 24]. As a matter of fact, these approaches are complementary. Namely, one focuses on the construction of physically secure objects while the other focuses on their exploitation in the construction of advanced functionalities.

3 Unpredictability vs. Indistinguishability

Although protecting leaking devices against key recovery attacks is already difficult, ensuring security against this type of attacks is unfortunately not sufficient. One generally expects pseudorandom generators, functions, or permutations to be indistinguishable from truly random. For example, we can refer to the formal definitions of security for symmetric encryption schemes in [4]. As an illustration, we use the real or random indistinguishability. In this setting, the adversary has access to an oracle $\mathsf{Enc}_k(\mathrm{RR}(\cdot, b))$ that takes an input message x and does the following: toss a coin b, if $b = 0$, return $\mathsf{Enc}_k(x)$, else return $\mathsf{Enc}_k(r)$, where $\mathsf{Enc}_k(x)$ is an encryption scheme (i.e., typically, our block cipher $\mathsf{E}_k(x)$ put into a certain mode of operation) and $r \xleftarrow{R} \mathcal{X}$ is a uniformly distributed random message.

Definition 1 Let $\mathsf{Enc} : \mathcal{K} \times \mathcal{X} \to \mathcal{X}$ be an encryption scheme and A be an algorithm that has access to the oracle $\mathsf{Enc}_k(\mathrm{RR}(\cdot, b))$. We consider

---

[1] In [31, 42], an implementation is defined as the combination of a target device and a measurement setup. We use the same definition in this chapter.

$$\mathrm{Succ}^{\mathrm{ror-ind}-b}_{\mathrm{Enc},\mathsf{A}} = \Pr[k \xleftarrow{R} \mathcal{K} : \mathsf{A}(\mathsf{Enc}_k(\mathrm{RR}(\cdot,b)))) = 1]$$

The *ror-advantage* of a chosen plaintext adversary A against Enc is:

$$\mathrm{Adv}^{\mathrm{ror-cpa}}_{\mathrm{Enc},\mathsf{A}} = \left| \mathrm{Succ}^{\mathrm{ror-ind}-1}_{\mathrm{Enc},\mathsf{A}} - \mathrm{Succ}^{\mathrm{ror-ind}-0}_{\mathrm{Enc},\mathsf{A}} \right|$$

We say that an encryption scheme Enc is ror-indistinguishable if this ror-advantage is negligible for any polynomial time adversary.

The central issue when trying to adapt this definition to a physically observable device is that in general, a leakage trace easily allows distinguishing real inputs/outputs from random ones. This can be intuitively understood by looking at Fig. 3 where the leakage trace corresponding to an encryption $y_i = \mathsf{Enc}_k(x_i)$ is plotted. In this trace, we see that different dependencies can be observed and exploited: the beginning of the trace mainly leaks about the input x_i; the core of the trace leaks jointly about the input x_i and the secret key k; finally, the end of the trace mainly leaks about the output y_i. In most practical side-channel attacks, the leakage is in fact sufficient to recover the secret key k, provided a sufficient amount of (different) encrypted plaintexts can be observed.

Say now that the adversary does not have to recover the key, but to distinguish a real output $\mathsf{Enc}_k(x_i)$ from a random one $\mathsf{Enc}_k(r)$, given the input x_i and the leakage trace corresponding to the encryption of x_i. In fact, the leakage trace will easily allow doing this distinction. For example, imagine that some samples in the leakage trace depend on the Hamming weight $H_W(x_i)$, a frequently considered model in practice. With high probability, this Hamming weight will not be equal to $H_W(r)$. In

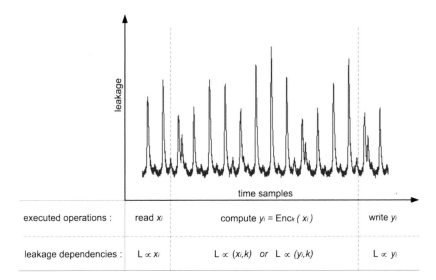

Fig. 3 Impossibility to assume indistinguishability for block ciphers

other words, since we can hardly achieve implementations secure against key recovery, it is even harder to achieve implementations providing an indistinguishability notion. At least, this is definitely not something that we can take for granted. Note that for exactly the same reasons, the existence of durable functions and maximal one-way functions as assumed in [31] cannot be considered as reasonable foundations for a leakage resilient cryptography.

The previous discussion suggests that "protecting" the inputs/outputs of a cryptographic algorithm against distinguishing attacks enhanced with physical leakages is a difficult task. Following this observation, a natural idea is to try analyzing simpler security notions for simpler primitives first. For example, [22, 31, 33, 34] consider PRGs. In this context, two security definitions can be considered, namely the next bit unpredictability introduced in [6] and the current output indistinguishability introduced in [49]. In a black box setting, these two notions are equivalent. But as pointed out in [31], this equivalence does not hold anymore with physically observable devices. The reason of this difference can be easily understood from the example in Fig. 3. Intuitively, what essentially matters from a side-channel point of view is the word "next" in these definitions. That is, distinguishing current outputs from random ones is trivial if one can access the leakage corresponding to this output (as in the Hamming weight example). But predicting the next output bit may still be difficult in this case. Therefore, the following sections will consider a definition of security that corresponds to the next output indistinguishability (or equivalently to the next bit unpredictability). We denote this security notion as *physical indistinguishability*.

Definition 2 Let $G^q : \mathcal{K} \rightarrow \mathcal{K} \times \mathcal{X}^q$ be an iterative pseudorandom generator, with q iterations denoted as $[k_1, x_1] = G(x_0), [k_2, x_2] = G(k_1), \ldots, [k_q, x_q] = G(k_{q-1})$. Let $\mathbf{l}_q = [L(k_1, x_1), L(k_2, x_2), \ldots, L(k_q, x_q)]$ be the leakage vector corresponding to these q iterations. Let $P^q = (G^q, L)$ be the physical implementation corresponding to the combination of the pseudorandom generator G^q with the leakage function L. Let finally A be an algorithm that takes the plaintexts $\mathbf{x}_q = [x_1, x_2, \ldots, x_q]$ and leakages $\mathbf{l}_q$ as input and returns a bit. We consider

$$\mathrm{Succ}_{P^q,A}^{\mathrm{prg-ind}-0} = \Pr[k_0 \xleftarrow{R} \mathcal{K}, [\mathbf{x}_{q+1}, k_{q+1}] = P^{q+1}(k_0) \rightsquigarrow \mathbf{l}_{q+1} : A(\mathbf{x}_{q+1}, \mathbf{l}_q) = 1),$$

$$\mathrm{Succ}_{P^q,A}^{\mathrm{prg-ind}-1} = \Pr[k_0 \xleftarrow{R} \mathcal{K}, [\mathbf{x}_q, k_q] = P^q(k_0) \rightsquigarrow \mathbf{l}_q; x_{q+1} \xleftarrow{R} \mathcal{X} : A(\mathbf{x}_{q+1}, \mathbf{l}_q) = 1).$$

The ind-advantage of A against P^q is defined as:

$$\mathrm{Adv}_{P^q,A}^{\mathrm{prg-ind}} = \left| \mathrm{Succ}_{P^q,A}^{\mathrm{prg-ind}-0} - \mathrm{Succ}_{P^q,A}^{\mathrm{prg-ind}-1} \right|$$

The implementation of a PRG is physically indistinguishable if the ind-advantage of any polynomial time adversary against this implementation is negligible.

We observe that, contrary to the definitions of Dziembowski and Pietrzak [12, 34], our definition is not adaptive in the sense that the leakage function is the predefined before the q iterations of the PRG. We believe that this definition is realistic since the information leaked is essentially a function of the targeted device rather than a choice of the adversary. Besides, letting the adversary select different leakage functions for different runs of the circuit results in an overly strong definition in many cases, as we will further discuss in Sects. 5.1 and 6.4.

4 Physical Assumptions: Local vs. Global Approach

Since the apparition of power analysis attacks in the late 1990s, various solutions have been proposed to counteract them. Among these countermeasures, a first category aims to provide design guidelines for the implementation of cryptographic primitives. That is, they study how to implement (e.g.,) a block cipher is in such a way that it leaks as little as possible. Such local countermeasures have been intensively analyzed in the last 10 years and typically include hiding [47] or masking [19]. Their limitation is that up to now, no solution allows to completely get rid of the leakages. For example, masking schemes have been shown vulnerable to higher order attacks [32] and hiding countermeasures generally give rise to small data-dependent leakages that can be exploited by an adversary. As a consequence, a complementary approach is to accept that cryptographic implementations leak a certain amount of information and try to use them in such a way that these leakages do not lead to complete security failures. In this global approach, one essentially assumes that a single iteration of the target cryptographic primitive "does not leak too much" in some sense.

In the following of this chapter, we investigate this second option. It implies the need to define what is meant by "bounded leakage." For this purpose, the proofs in [34] assume a leakage of λ bits on a key K if this key is (computationally) indistinguishable from a distribution Y having an average min entropy conditioned on the leakage of $n - \lambda$ bits. This is formalized by the notion of HILL pseudoentropy. Here, we denote with $\delta^{\mathsf{D}}(K; Y)$ the advantage of a circuit D in distinguishing the random variables K, Y, i.e., $\delta^{\mathsf{D}}(K; Y) = | \Pr[\mathsf{D}(K) = 1] - \Pr[\mathsf{D}(Y) = 1]|$. We also define $\delta_s(K; Y)$ as the maximum of $\delta^{\mathsf{D}}(K; Y)$ taken over all circuits D of size s. We finally use the standard definitions:

Definition 3 The min entropy of a random variable K is defined as:

$$H_\infty(K) = - \log \max_{k \in \mathcal{K}} \Pr[K = k],$$

and the average min entropy of K conditioned on L is defined as:

$$H_\infty(K|L) = - \log \left(\mathbf{E}_{l \leftarrow L} \left[\max_{k \in \mathcal{K}} \Pr[K = k | L = l] \right] \right),$$

Then, we define the following computational analogues:

Definition 4 K has HILL pseudoentropy n, denoted by $H_{\varepsilon,s}^{\mathsf{HILL}}(K) \geq n$, if there exist a distribution Y with min entropy $H_\infty(Y) \geq n$ where $\delta_s(K; Y) \leq \varepsilon$.

Definition 5 K has HILL pseudoentropy n conditioned on L, denoted by $H_{\varepsilon,s}^{HILL}$ $(K|L) \geq n$, if there exists a collection of distributions Y_l (giving rise to a joint distribution (Y,L)), such that $H_\infty(Y|L) \geq n$ and $\delta_s((K, L); (Y, L)) \leq \varepsilon$.

From such definitions, there are two possible research directions. First, one can investigate how to best exploit a λ-bit leakage, e.g., in the design of PRGs (this will be analyzed in Sect. 5). Second, it is also required to determine what is a reasonable value for λ, in practical settings. As already mentioned, this amount of leakage highly depends on the target device and adversarial strategy. Leakage traces can be made of gigabytes of data and the selection of a good decoding algorithm to extract λ bits of information is a quite challenging issue. As for classical cryptanalysis concerns, this is where reasonable assumptions have to be introduced in accordance with practical works in the area of side-channel attacks. We show in this section that the framework of [42] can be used for this purpose. Without entering into details that are out of the scope of this chapter, the goal of this framework is to provide tools allowing one to evaluate a leaking implementation and a side-channel adversary. As summarized in Fig. 4, it can be seen as an interface between practical and theoretical concerns raised by physically observable devices. When designing new attacks or local countermeasures, it allows determining their effectiveness with a combination of information theoretic and security metrics. For example, it is shown in [43] that an information theoretic metric nicely captures the impact of a local countermeasure such as masking. When building new cryptographic primitives, it allows to quantify the λ-bit leakage assumed in the proofs of constructions such as [12, 34].

Literally, the min entropy $H_\infty(X)$ is the negative logarithm of the probability that a variable X is determined correctly with only one guess of the form "is X equal to x?" with an optimal strategy [7]. Since an exact evaluation of the min entropy (and its computational analogues such as the HILL pseudoentropy) is generally hard to obtain, a simple assumption is to approximate it with the success rate of a

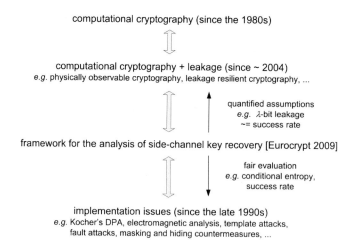

Fig. 4 Interfacing the theory and practice of side-channel attacks

side-channel key recovery attack (briefly recalled in Sect. 10.1), as can be estimated in practical settings:

Assumption. *The HILL pseudoentropy remaining after a q-query side-channel key recovery attack has been performed against a key variable K is approximated by* $H^{HILL}_{\varepsilon,s}(K|L_q) \simeq -\log_2(\mathrm{Succ}^{sc-kr-1,K}_{A_{E_K,L}}(\tau, m, q))$, *where* $A_{E_K,L}$ *is a "best available" adversary with time, memory, and data complexity τ, m, and q.*

Otherwise said, we have: $\lambda(\tau, m, q) \simeq n + \log_2(\mathrm{Succ}^{sc-kr-1,K}_{A_{E_K,L}}(\tau, m, q))$. Exemplary values for λ in different contexts will be discussed in the next section. We mention that considering the success rate of specific attacks may appear as weak from a theoretical point of view. But this assumption is in fact imposed to some extent by the computational difficulty of perfectly estimating the side-channel leakage, as we now argue. That is, ideally, estimating λ would require, for each key value (that is the only unknown variable in a side-channel attack):

- to estimate the probability distribution of the leakage conditionally to this key,
- to compute the (e.g., average min) entropy by integrating this distribution.

For practical ciphers (e.g., the AES Rijndael), this would mean the estimation of 2^{128} distributions, which is unfeasible. Note that for large leakage traces (as usually observed in practice, see [44]), even the estimation of one distribution may be computationally hard (e.g., assuming that the leakages are normally distributed, it requires to estimate a covariance matrix of size $N_l \times N_l$). In order to avoid such limitations, actual side-channel attacks generally exploit the approximated probability distribution of a reduced set of leakage samples and for enumerable subkeys, i.e., the generic template attacks detailed in [42].

As a consequence, there is a hardly avoidable gap between the λ-bit leakage assumed in the proofs and the λ-bit leakage that can be approximated in practice. Solutions to reduce this gap can also be devised in two directions. On the one hand, the approximations of λ should consider reasonable security margins, as in the practical security approach when designing block ciphers, e.g., in [25]. On the other hand, proofs could be based on weaker assumptions than $n - \lambda$ bits of HILL pseudoentropy. For example, one could try to exploit the unpredictability entropy defined in [36] (that is implied by HILL pseudoentropy):

Definition 6 K has unpredictability entropy n conditioned on L, denoted by $H^{unp}_s(K|L) = n$, if for any A of size s it holds that $\Pr[A(L) = K] \leq 2^{-n}$.

Or, alternatively, one could assume that the leakage function is hard to invert (which seems a very minimal assumption), as in the work of Dodis et al. on cryptography with auxiliary input. We will use this assumption in Sect. 5.3.

Note finally that the parameters τ and m correspond to the circuit size s in our definitions of computational entropy. They allow considering the effectiveness of techniques combining side-channel leakages with classical cryptanalysis such as the collision or algebraic side-channel attacks introduced in [40] and [36, 37], i.e., attacks in which the time (and memory) complexity are non-negligible.

4.1 Analogy with Classical Cryptanalysis

Before moving to the security analysis of different PRGs, it is worth emphasizing that this situation, although not satisfying, is not very different than the one in classical (e.g., linear, differential) cryptanalysis. In this case, one also considers the best available attacks to evaluate the security of a cipher. Hence, local and global countermeasures are not contradictory. It is both required to know how to design implementations that do not leak too much and how to exploit such implementations in order to provide good cryptographic properties. By analogy, there is no contradiction between the wide trail strategy [9], that has been used in the design of the AES Rijndael (i.e., a type of local countermeasure against linear/differential attacks), and the proof that an encryption mode is secure if its underlying block cipher is indistinguishable from a pseudorandom permutation [4] (i.e., a more global approach). Similarly, countermeasures such as masking and hiding (or more formal solutions like [8, 16, 22]) can be used to design implementations with limited leakages that can then be used in secure PRG (or other) constructions.

5 Leakage Resilient PRGs

5.1 On the Difficulty of Modeling a Leakage Function

An appealing solution to build PRGs secure against side-channel attacks is to consider forward secure primitives, e.g., as introduced in [5] for symmetric encryption. In this section, we analyze the "future computation attack" to illustrate the need of a new type of construction in [34]. This attack can be explained from Fig. 5, in which a length doubling PRG is denoted as 2PRG and the l states corresponding to l iterations of the arbitrary length PRG are denoted as S_i.

In [34], the leakage of each iteration of the 2PRG is bounded to λ bits. As discussed in the previous section, this is a reasonable abstraction. But the λ bits leaked by the computation of a state S_i can be selected adaptively by the adversary, as long as they are a polynomial time function of this state. Therefore, a straightforward attack depicted on the figure is to select λ bits of S_l during the computation of S_1, another λ bits of S_l during the computation of S_2, etc., until the state S_l

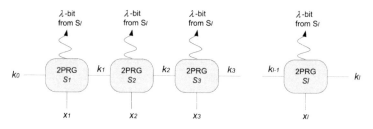

Fig. 5 The "future computation attack"

is completely revealed which trivially (and completely) breaks the security of the scheme. Looking at the physical reality, this attack is obviously an overestimation of what a side-channel adversary could achieve. In general, the leakage function is not selected by the adversary, at least not arbitrarily. It is rather a feature of the target device. In certain settings (e.g., electromagnetic leakages), one could imagine that the antenna is moved adaptively during the attack. But it at least remains that a leakage function never leak about future computations. In reaction to this type of (unrealistic) attacks, three different positions could be adopted that we now detail:

1. One can consider stronger assumptions for the 2PRG. For example, in the ideal cipher model, the outputs of any iteration of the PRG are uniformly random and the computation of a state S_i cannot leak about any state S_j with $j > i$. This is typically the solution considered in [33].
2. Another solution is to keep the model as it is (i.e.,unrealistic with respect to the physics) and to build constructions that can even cope with this type of leakage. This is typically the solution considered in [34].
3. Eventually, a more elegant solution is to restrict the leakage function in a meaningful way. A natural direction with this respect would be to limit the complexity of this function (e.g., in terms of circuit size).

The goal of the next section is to analyze the security of different PRGs against side-channel attacks. For simplicity, we selected the forward secure PRG from [5] and the leakage resilient PRG from [34], pictured in Fig. 6.

Note that the block cipher-based PRG from [33] could be similarly investigated (i.e., it has the same properties as [5] in terms of leakage resilience). Following these chapters, we aim to work out the question: "Is the alternating structure of [34]

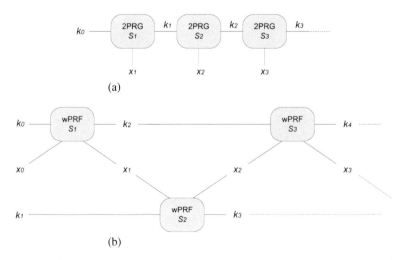

Fig. 6 Two PRG constructions. (**a**) Forward secure PRG from [5], (**b**) Leakage resilient PRG from [34]

a requirement for leakage resilient PRGs or is a forward secure primitive such as [5, 33] sufficient to withstand side-channel attacks?"

5.2 Theoretical Security Analysis and Limitations

We start with an intuitive description of the two approaches.

5.2.1 Alternating Structure

As previously mentioned, a central difficulty when modeling a leakage function is the fact that polynomial time computations from a state S_i may potentially leak information about future states S_j with $j > i$. The solution proposed in [34] may be summarized as follows:

1. Double the key size and use an "alternating structure" such as in the lower part of Fig. 6 in which wPRF is a weak pseudorandom function (i.e., a PRF in which the inputs are not chosen by the adversary but at random).
2. Assume that when computing the odd states (i.e., $S_1, S_3, \ldots$), the even states (i.e., $S_2, S_4, \ldots$) are not manipulated by the device and therefore do not leak.

As a result, the sequence of keys $k_0, k_2, k_4, \ldots$ cannot be determined from the sequence $k_1, k_3, k_5, \ldots$ It implies that, e.g., when leaking λ bits about S_1, the states S_i for any $i > 1$ are not polynomial time computable from S_1 because k_1 is still safe. This prevents the future computation attack. The main limitation of this proposal is that it is not clear if the alternating structure is only motivated by the proof technique (i.e., the willing to be in the standard model) or if there is also a physical concern that makes it necessary. Its main advantage is the security analysis combining physical and classical security notions (whereas, e.g., [33] was dealing with both issues separately, in a more specific scenario).

5.2.2 Forward Secure PRGs

Alternatively, one can prevent side-channel attacks with a forward secure primitive without alternating structure. But if no additional restrictions are considered for the leakage function, proving the security against the future computation attack then requires the assumption that the 2PRGs in the construction of Fig. 6a behave as random oracles that the leakage function cannot query. Interestingly, this solution also allows to get rid of certain physical assumptions, as will be detailed in the next section.

The main security features of these two approaches are listed in Table 1. In summary, we have an efficient construction that requires strong black box assumptions on the one hand. And on the other hand, we have a less efficient construction proven in the standard model. It is worth to emphasize that the random oracle is only used to prove the leakage resilience of [5]. Yet, in the black box setting, this construction remains secure in the standard model. In other words, the random oracle essentially

Table 1 Summary of the security features for two leakage resilient PRGs

	Forward secure PRG [5]	Alternating structure [34]
Black box assumptions	Random oracles	Standard model
Leakage assumptions	Non-invertibility [10] + non-adaptivity	HILL pseudoentropy + "independent leakages"
Key size	n	$2n$
Output bits per round	n	n
Exhaustive key search	$\approx 2^n$	$\approx 2^n$
Construction	2PRG $\leadsto$ PRG	wPRF $\leadsto$ PRG
Tolerated leakage	$\lambda = \alpha n$ (with $\alpha \in [0, 1]$)	$\lambda \approx \log_2(n)$ (if wPRF secure against poly. size adversaries) $\lambda = \alpha n$ (if wPRF secure against exp. size adversaries)

prevents the "future computation attack," which seems a reasonable abstraction, while keeping the proofs simpler.

5.3 Proving Leakage Resilience with Random Oracles

In this section, we provide a proof of security for the PRG depicted in Fig. 6a. For this purpose, we first detail the three assumptions that we require for our proof to hold and discuss their pros and cons compared to the approach in [34].

a. *Random oracles.* In order to ensure the locality of the leakages, we model 2PRG : $k_i \rightarrow (k_{i+1}, x_{i+1})$ as a random oracle mapping values from $\{0, 1\}^n$ to values in $\{0, 1\}^{2n}$. Then, during the ith invocation of 2PRG , the attacker receives two leakages associated to this evaluation, $L_i^i(k_{i-1})$ and $L_i^o(k_i, x_i)$, together with the output x_i of the execution of 2PRG. Most importantly, we assume that the leakage functions cannot query 2PRG, which makes it impossible to use them to obtain information about previous or future invocations of 2PRG.

The fact that we model 2PRG as a random oracle that the leakage function cannot query corresponds to the experimental observation that any useful information obtained through the side channels of a circuit implementation is related to simple functions of the inputs and outputs of that circuit, but will not provide any sophisticated function that is not present in the circuit state, e.g., a future output of the PRG. This, in particular, appears to be realistic for circuits implementing block ciphers where the measured information can be interpreted as a simple function of the block cipher input and key during the first few rounds of the block cipher computation and/or as a simple function of the block cipher output and key during the last few rounds of the block cipher computation, but where any useful function of the block cipher input and output remains elusive (or would be the sign of a completely broken implementation, e.g., as in [36, 37]).

b. *Bounded leakage per iteration.* Formally, we require that the leakages given to the adversary preserve the secrecy of the PRG seed in the following sense: the probability that an adversary recovers the seed used as input or provided as output during two iterations of the PRG construction should be small. Considering two iterations is minimal here since half of the output of an iteration is the input of the next iteration, and there are therefore two leakages taken on each secret variable. This is formalized through the following definition.

Definition 7 Let (L^o, L^i) be a pair of functions, A^{2PRG} an algorithm representing the side-channel adversary with oracle access to 2PRG, n a fixed integer, and $Pr_{Guess}(n)$ the following probability: $Pr[A^{2PRG}(L^o(k_1, x_1), x_1, L^i(k_1)) = k_1 : k_0 \leftarrow \{0, 1\}^n; (k_1, x_1) := 2PRG(k_0)]$. The pair (L^o, L^i) is said to be ε-*seed preserving* for security parameter n and A^{2PRG} if $Pr_{Guess}(n) \leq \varepsilon$. A pair of functions (L^o, L^i) is said to be *seed preserving* if, for every PPT A^{2PRG}, there is a negligible function ε such that (L^o, L^i) is $\varepsilon(n)$-seed preserving for every security parameter n and A^{2PRG} running on input 1^n. A sequence of pairs of functions $(L^o_1, L^i_1), \ldots, (L^o_l, L^i_l)$ is said to be *uniformly seed preserving* if, for every PPT A^{2PRG}, there is a negligible function ε such that each pair of this sequence is $\varepsilon(n)$-seed preserving for every security parameter n and A^{2PRG} running on input 1^n.

Assuming that adversaries only receive outputs of seed-preserving functions is in fact similar to the assumptions made in the cryptography with auxiliary input setting [10]. We believe it captures physical leakages particularly well in the sense that we do not put any constraint in the form of the leakage: it can be a simple computation time, a huge sequence of power consumption measurements, a map of electromagnetic radiations, or anything else. Moreover, it might very well be the case that the full information about the 2PRG seeds is included in these leakages. We only require that the leakage functions cannot be inverted efficiently. Overall, this is a weaker assumption than requiring a high HILL pseudoentropy as in [12, 34]. But when to be quantified in practice, it also suffers from the gap described in Sect. 4. That is, we can hardly evaluate the performance of every possible adversary and need to rely on specific attacks.

Stronger versions of these notions of seed preservation would provide the extra-ability to the adversary to check whether a candidate key k_1 is the correct one. This can happen in different contexts in practice: it might be the case that half of $2PRG(k_1)$ is available as public output of a next round, enabling the adversary to perform some comparisons; it might also be the case that the adversary is able to re-initialize the circuit to a value of his choice and to compare the leakages observed in the targeted execution to the leakage occurring in an execution triggered after re-initialization. The security of the PRG and PRF construction, as claimed next in Theorems 1 and 2, could be rephrased in terms of this strengthened notion of seed preservation. The proofs would remain the same, except that the reductions would become tighter by a factor corresponding to the number of random oracle queries made by the adversary.

c. *Only the current iteration leaks.* We assume that the leakage of each state S_i only depends on its current inputs/outputs and is independent of the previous states. This is a reasonable restriction in regard to most experimental side-channel attacks. But in fact, even if history effects were observed in the leakages (e.g., if the leakage in state S_i was dependent on k_{i-2}), it would be possible to relax this assumption by simply generalizing the seed-preserving property to more than two PRG iterations. We then would require, for instance, that a triple of leakage functions (L_1, L_2, L_3) with appropriate inputs does not give any PPT adversary a non-negligible advantage in guessing any of the involved secrets.

Finally, it is worth emphasizing that we do not need the "only computation leaks" assumption that was first introduced in [31]. This is interesting since in advanced CMOS technologies (65 nm or smaller), the power consumption is not dominated by its dynamic part anymore and the so-called static leakages start to play a significant role. This means that leakages happen independently of the fact that computations are performed. In our setting, these non-computational leakages can simply be given to the adversary in Definition 7, as long as they provide a negligible advantage in guessing the seeds.

Similarly, we do not need the condition of independent leakages as in [34]. In this respect, and contrary to what is claimed in [35], this independence requirement is not a completely mild assumption and it may be contradicted by coupling effects in microelectronic devices. As an illustration, [2] suggests different leakage models that can be observed, among which linear and quadratic ones. If we denote by $S(i)$ the ith bit of a state S, it yields

$$L_{quad}(S) = \underbrace{\sum_i \alpha_i \times S(i)}_{L_{lin}(S)} + \sum_{i,j} \beta_{i,j} \times S(i) \times S(j) + \cdots$$

Clearly, if the quadratic term (that typically captures coupling effects) is non-negligible and S contains two consecutive states of the alternating structure, their leakages are not independent (i.e., depend on the combination of both parts). Note that even if partially unverified, none of these assumption has the potential to break the constructions in practice. But quantifying them empirically and reflecting non-computational or dependent leakages in the proofs would increase their relevance. Summarizing, the combination of an alternating structure and the assumption of independent leakages can be seen as the counterparts that one has to pay in order to get rid of the random oracles in the proofs of [34].

d. *Security of the forward secure PRG of Fig. 6a.* We define the security for the PRG construction of Fig. 6a through the experiment $\mathsf{Pred}_{A^{2PRG}, \overline{L}}(n)$ during which the adversary A^{2PRG} tries to predict something about the next round's output of PRG given the output and leakages from the past rounds, where the leakages are taken from a family of leakage functions $\overline{L} := \langle L_1^i, L_1^o, L_2^i, L_2^o, \ldots \rangle$.

Experiment $\mathsf{Pred}_{\mathsf{A}^{2\mathsf{PRG}},\mathsf{L}}(n)$:

1. During initialization, a key k_0 is selected uniformly at random in the set $\{0, 1\}^n$, and a counter i is set to 0.
2. On input 1^n, adversary $\mathsf{A}^{2\mathsf{PRG}}$ starts sending request queries. On each request query, the counter i is incremented, the pair (k_i, x_i) is computed as $2\mathsf{PRG}(k_{i-1})$, and the leakages $\mathsf{L}_i^i(k_{i-1})$ and $\mathsf{L}_i^o(k_i, x_i)$ are returned to $\mathsf{A}^{2\mathsf{PRG}}$, together with x_i.
3. When $\mathsf{A}^{2\mathsf{PRG}}$ outputs a test query, a bit $b \in \{0, 1\}$ and a value $r \in \{0, 1\}^n$ are chosen uniformly at random, and r is given to $\mathsf{A}^{2\mathsf{PRG}}$ if $b = 0$ or x_{i+1} is given otherwise, computed as $(k_{i+1}, x_{i+1}) := 2\mathsf{PRG}(k_i)$.
4. $\mathsf{A}^{2\mathsf{PRG}}$ outputs a bit b', and $\mathsf{Pred}_{\mathsf{A}^{2\mathsf{PRG}},\mathsf{L}}(n)$ is set to 1 iff $b = b'$.

We show that as long as the pairs of leakage functions that are evaluated on the same keys are uniformly seed preserving and can be evaluated efficiently, no efficient adversary can efficiently predict the output of the next round of the PRG from the outputs and leakages of the past rounds. That is

Theorem 1 *Let* $\mathsf{A}^{2\mathsf{PRG}}$ *be a PPT adversary playing the* $\mathsf{Pred}_{\mathsf{A}^{2\mathsf{PRG}},\mathsf{L}}(n)$ *experiment. Then, we have* $\Pr[\mathsf{Pred}_{\mathsf{A}^{2\mathsf{PRG}},\mathsf{L}}(n) = 1] = \frac{1}{2} + \mathsf{negl}(n)$, *provided that the family of pairs of leakage functions* $(\bot, \mathsf{L}_1^i)$, $(\mathsf{L}_1^o, \mathsf{L}_2^i)$, ... *is uniformly seed preserving and that all leakage functions can be evaluated in probabilistic polynomial time.*

In other words, the implementation of the PRG in Fig. 6a is physically indistinguishable (as in Definition 2). The proof is given in Sect. 10.2.

5.4 Practical Security Analysis

The previous section provided an overview of the theoretical security analysis for two PRGs. Given a "small enough" leakage, these two constructions are expected to be secure against side-channel attacks. In order to observe these results for real devices, it then remains to evaluate exactly how much information is leaked in practice. For this purpose, a first step is to instantiate the 2PRGs and wPRFs that are necessary to implement the leakage resilient PRGs. For illustration and because they are usual targets of practical works in the area of side-channel attacks, we used the block cipher-based constructions in Fig. 7, following the suggestion of Pietrzak [34] for his wPRF implementation.

As detailed in Sect. 4, a reasonable estimation of the leakage in one iteration of a PRG is given by the success rate of the "best available adversary" for which the most important parameter is the data complexity q. A consequence is that the practical security of a construction is mainly determined by the number of different traces that can be exploited to identify each secret key. In other words, the practical security of a construction can be related to the notion of q-limited adversary that has been formalized by Vaudenay in his decorrelation theory [48]. For example, in the context of block ciphers, it yields

Definition 8 An adversary against a block cipher $\mathsf{E}_k(x)$ is q-limited for a key k if he can only observe the encryption of q different plaintexts under this key.

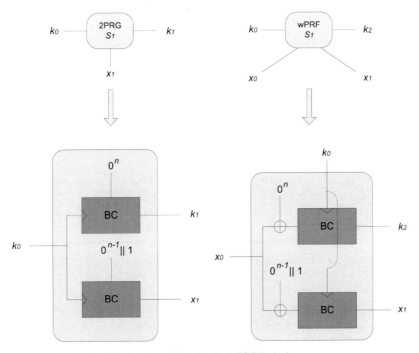

Fig. 7 Instantiation of a 2PRG and a wPRF with the AES Rijndael

Following this definition, it is clear that the 2PRG and wPRF of Fig. 7 are similar in terms of practical security. For both constructions, one iteration limits the side-channel adversaries to two queries. We can then refine the definition:

Definition 9 A block cipher-based construction is q-limiting for a side-channel adversary A if this construction limits the number of different encryptions performed under a single key that A can observe (i.e., the data complexity) to q.

Of course, having a q-limiting construction is not a sufficient condition to be secure against side-channel attacks. Overall, we need a combination of theoretical security (i.e., the physical indistinguishability discussed in Sects. 3, 5.2 and 5.3) and practical security. But the q-limit allows hardware designers to know how much leakage they need to face. If a construction is 2-limiting, their goal will be to limit the success rate of the best available adversary for $q = 2$ queries. As an illustration, Table 2 provides the estimated success rates for different attacks against implementations of the DES and AES Rijndael.

These results illustrate that small differences in data complexity can lead to largely different success rates. Recall that side-channel attacks can be performed in various contexts (e.g., allowing the profiling of a device or not) which explains this large variance. Hence, the table makes a case for evaluating the metrics of [42] in various experimental settings, e.g., for protected circuits. Initiatives such as the DPA Contest [11] could typically be used for this purpose.

Table 2 Approximated success rates for different attacks

Algorithm	Device	Attack	References	$q = 2$	$q = 10$	$q = 100$
AES	PIC 8-bit controller	Algebraic	[37]	≈ 1	≈ 1	≈ 1
AES	Atmel 8-bit controller	Hexavariate templates	[45]	$\approx 2^{-16}$	≈ 1	≈ 1
AES	Atmel 8-bit controller	Correlation	[45]	$\approx 2^{-128}$	$\approx 2^{-37}$	≈ 1
DES	64-bit ASIC	DoM test	[46]	$\approx 2^{-56}$	$\approx 2^{-56}$	$\approx 2^{-12}$

Note that the table already suggests that, for small devices (like 8-bit controllers), a reasonable λ-bit leakage can hardly be achieved if no additional (local) counter-measures are considered. For example and when applicable, the 1-limited algebraic side-channel attacks in [37] leak the complete AES key.

We also mention that it is important to distinguish the data complexity of a side-channel attack from the number of measurements performed by the adversary. Indeed, for a given data complexity q, the adversary may wish to repeat the measurement of his q different traces several (say n_r) times. For example, the left part of Fig. 8 illustrates the impact of measuring the same leakage trace l_1 three different times (giving rise to measurements l_1, l_1' and l_1''). Intuitively, the combination of these measurements can be used to remove the noise from the traces, e.g., in order to obtain a more precise information about an intermediate value Y in the target cryptographic computation. This information can then be translated into information about a subkey S. Here the data complexity is $q = 1$ and we have $n_r = 3$. By contrast, standard DPA attacks generally try to combine the leakage corresponding to different plaintexts. As illustrated in the right part of Fig. 8, each trace then brings a different information about the target intermediate value Y. And by translating these leakages into subkey information, one can have the intersection between the set of possible subkeys arbitrary small. Here, the data complexity is $q = 2$ and we have $n_r = 1$.

Following this observation, the expectation when using q-limiting constructions is that even if a polynomial time adversary can remove a significant part of the noise from his side-channel traces (e.g., the ones in Fig. 2), there remains some uncertainty

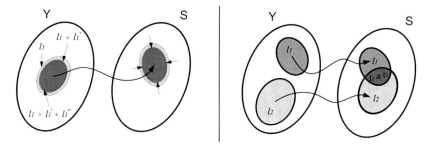

Fig. 8 Repeating a measurement vs. measuring a new trace

on the key because of the bounded data complexity. At least, one can hope that it is easier for hardware designers to guarantee this condition than to bound the success rate for a non-limiting construction.

We conclude this section with two remarks.

5.4.1 Remark 1

It is interesting to mention that, depending on the block cipher used in the 2PRG construction of Fig. 7, such an instantiation may introduce a gap with the assumptions of Sect. 5.3. Just observe that we consider the leakage on the output of a 2PRG $L^o(k_i, x_i)$ and the one on its input $L^i(k_{i-1})$ as independent. But if a block cipher with an invertible key scheduling algorithm (e.g., the AES Rijndael) is used in Fig. 7, the output leakage may potentially leak on the key that was used to produce this output. This observation is not expected to modify the practical security of the resulting PRG (a similar leakage actually occurs for the wPRF construction too), but suggests that carefully instantiating block cipher-based constructions may be important to prevent side-channel attacks. It also recalls that non-invertible key scheduling algorithms (e.g., as in the FOX block cipher [23]) are desirable in the context of leaking devices. Alternatively, strictly respecting the assumption of Sect. 5.3 may require some performance overheads, e.g., by replacing $2\mathsf{PRG}(x) := \big(\mathsf{BC}_x(0^n), \mathsf{BC}_x(1||0^{n-1})\big)$ by a slightly more expensive one, e.g., $2\mathsf{PRG}(x) := \big(\mathsf{BC}_{\mathsf{BC}_x(0)}(0^n), \mathsf{BC}_{\mathsf{BC}_x(0)}(1||0^{n-1})\big)$.

5.4.2 Remark 2

It is also worth noticing that by adding more block ciphers to the 2PRG example of Fig. 7, one can easily extend it toward a 3PRG, 4PRG, ... This leads to a simple tradeoff between security and performance. Namely, for a given q-limit, one can use a qPRG in Fig. 6a and consequently generate $\frac{n \cdot (q-1)}{q}$ pseudorandom bits per block cipher execution.

6 Initialization Issues

6.1 Breaking [34] with a Standard DPA

The claim in [34] is that the alternating structure in Fig. 6b can be used as a leakage resilient stream cipher. This implies the need of an initialization process with a public initialization vector. For this purpose, the authors suggest to pick up the keys k_0 and k_1 as well as the public x_0 at random. But this is in fact in contradiction with the practical requirements of a stream cipher. It is generally expected that one can re-initialize or re-synchronize a stream cipher without picking up new keys at random (e.g., see [15]). This is important, e.g., if the stream cipher has to be used in a challenge–response protocol. Unfortunately, using x_0 as an initialization vector without changing the keys k_0 and k_1 in the same time leads to straightforward DPA attacks that break the stream cipher of [34] in practice. Just observe that, because of the AES-based construction in Fig. 7, the adversary targeting the alternating structure for a given x_0 is 2-limited. Say now the adversary re-initializes the PRG with multiple random x_0 values, e.g., say he does it t times with the same key (k_0, k_1). Then, the first iteration of the PRG is not 2-limited anymore but $2 \cdot t$-limited, where t can be arbitrarily large. As a consequence, the construction is no more leakage resilient. A standard DPA attack such as in the right part of Fig. 8 can be applied.

Summarizing, the alternating structure from [34] is limited to a fixed x_0 and its proof holds – but then, the resulting stream cipher cannot be efficiently re-initialized, re-synchronized, used in a challenge–response protocol, ... Or it allows the adversary to observe multiple x_0 values – but then it is insecure. In other words, there is a component missing in this construction.

6.2 Secure Initialization Process

In order to avoid the previous issue, [33] proposed a secure initialization process of which different variants can be designed. Essentially, the idea is to bound the increase of the q-limit during initialization at the cost of some performance overheads. This can be simply achieved by adding multiplexors in the implementation and using the selection bit(s) to insert the random IV progressively in the PRG iterations. In its original form, pictured in Fig. 9, this process was made of two phases (we refer to the original publication for the details):

1. In a first (initialization) phase, n iterations of the PRG are executed with inputs z_i's that are obtained as follows: $z_i = C_0$ if $IV(i) = 0$, $z_i = C_1$ if $IV(i) = 1$, where $IV(i)$ is the ith bit of the initialization vector and C_0, C_1 are two constants. The n first y_i values are not transmitted as output.
2. In a second (generation) phase, the PRG goes on iterating with a fixed input C_0 and now transmits the pseudorandom blocks y_i as outputs.

In this basic proposal and if a n-bit IV is considered, the secure initialization requires n iterations of the PRG before pseudorandom blocks can be produced. The advantage of this constructions is that it is still 2-limiting. Trading performances for security is again possible by adapting the q-limit and using larger multiplexors (i.e., incorporating more IV bits in each iteration).

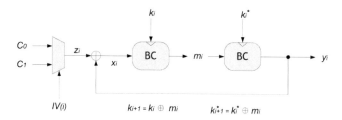

Fig. 9 Secure initialization process from ASIACCS 2008

Applying this idea to the PRGs of [5, 34] can then be done in different ways. The simplest one is to use the IV bits to select which of the 2PRG and wPRF outputs is used as a key in the next round (i.e., if $IV(i) = 0$, use the lower outputs in Fig. 7, if $IV(i) = 1$, use the upper one) – again without generating any pseudorandom block during initialization. Adding a multiplexor as in Fig. 9 and XORing its output with the plaintexts (or keys) in the PRG constructions of Fig. 6 is another solution that also allows various tradeoffs (but may requires a more careful analysis). The next section details yet another simple solution that we only apply to the forward secure PRG of [5] for convenience.

6.3 A More Elegant (and Standard) Construction

In fact, the initialization of [33] can be directly related to the construction for pseudorandom functions introduced by Goldreich, Goldwasser, and Micali in 1986 [18]. This leads to the simple process represented in Fig. 10. In its basic form (in the left part of the figure), this construction can be viewed as a binary tree of depth n. The value of the root equals k_0. Then, depending on the IV bits, the left or right outputs of the 2PRG are selected. After n iterations of the 2PRG, we obtain a key k'_0 that has been initialized with a public IV and can be used as a seed in PRG of Fig. 6a. Again, it is possible to reduce the depth of the tree (i.e., to increase performances)

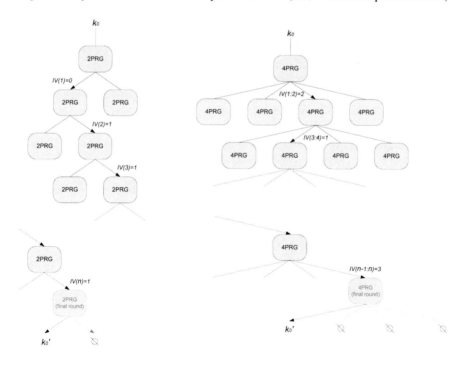

Fig. 10 Leakage resilient PRF. *Left*: n-iteration, *Right*: $\frac{n}{2}$-iteration

by increasing the q-limit (e.g., the right part of Fig. 10 uses 4PRGs with $q = 4$ and $n/2$ iterations), leading to the tradeoff:

- Number of block cipher executions for initialization: n/β,
- q-limit: 2^β, where β is an integer $\in [1, n]$.

The security of such an initialization process will be discussed in Sect. 7.

6.3.1 Remark 3

Combining the PRG in the upper part of Fig. 6 with the initialization process of Fig. 10 leads to an overall tradeoff between security and performances. As an illustration, two extreme solutions are given in section, Fig. 11. In the upper part, a completely trusted block cipher is exploited. Hence, a 2^n-limited construction is tolerated, with an initialization process in one block cipher execution and producing n pseudorandom bits per block cipher execution. In the lower part, the block cipher has significant leakages. Hence, a 2-limited construction is used, with an initialization process in n block cipher executions and producing $n/2$ pseudorandom bits per block cipher execution.

6.4 Remark on the Impossibility of a Secure Initialization Process with an Adaptive Selection of the Leakages

It is interesting to highlight that a secure and efficient initialization process (i.e., where re-synchronization does not imply a new key exchange) is incompatible with a fully adaptive selection of the leakage function for stateless devices (i.e., devices that do not save any part of their state between two reinitializations). Even the previous proposals can be broken in this context. Indeed, if a new leakage function can be chosen anytime the PRG is initialized with the same input, then a "future computation attack" exploiting only the first iteration of the initialization process can be mounted. This can be intuitively related to Fig. 8. In fact, the goal of a secure initialization process is to keep the q-limit as small as possible. Say for illustration that we have a 1-limiting process. Then, the expectation is that repeating the measurement of the same trace l_1 can only be used to remove noise, as in the left part of the figure. Good implementations should be such that this is not sufficient to leak the whole subkey S. But say now the adversary can adaptively choose his leakage function anytime he measures the trace corresponding to the same input plaintext x_1. Then, DPA attacks can again be mounted as in the right part of the figure. Quite problematically, such an attack would imply that any leaking implementation can be broken in linear number of measurements, making it impossible to build a secure device.

One solution to prevent the previous issue could be to limit the adaptivity of the leakage function to different inputs. That is, one could require that the same inputs to the target device should always give rise to the same leakage function. But from an operational point of view, the adaptivity of the leakage function relates to the

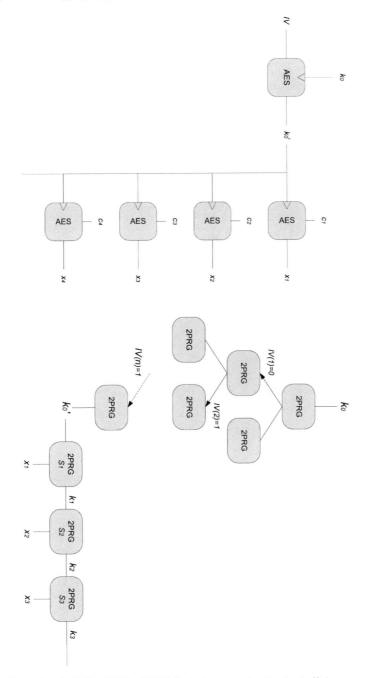

Fig. 11 Securely re-initialized PRGs. *UP*: 2^n-limited construction (i.e., best efficiency, worst security); *Down*: 2-limited construction (best security, worst efficiency)

possibility to change the measurement conditions during a side-channel attack (e.g., the antenna's position in an electromagnetic analysis). Hence, whatever modification of the setup that an adversary can do when changing the inputs can also be done when these inputs are kept constant. Therefore, if we assume that the leakage function cannot be chosen adaptively by the adversary when inputs are kept constant (which is mandatory in order to avoid trivial attacks as just described), there is no reason to allow it for variable inputs. We conclude that the leakage function should not be considered as adaptively selected by the adversary. In fact, a better solution to reflect the possible adaptivity of the measurement conditions is to include this adaptivity in the adversary's abilities and to quantify it directly in the λ-bit leakage bound.

7 Generalization to PRFs

In addition to its interesting features for initializing a PRG, the construction of Fig. 10 can also be used as a PRF, if the IV is replaced by an input x. Proving the security of these constructions can be done using essentially the same technique as in Sects. 5.3 and 10.2. Namely, each IV determines one trail through the tree of Fig. 10. And each of these trails is similar to one iteration of a PRG. As in Sect. 3, we need to extend the black box security definition of a PRF (against adaptive queries) from [18] to physical indistinguishability in presence of all leakages except for the challenge (here meaning that the output to be distinguished from random does not come with a leakage). That is

Definition 10 Let $F : \mathcal{X} \times \mathcal{K} \to \mathcal{X}$ be a pseudorandom function and $P = (F, L)$ be the physical implementations corresponding to the combination of F with a leakage function L. Let finally $A = (A_1, A_2)$ be a pair of PPT algorithms where A_1 takes a set of plaintexts S and some information $\mathcal{I}$ as input and outputs a plaintext, while A_2 takes a plaintext, a ciphertext, and some information $\mathcal{I}$ as input and outputs a bit. We consider the following experiments:

$\mathrm{Exp}_{P,A}^{\mathrm{prf-ind-0}}$

$S = \varnothing;$
$\mathcal{I} = \varnothing;$
$k \xleftarrow{R} \{0, 1\}^n;$
for $i = 1 : q$
    $x_i \leftarrow A_1(S, \mathcal{I});$
    $S = S \cup \{x_i\};$
    $\mathcal{I} = \mathcal{I} \cup \{x_i, F_k(x_i) \rightsquigarrow l_i\};$
end;
$x_{q+1} \notin S \leftarrow A_1(S, \mathcal{I});$

$-$

$b \leftarrow A_2(x_{q+1}, F_k(x_{q+1}), \mathcal{I});$

$\mathrm{Exp}_{P,A}^{\mathrm{prf-ind-1}}$

$S = \varnothing;$
$\mathcal{I} = \varnothing;$
$k \xleftarrow{R} \{0, 1\}^n;$
for $i = 1 : q$
    $x_i \leftarrow A_1(S, \mathcal{I});$
    $S = S \cup \{x_i\};$
    $\mathcal{I} = \mathcal{I} \cup \{x_i, F_k(x_i) \rightsquigarrow l_i\};$
end;
$x_{q+1} \notin S \leftarrow A_1(S, \mathcal{I});$
$y \xleftarrow{R} \{0, 1\}^n;$
$b \leftarrow A_2(x_{q+1}, y, \mathcal{I});$

The ind-advantage of A against P is defined as:

$$\text{Adv}_{\mathsf{P},\mathsf{A}}^{\text{prf}-\text{ind}} = |\Pr[\text{Exp}_{\mathsf{P},\mathsf{A}}^{\text{prf}-\text{ind}-0} = 1] - \Pr[\text{Exp}_{\mathsf{P},\mathsf{A}}^{\text{prf}-\text{ind}-1} = 1]|$$

The implementation of a PRF is physically indistinguishable if the ind-advantage of any polynomial time adversary against this implementation is negligible.

In other words, the leakage is provided for the q first queries but not for the $q + 1$th one for which the indistinguishability game is played.[2] Also, the plaintexts can be selected adaptively by the adversary, but not the leakage function.

We show in this section that if the leakage functions corresponding to the different rounds of the PRF in Fig. 10 are seed preserving, then the implementation of the PRF is physically indistinguishable (satisfying Definition 10). But we need a slight modification of the seed-preserving notion for this purpose. Just observe that Definition 7 has no guarantee for preserving the right-hand output of 2PRG. For example, a leakage function such that $L^o(k_1, x_1) = x_1$ satisfies this definition but trivially leaks all PRF outputs rooted from x_1 to adversary A^{2PRG}. Hence, we introduce symmetric seed-preserving leakage functions in Definition 11, and then state the main result for our PRF construction in Theorem 2.

Definition 11 Let (L^o, L^i) be a pair of functions, A^{2PRG} an algorithms representing the side-channel adversary with oracle access to 2PRG, n a fixed integer, and $\Pr_{\text{Guess}}(n, b)$ the following probability: $\Pr[\mathsf{A}^{\text{2PRG}}(L^o(k_0, k_1), L^i(k_b), k_{\bar{b}}) = k_b :$ $k \leftarrow \{0, 1\}^n; (k_0, k_1) := \text{2PRG}(k)]$. The pair (L^o, L^i) is said to be ε-symmetric seed preserving for parameter n and A^{2PRG} if $\Pr_{\text{Guess}}(n, b) \leq \varepsilon$ for any $b \in \{0, 1\}$. A pair of functions (L^o, L^i) is said to be *symmetric seed preserving* if, for every PPT A^{2PRG}, there is a negligible function $\varepsilon(n)$ such that (L^o, L^i) is $\varepsilon(n)$-symmetric seed preserving for every security parameter n and A^{2PRG} running on input 1^n. A sequence of pairs of functions $(L_1^o, L_1^i), \ldots, (L_l^o, L_l^i)$ is said to be *uniformly symmetric seed preserving* if, for every PPT A^{2PRG}, there is a negligible function $\varepsilon(n)$ such that each pair of this sequence is $\varepsilon(n)$-symmetric seed preserving for every security parameter n and A^{2PRG} running on input 1^n.

In order to state our theorem, let us assume $\mathcal{X} = \mathcal{K} = \{0, 1\}^n$ and denote the input and key of the PRF in Fig. 10 as $x = b_1 \cdots b_n \in \{0, 1\}^n$ and $k \in \{0, 1\}^n$, respectively. Let us also define the following notations:

$$G_0(k) \| G_1(k) \overset{\text{def}}{=} \text{2PRG}(k), \text{ with } |G_0(k)| = |G_1(k)|,$$
$$k_{b_1 \cdots b_i} \overset{\text{def}}{=} G_{b_i}(\cdots G_{b_2}(G_{b_1}(k)) \cdots) \text{ for } 1 \leq i \leq n - 1,$$
$$k_{b_1 \cdots b_n} \overset{\text{def}}{=} G_0(G_{b_n}(k_{b_1 \cdots b_{n-1}})),$$

where subscripts $b_1 \cdots b_i$ identify the invocation path from root node k with one more invocation of G_0 appended to the last level, and thus the PRF output $F_k(x)$

---

[2] In fact, for the PRF construction of Fig. 10, we can prove a slightly stronger result. Namely, we only need that the leakage of the last PRF round (i.e., the last 2PRG invocation) of the last query x_{q+1} is not provided to the adversary.

is simply given by k_x. We observe that this construction is essentially the GGM one [18], except that an extra round has been added in the end. This extra round has been introduced in order to make sure that the leakage occurring during the evaluation of F_k on one input does not trivially provide information on the output of F_k evaluated on an input that would only differ from the first one by the last bit. Note that, if 2PRG is implemented using two AES, as depicted in Fig. 7, then it is possible to evaluate only one of the two output halves, and this extra round is not needed anymore. As previously, we model the leakage function L with $n + 1$ pairs of functions, $\overline{L} := \langle (L_1^i, L_1^o), \ldots, (L_{n+1}^i, L_{n+1}^o) \rangle$, such that for any jth level PRG invocation $(k_{b_1 \cdots b_{j-1} 0}, k_{b_1 \cdots b_{j-1} 1}) := 2PRG(k_{b_1 \cdots b_{j-1}})$, the leakage is given by $(L_j^i(k_{b_1 \cdots b_j}), L_j^o(k_{b_1 \cdots b_j 0}, k_{b_1 \cdots b_j 1}))$. It directly yields:

Theorem 2 *The PRF construction given by* $F_k(x) = G_0(G_{b_n}(\cdots G_{b_1}(k) \cdots))$ *is physically indistinguishable provided that the family of pairs of leakage functions* $\{(\bot, L_1^i), (L_1^o, L_2^i), \ldots, (L_n^o, L_{n+1}^i)(L_{n+1}^o, \bot)\}_{n \in \mathbb{N}}$, *is uniformly symmetric seed preserving and can be evaluated in probabilistic polynomial time.*

The proof of Theorem 2 is given in Sect. 10.3.

8 Remark on the Impossibility of Proving the Leakage Resilience for the Forward Secure PRG of Fig. 6a in the Standard Model

Before to conclude this chapter, we finally note that restricting the leakage function only will not be sufficient to prove the leakage resilience for the forward secure PRG of Fig. 6a in the standard model. For example, say the 2PRG used in this construction can be written as follows:

$$G(x_1, x_2, \ldots, x_n) = (x_1, G^*(x_2, x_3, \ldots, x_n)) = (y_1, y_2, \ldots, y_{2n}),$$

with $G^* : \{0, 1\}^{n-1} \to \{0, 1\}^{2n-1}$ a secure PRG. Using this 2PRG in the upper scheme of Figure 6 gives rise to a secure PRG in the black box setting. But if the first n bits of $G(x_1, x_2, \ldots, x_n)$ are used as intermediate key and the second n bits as output, then a simple leakage function that only leaks x_1 will straightforwardly allow breaking the scheme. Importantly, this attack assumes a non-adaptive leakage function of which the computational complexity is very low (it just leaks one bit per iteration). This counterexample shows that proving the leakage resilience of a PRG such as [5] in the standard model also requires stronger black box assumptions than traditionally considered for 2PRGs.

9 Open Problems

This report implies two important scopes for further research. A first one is to better investigate side-channel resilient constructions of PRGs and PRFs (and their possible extension to PRPs). This requires to work on the minimum black box and physical assumptions that can be used to analyze the security of cryptographic devices in a formal setting. For example, important assumptions include the following:

1. The adaptivity of the leakage function. As discussed in Sect. 6.4, it is reasonable to remove this ability from the adversary's power and to reflect the possible adaptivity of the measurement setup in the λ-bit leakage bound.
2. "Only computations leak." As discussed in Sect. 5.3, this (and related) assumptions should ideally be avoided or reflected in the proofs quantitatively, in order to capture the behavior of advanced circuit technologies where static leakages and coupling effects are not negligible. Or, alternatively, one could investigate the exact type of dependencies that should be avoided (and translate them as requirement for hardware designers) to keep sound proofs.
3. The computational limits of the leakage function. In [31, 34], the leakage function is assumed to be a polynomial time computable function of its inputs. As discussed in Sect. 5.1, this incorporates attacks that exceed the power of actual adversaries. Hence, an interesting direction would be to further restrict this function. Promising candidates can be found in the early literature on side-channel attacks such as [2]. For example, considering linear functions of a device's internal configuration (or quadratic functions in order to capture possible coupling effects in the circuits) appears as a good starting point. At least, those type of functions should be captured by theoretical analysis, in view of their close connection to practice.

These physical assumptions then have to be combined with minimum black box requirements. We note that considering the black box security and the physical security with different assumptions may be an interesting alternative to demonstrate the leakage resilience while keeping proofs simple. For example, one can show the security of a construction in the standard model without leakage and then use a random oracle (that cannot be queried by the leakage function) to prove side-channel resilience, as in this report. Since the random oracle is then mainly used to capture the idea that "side-channel leakages do not leak about the future," it seems a reasonable abstraction in this context. Overall, this discussion shows that there are interesting tradeoffs to consider between the black box and physical requirements that one imposes to algorithms and implementations.

Second, we need to move from the abstract description of leakage resilient primitives toward their concrete implementations, in order to quantify the actual information leakages that they provide, in function of the adversary's abilities (i.e., data, time, and memory complexity). It implies extending Table 2 in this chapter and evaluating the security of more algorithms and devices against various attacks. This situation is in fact analogical to the classical cryptanalysis of block ciphers, where it is essential to determine the best known attacks against various ciphers. In a physical setting, the same question arises for every combination of algorithm and device – with the additional difficulty of exactly specifying the adversary's power, e.g., in terms of profiling and a priori knowledge of the underlying hardware. Concentrating the community's experimental efforts on a few implementations (e.g., by standardizing measurement boards [38]) would be very useful in this respect. Initiatives such as [11] could also be adapted for this purpose. We note that the large variability of the success rates in Table 2 suggests that keeping λ as small as possible in practical devices is certainly as challenging as properly exploiting small λ's in secure PRG (or other) constructions.

Eventually, physical security is a young topic and its proper formalization is still a scope for further research. Hence, it is important to question the validity of the models used to analyze the security of leaking devices first. Because of the physical origin of, e.g., side-channel attacks, it implies the need of experimental evaluations. And the implementation cost also has to be part of this analysis since overall (and in particular for small embedded devices), what matters to hardware designers is to obtain the best security at the lowest cost.

10 Further Details

10.1 Security Metric

We consider a side-channel key recovery adversary $\mathsf{A}_{\mathsf{E}_K,\mathsf{L}}$ with time complexity τ, memory complexity m and q queries to the target physical computer. His goal is to guess a key class s with non-negligible probability. For this purpose and for each candidate s^*, he compares the actual observation of a leaking device l_q with some key-dependent model for these leakages $\mathsf{M}(s^*, .)$. Let $\mathsf{T}(\mathsf{l}_q, \mathsf{M}(s^*, .))$ be the statistical test used in the comparison. We assume that the highest value of the statistic corresponds to the most likely key candidate. For each observation l_q, we store the result of the statistical test T in a vector $\mathbf{g}_q = \mathsf{T}(\mathsf{l}_q, \mathsf{M}(s^*, .))$ containing the key candidates sorted according to their likelihood: $\mathbf{g}_q := [g_1, g_2, \ldots, g_{|\mathcal{K}|}]$ (e.g., $|\mathcal{S}|=256$ for key bytes). Then, for any side-channel attack exploiting a leakage vector l_q and giving rise to a result $\mathbf{g}_q$, we define the success function of order o against a key byte s as: $\mathsf{S}_k^o(\mathbf{g}_q)=1$ if $k \in [g_1, \ldots, g_o]$, else $\mathsf{S}_k^o(\mathbf{g}_q)=0$. It leads to the oth-order success rate:

$$\mathrm{Succ}_{\mathsf{A}_{\mathsf{E}_K,\mathsf{L}}}^{sc-kr-o,S}(\tau, m, q) = \mathop{\mathbf{E}}_{k} \mathop{\mathbf{E}}_{\mathsf{l}_q} \mathsf{S}_k^o(\mathbf{g}_q). \tag{2}$$

Intuitively, a success rate of order 1 (*resp.* 2) relates to the probability that the correct key byte is sorted first (*resp.* among the two first ones) by the adversary.

10.2 Proof of Theorem 1

Proof (Theorem 1) Let $\mathsf{A}^{\mathsf{2PRG}}(1^n)$ be an adversary who wins the $\mathsf{Pred}_{\mathsf{A}^{\mathsf{2PRG}},\mathsf{\Gamma}}(n)$ game with probability $\frac{1}{2} + \eta(n)$, and let p be a polynomial such that $p(n)$ is an upper bound on the number of **request** queries made by $\mathsf{A}^{\mathsf{2PRG}}(1^n)$. Let Query_l (resp.

Query_a) be the event that $A^{\text{2PRG}}(1^n)$ makes a query to 2PRG on the last key k_i (resp. any key) computed by the challenger before the test query is made.

We distinguish between the cases where the Query_l event happens or not: $\Pr[\text{Pred}_{A^{\text{2PRG}},\overline{L}}(n) = 1] \leq \Pr[\text{Pred}_{A^{\text{2PRG}},\overline{L}}(n) = 1 \land \neg \text{Query}_l] + \Pr[\text{Query}_l]$.

The probability $\Pr[\text{Pred}_{A^{\text{2PRG}},\overline{L}}(n) = 1 \land \neg \text{Query}_l]$ is bounded by $\frac{1}{2} + \frac{p(n)^2}{2^n}$, which is the sum of the probability of a pure guess and an upper bound on the probability that a collision happens between PRG's last output and an output of a previous round.

We now show that $\Pr[\text{Query}_l]$ is negligible. To this purpose we build an adversary A'^{2PRG} as follows.:

Adversary A'^{2PRG}:

1. On input 1^n, start an instance of A^{2PRG} with input 1^n, and record all interactions between A^{2PRG} and the 2PRG oracle.
2. Pick $j \leftarrow [0, p(n)]$ and $r_0 \leftarrow \{0, 1\}^n$ uniformly at random, and set a counter i to 0.
3. Ask a challenger to pick $k_0 \in \{0, 1\}^n$ uniformly at random, to compute $(k_1, x_1) := \text{2PRG}(k_0)$ and to provide $(\mathsf{L}_j^o(k_1, x_1), x_1, \mathsf{L}_{j+1}^i(x_1))$.
4. On each request query from A^{2PRG}, proceed as follows: increment the counter i, select $(r_i, x_i) \leftarrow (\{0, 1\}^n)^2$ uniformly at random, and submit $\mathsf{L}_i^i(r_{i-1})$, y_i and $\mathsf{L}_i^o(r_i, y_i)$ to A^{2PRG}, unless $i = j$ in which case $\mathsf{L}_j^o(k_1, x_1)$, x_1 and $\mathsf{L}_{j+1}^i(x_1)$ are submitted instead.
5. On the test query from A^{2PRG}, pick $y_{i+1} \leftarrow \{0, 1\}^n$ uniformly at random and submit that value to A^{2PRG}.
6. Let $\{z_1, \ldots, z_q\}$ be the set of requests made by A^{2PRG} to 2PRG until it halts. Output an element z selected uniformly at random into that set.

The strategy of adversary A'^{2PRG} is based on the assumption that, in a normal run of the $\text{Pred}_{A^{\text{2PRG}},\overline{L}}(n)$ experiment, A^{2PRG} would make a query on (at least) one of the keys involved in the experiment. So, A'^{2PRG} makes a uniform guess on the index of the first key on which such a query is made; guessing the first queried key ensuring that key will only be correlated to one thing: the corresponding leakages (and not any previous call on 2PRG). This guess will be correct with probability $\frac{1}{p(n)+1}$. Then, A'^{2PRG} provides leakages to A^{2PRG} computed from random values of its own choice, except for the j index, for which the leakages and PRG output are replaced by those obtained from a challenger for the seed-preserving property. A'^{2PRG} also provides a random value y_{l+1} as final input to A^{2PRG}. If the guess on the index j is correct, all the inputs sent to A^{2PRG} are distributed exactly as in the $\text{Pred}_{A^{\text{2PRG}},\overline{L}}(n)$ experiment, as long as A^{2PRG} does not make a query on the value k_1 computed by the challenger. Therefore, when A^{2PRG} halts, A'^{2PRG} can select one of the inputs of the q queries made by A^{2PRG} and, if A^{2PRG} made a query on k_1, that guess will be correct with probability $\frac{1}{q}$. So, eventually, we have that $\Pr[z = k_1 | \text{Query}_a] = \frac{1}{q(p(n)+1)}$.

Now, we observe that $\Pr[z = k_1 | \text{Query}_a] \leq \frac{\Pr[z=k_1]}{\Pr[\text{Query}_a]}$, and that $\Pr[\text{Query}_l] \leq \Pr[\text{Query}_a]$, which implies that $\Pr[\text{Query}_l] \leq q(p(n) + 1)\Pr[z = k_1]$.

Eventually, we observe that A'^{2PRG} runs in PPT: A^{2PRG} runs in PPT, and the leakage functions can be evaluated in PPT too. Therefore, since the leakage function family $\overline{L}$ is uniformly seed preserving, there is a negligible function ε such that $\Pr[x = k_1] \leq \varepsilon(n)$. As a result, $\Pr[\mathsf{Query}_l] \leq q(p(n)+1)\varepsilon(n)$, which is negligible.

So, we have that $\Pr[\mathsf{Pred}_{A^{2PRG},\overline{L}}(n) = 1] \leq \frac{1}{2} + \frac{p(n)^2}{2^n} + q(p(n)+1)\varepsilon(n)$, as desired.

10.3 Proof of Theorem 2

Proof (Theorem 2) In the context of this construction, $\mathsf{Exp}_{P,A}^{prf-ind-0}$ (as in Definition 10) is the output of A^{2PRG} with the first q adaptive queries to $(F, \overline{L})$ and the $(q + 1)^{th}$ query to F alone, which we visually write as $A^{2PRG,\langle (F,\overline{L})_{[1,q]},(F,\perp)_{[q+1]}\rangle}$ and analogously denote $\mathsf{Exp}_{P,A}^{prf-ind-1}$ as $A^{2PRG,\langle (F,\overline{L})_{[1,q]},(R,\perp)_{[q+1]}\rangle}$, where random function R is constructed using the same tree structure except that all tree nodes are fresh randomness (rather than being generating by invoking 2PRG on their parent node as in the F), the output of R on input x is the leaf node reached by taking path x from the root. Then, by triangle inequality we have

$$|\Pr[\mathsf{Exp}_{P,A}^{prf-ind-0} = 1] - \Pr[\mathsf{Exp}_{P,A}^{prf-ind-1} = 1]|$$

$$\leq |\Pr[A^{2PRG,\langle (F,\overline{L})_{[1,q]},(F,\perp)_{[q+1]}\rangle} = 1] - \Pr[A^{2PRG,\langle (R,\overline{L})_{[1,q]},(R,\perp)_{[q+1]}\rangle} = 1]|$$

$$+ |\Pr[A^{2PRG,\langle (R,\overline{L})_{[1,q]},(R,\perp)_{[q+1]}\rangle} = 1] - \Pr[A^{2PRG,\langle (F,\overline{L})_{[1,q]},(R,\perp)_{[q+1]}\rangle} = 1]|$$

$$\leq |\Pr[A^{2PRG,(F,\overline{L})_{[1,q+1]}} = 1] - \Pr[A^{2PRG,(R,\overline{L})_{[1,q+1]}} = 1]|$$

$$+ |\Pr[A^{2PRG,(R,\overline{L})_{[1,q]}} = 1] - \Pr[A^{2PRG,(F,\overline{L})_{[1,q]}} = 1]|.$$

Therefore, we reduce the problem to showing the oracle indistinguishability between $(F, \overline{L})$ and $(R, \overline{L})$.

The main idea of the rest proof is that the ability to distinguish $(F, \overline{L})$ and $(R, \overline{L})$ using p queries to 2PRG and q queries to the above oracle pair (by a hybrid argument and efficient simulation of the GGM tree [18]) implies an efficient algorithm (with only slightly more complexity than the distinguisher) to invert one of $2q$ independent instances of symmetric seed-preserving functions with probability of $\varepsilon(n)/(2pn)$, or equivalently, solve one such instance with probability at least $\varepsilon(n)/(4pqn)$, and thus a contradiction to the symmetric seed-preserving property in Definition 11.

Following Sect. 7, we use $k_{b_1\cdots b_j}$ to denote the string on the jth-level tree node by taking invocation path $b_1, \cdots, b_j$ from the 0th level (root) node k. We consider hybrids $(H_0, L_0), \cdots, (H_{n+1}, L_{n+1})$, where each (H_j, L_j) is constructed using the tree structure with all nodes of up to level j are randomly chosen, and the rest nodes (of higher levels) obtained by invoking 2PRG on their parent node, and let $H_j(x)$ be the leaf node reached by taking path x from the root, and define L_j by invoking $\overline{L}$

on the corresponding nodes. It is not hard to see that (F, L_F) is identical to (H_0, L_0) and that (R, L_R) is identical to (H_{n+1}, L_{n+1}).

Suppose to the contrary that there exists A^{2PRG} that distinguishes (F, L_F) and (R, L_R) with advantage $\varepsilon(n)$ by making p queries to 2PRG and q queries to $(F, L_F)/(R, L_R)$, then A^{2PRG} can distinguish at least one neighboring hybrids (H_j, L_j) and (H_{j+1}, L_{j+1}) with advantage at least $\varepsilon(n)/(n+1)$ by the same number of queries. We stress that (H_j, L_j) and (H_{j+1}, L_{j+1}) are identically distributed to anyone without access to 2PRG, or otherwise said, A^{2PRG} must obtain the $\varepsilon(n)/(n+1)$ advantage by recovering of one of the jth level on-path (with respect to the q queries) nodes of (H_j, L_j) and then run 2PRG on it iteratively to verify whether the final output corresponds to the output of H_j. Note that the ability to recover nodes on any other level does not gives A^{2PRG} any advantage in distinguishing (H_j, L_j) and (H_{j+1}, L_{j+1}), which can be seen by a comparison between the two. We show in the following an inverting algorithm that uses the above distinguisher to solve one of $2q$ independent instances of pair of functions (L^o_{j-1}, L^i_j) with probability at least $\varepsilon(n)/(2p(n+1))$.

The inverting algorithm Inv^{2PRG} that runs A^{2PRG} as subroutine, simulates (H_j, L_j), and inverts $2q$ independent instances of (L^o_{j-1}, L^i_j) works as follows. We recall (see Definition 11) that each problem instance is in the format of $(L^o_{j-1}(k_0, k_1), L^i_j(k_b), k_{\bar{b}})$, and the challenge is to recover k_b.

Adversary Inv^{2PRG}:

1. Inv^{2PRG} takes as input 1^n, q independent instances of (L^o_{j-1}, L^i_j) with $b = 0$, and another q independent instances with $b = 1$. On startup, Inv^{2PRG} starts an instance of A^{2PRG} with input 1^n.

2. For each query $b_1, \cdots, b_n$ from A^{2PRG} to (H_j, L_j), Inv^{2PRG} simulates (H_j, L_j) as follows: for the first query (otherwise skip the common prefix that matches any of the previous ones), sample and record random strings k', $(k'_{b_1}, k'_{\bar{b_1}})$, $(k'_{b_1 b_2}, k'_{b_1 \bar{b_2}})$, $\cdots$, $(k'_{b_1 \cdots b_{j-2} b_{j-1}}, k'_{b_1 \cdots b_{j-2} \bar{b_{j-1}}})$ of level up to $j - 1$, and compute corresponding leakages using the leakage functions from $\bar{L}$. At the jth level, if $b_1, \cdots, b_{j-1}$ does not match the prefix of previous queries, toss a random bit $a_t \xleftarrow{R} \{0, 1\}$ $(1 \le t \le q)$ and install a new instance of (L^o_{j-1}, L^i_j) with $b = b_j \oplus a_t$ on it, i.e., $(L^o_{j-1}(k_0, k_1), L^i_j(k_b), k_{\bar{b}})$ as the output-layer leakage of level j, input-layer leakage on input $k'_{b_1 \cdots b_{j-1} b}$, and $k'_{b_1 \cdots b_{j-1} \bar{b}}$, respectively. In case that $a_t = 1$ (i.e., $b_j = b$), the strings and leakages on level j and above can be simulated by invoking 2PRG (on $k_{\bar{b}}$ and onward) and computing the leakage functions, and in the other case that $a_t = 0$, the simulation is identical except that $k'_{b_1 \cdots b_j 0}$ and $k'_{b_1 \cdots b_j 1}$ are sampled randomly instead of being obtained by invoking 2PRG on $k_{\bar{b}}$.

3. Inv^{2PRG} records all A^{2PRG}'s queries to 2PRG. When A^{2PRG} halts, Inv^{2PRG} randomly selects one of the recorded p queries and produces it as output, and then halts.

As discussed, A^{2PRG} can invert one (say the tth) of the $2q$ instances with chance at least $\varepsilon(n)/(n+1)$ on condition that it is selected from the p candidates (with probability $1/p$) and that $a_t = 0$ (with probability $1/2$), and thus the overall probability is $\varepsilon(n)/(2p(n+1))$, as desired.

References

1. A. Akavia, S. Goldwasser, V. Vaikuntanathan, in *Simultaneous Hardcore Bits and Cryptography Against Memory Attacks*. Proceedings of TCC 2009, San Francisco, CA, USA. Lecture Notes in Computer Science, vol. 5444 (Springer, Berlin, Heidelberg, Mar 2009), pp. 474–495
2. M.L. Akkar, R. Bévan, P. Dischamp, D. Moyart, in *Power Analysis, What Is Now Possible....* Proceedings of ASIACRYPT 2001, Kyoto, Japan. Lecture Notes in Computer Science, vol. 1976 (Springer, Berlin, Heidelberg, New York, Dec 2001), pp. 489–502
3. R. Anderson, M. Kuhn, in *Tamper Resistance - A Cautionary Note*, USENIX Workshop on Electronic Commerce, Oakland, CA, USA, Nov 1996, pp 1–11
4. M. Bellare, A. Desai, E. Jokipii, P. Rogaway, in *A Concrete Security Treatment of Symmetric Encryption*. Proceedings of FOCS 1997, Miami, FL, USA, Oct 1997, pp. 394–403.
5. M. Bellare, B. Yee, in *Forward-Security in Private-Key Cryptography*. Proceedings of CT-RSA 03, San Francisco, CA, USA. Lecture Notes in Computer Science, vol. 2612 (Springer, Heidelberg, Apr 2003), pp. 1–18
6. M. Blum, S. Micali, How to generate cryptographically strong sequences of pseudo-random bits. SIAM J. Comput. **13**(4), 850–863 (1984)
7. C. Cachin, *Entropy Measures and Unconditional Security in Cryptography*. Ph.D. thesis, ETH Dissertation, num 12187, 1997
8. S. Chari, C.S. Jutla, J.R. Rao, P. Rohatgi, in *Towards Sound Approaches to Counteract Power-Analysis Attacks*. Proceedings of CRYPTO 1999, Santa Barbara, CA, USA. Lecture Notes in Computer Science, vol. 1666 (Springer, Berlin, Heidelberg, New York, Aug 1999), pp. 398–412
9. J. Daemen, V. Rijmen, in *The Wide Trail Design Strategy*. Proceedings of Cryptography and Coding, 8th IMA International Conference, Cirencester, UK. Lecture Notes in Computer Science, vol. 2260 (Springer, Berlin, Dec 2001), pp. 222–238
10. Y. Dodis, Y. Tauman Kalai, S. Lovett, *On Cryptography with Auxiliary Input*, in the proceedings of STOC 2009, pp 621-630, Bethesda, Maryland, USA, June 2009.
11. Télécom ParisTech, *The DPA Contest*, http://www.dpacontest.org/
12. S. Dziembowski, K. Pietrzak, in *Leakage-Resilient Cryptography*. Proceedings of FOCS 2008, Washington, DC, USA, Oct 2008, pp. 293–302
13. T. Eisenbarth, T. Kasper, A. Moradi, C. Paar, M. Salmasizadeh, M.T. Manzuri Shalmani, in *On the Power of Power Analysis in the Real World: A Complete Break of the KeeLoqCode Hopping Scheme*. Proceedings of CRYPTO 2008, Santa Barbara, CA, USA. Lecture Notes in Computer Science, vol. 5157 (Springer, Berlin, Heidelberg, Aug 2008), pp. 203–220
14. ECRYPT Network of Excellence in Cryptology, *The Side-Channel Cryptanalysis Lounge*, http://www.crypto.ruhr-uni-bochum.de/ensclounge.html
15. ECRYPT Network of Excellence in Cryptology, *The eSTREAM Project*, http://www.ecrypt.eu.org/stream/, http://www.ecrypt.eu.org/stream/call/
16. S. Faust, L. Reyzin, E. Tromer, *Protecting Circuits from Computationally-Bounded Leakage*, Cryptology ePrint Archive, Report 2009/379
17. R. Gennaro, A. Lysyanskaya, T. Malkin, S. Micali, T. Rabin, in *Algorithmic Tamper-Proof (ATP) Security: Theoretical Foundations for Security against Hardware Tampering*. Proceedings of TCC 2004, Cambridge, MA, USA. Lecture Notes in Computer Science, vol. 2951 (Springer, Berlin, Feb 2004), pp. 258–277
18. O. Goldreich, S. Goldwasser, S. Micali, How to construct random functions. J. ACM **33**(4), 792–807 (1986)
19. L. Goubin, J. Patarin, in *DES and Differential Power Analysis*. Proceedings of CHES 1999, Worcester, MA, USA. Lecture Notes in Computer Science, vol. 1717 (Springer, Berlin, Aug 1999), pp. 158–172
20. J.A Halderman, S.D. Schoen, N. Heninger, W. Clarkson, J.A. Calandrino, A.J. Feldman, J. Appelbaum, E.W. Felten, in *Lest We Remember: Cold Boot Attacks on Encryption Keys*. Proceedings of the USENIX Security Symposium 2008, San Jose, CA, USA, Aug 2008, pp. 45–60

21. C. Hsiao, C. Lu, L. Reyzin, in *Conditional Computational Entropy, or Toward Separating Pseudoentropy from Compressibility*. Proceedings of EUROCRYPT 2007, Barcelona, Spain. Lecture Notes in Computer Science, vol. 4515 (Springer, Berlin, May 2007), pp. 169–186

22. Y. Ishai, A. Sahai, D. Wagner, in *Private Circuits: Securing Hardware against Probing Attacks*. Proceedings of Crypto 2003, Santa Barbara, CA, USA. Lecture Notes in Computer Science, vol. 2729 (Springer, Berlin, Aug 2003), pp. 463–481

23. P. Junod, S. Vaudenay, in *FOX: A New Family of Block Ciphers*. Proceedings of SAC 2004, Waterloo, Canada. Lecture Notes in Computer Science, vol. 3357, (Springer, Heidelberg, Aug 2004), pp. 114–129

24. J. Katz, in *Universally Composable Multi-Party Computation Using Tamper-Proof Hardware*. Proceedings of EUROCRYPT 2007, Barcelona, Spain. Lecture Notes in Computer Science, vol. 4515 (Springer, Berlin, Heidelberg, May 2007), pp. 115–128

25. L.R. Knudsen, in *Practically Secure Feistel Ciphers*. Proceedings FSE 1993, Cambridge, UK. Lecture Notes in Computer Science, vol. 809 (Springer, Berlin, Heidelberg, Dec 1993), pp. 211–221

26. P. Kocher, J. Jaffe, B. Jun, in *Differential Power Analysis*. Proceedings of Crypto 1999, Santa Barbara, CA, USA. Lecture Notes in Computer Science, vol. 1666 (Springer, Berlin, Heidelberg, New York, Aug 1999), pp. 398–412

27. P. Kocher, *Leak Resistant Cryptographic Indexed Key Update*, U.S. Patent 6539092, 2003

28. P. Kocher, in *Design and Validation Strategies for Obtaining Assurance in Countermeasures to Power Analysis and Related Attacks*. Proceedings of the NIST Physical Security Workshop, Honolulu, HI, USA, Sept 2005

29. B. Köpf, D. Basin, in *An Information Theoretic Model for Adaptive Side-Channel Attacks*. Proceedings of the ACM Conference on Computer and Communications Security 2007, Alexandria, VA, USA, Oct 2007, pp. 286–296

30. M. Luby, C. Rackoff, How to construct pseudorandom permutations from pseudorandom functions. SIAM J. Comput. **17**(2), 373–386 (1988)

31. S. Micali, L. Reyzin, in *Physically Observable Cryptography*. Proceedings of TCC 2004, Cambridge, MA, USA. Lecture Notes in Computer Science, vol. 2951 (Springer, Heidelberg, Feb 2004), pp. 278–296

32. T.S. Messerges, in *Using Second-Order Power Analysis to Attack DPA Resistant Software*. Proceedings of CHES 2000, Worcester, MA, USA. Lecture Notes in Computer Science, vol. 2523 (Springer, Berlin, Heidelberg, New York, Aug 2000), pp. 238–251

33. C. Petit, F.-X. Standaert, O. Pereira, T.G. Malkin, M. Yung, in *A Block Cipher Based PRNG Secure Against Side-Channel Key Recovery*. Proceedings of ASIACCS 2008, Tokyo, Japan, Mar 2008, pp. 56–65

34. K. Pietrzak, in *A Leakage-Resilient Mode of Operation*. Proceedings of Eurocrypt 2009, Cologne, Germany. Lecture Notes in Computer Science, vol. 5479 (Springer, Berlin, Heidelberg, Apr 2009), pp. 462–482

35. K. Pietrzak, in *Provable Security for Physical Cryptography*, invited talk. Proceedings of WEWORC 2009, Graz, Austria, July 2009

36. M. Renauld, F.-X. Standaert, *Algebraic Side-Channel Attacks*, Cryptology ePrint Archive: Report 2009/279

37. M. Renauld, F.-X. Standaert, N. Veyrat-Charvillon, in *Algebraic Side-Channel Attacks on the AES: Why Time Also Matters in DPA*. Proceedings of CHES 2009, Lausanne, Switzerland. Lecture Notes in Computer Science, vol. 5746 (Springer, Berlin, Sept 2009), pp. 97–111

38. RCIS (Research Center for Information Security), *SASEBO (Side-Channel Attack Standard Evaluation Boards)*, http://www.rcis.aist.go.jp/special/SASEBO/

39. W. Schindler, K. Lemke, C. Paar, in *A Stochastic Model for Differential Side-Channel Cryptanalysis*. Proceedings of CHES 2005, Edinburgh, Scotland. Lecture Notes in Computer Science, vol. 3659 (Springer, Berlin, Sept 2005), pp. 30–46

40. K. Schramm, T.J. Wollinger, C. Paar, in *A New Class of Collision Attacks and Its Application to DES*. Proceedings of FSE 2003, Lund, Sweden. Lecture Notes in Computer Science, vol. 2887 (Springer, Heidelberg, Feb 2003), pp. 206–222

41. N. Smart, D. Page, E. Oswald, Randomised representations. IET Inf. Secur. **2**(2), 19–27 (June 2008)
42. F.-X. Standaert, T.G. Malkin, M. Yung, in *A Unified Framework for the Analysis of Side-Channel Key Recovery Attacks*. Proceedings of Eurocrypt 2009, Cologne, Germany. Lecture Notes in Computer Science, vol. 5479 (Springer, Berlin, Heidelberg, Apr 2009), pp. 443–461, extended version available on the Cryptology ePrint Archive, Report 2006/139, http://eprint.iacr.org/2006/139
43. F.-X. Standaert, E. Peeters, C. Archambeau, J.-J. Quisquater, in *Towards Security Limits in Side-Channel Attacks*. Proceedings of CHES 2006, Yokohama, Japan, Oct 2006. Lecture Notes in Computer Science, vol. 4249 (Springer, Heidelberg, 2006), pp. 30–45, latest version available on the Cryptology ePrint Archive, Report 2007/222, http://eprint.iacr.org/2007/222.
44. F.-X. Standaert, C. Archambeau, in *Using Subspace-Based Template Attacks to Compare and Combine Power and Electromagnetic Information Leakages*. Proceedings of CHES 2008, Washington, DC, USA. Lecture Notes in Computer Science, vol. 5154 (Springer, Berlin, Heidelberg, Aug 2008), pp. 411–425
45. F.-X. Standaert, B. Gierlichs, I. Verbauwhede, in *Partition vs. Comparison Side-Channel Distinguishers: An Empirical Evaluation of Statistical Tests for Univariate Side-Channel Attacks*. Proceedings of ICISC 2008, Seoul, Korea. Lecture Notes in Computer Science, vol. 5461 (Springer, Berlin, Dec 2008), pp. 253–267
46. F.-X. Standaert, P. Bulens, G. de Meulenaer, N. Veyrat-Charvillon, *Improving the Rules of the DPA Contest*, Cryptology ePrint Archive, Report 2006/139, http://eprint.iacr.org/2006/139
47. K. Tiri, M. Akmal, I. Verbauwhede, *A Dynamic and Differential CMOS Logic with Signal Independent Power Consumption to Withstand Differential Power Analysis on Smart Cards*, ESSCIRC 2003, Estoril, Portugal, September 2003
48. S. Vaudenay, Decorrelation: a theory for block cipher security. J. Cryptol. **16**(4), 249–286 (2003)
49. A.C. Yao, in *Theory and Applications of Trapdoor Functions (Extended Abstract)*. Proceedings of FOCS 1982, Chicago, IL, USA, Nov 1982, pp. 80–91

Memory Leakage-Resilient Encryption Based on Physically Unclonable Functions

Frederik Armknecht, Roel Maes, Ahmad-Reza Sadeghi, Berk Sunar, and Pim Tuyls

1 Introduction

Modern cryptography provides a variety of tools and methodologies to analyze and to prove the security of cryptographic schemes such as in [6–9]. These proofs always start from a particular setting with a well-defined adversary model and security notion. The vast majority of these proofs assume a *black box* model: the attacker knows all details of the used algorithms and protocols but has no knowledge of or access to the secrets of the participants, nor can he observe any internal computations. The idealized model allows one to derive security guarantees and gain valuable insights.

However, as soon as this basic assumption fails most security guarantees are off and a new open field of study arises. In cryptographic implementations long-term secret keys are typically stored by configuring a non-volatile memory such as ROM, EEPROM, flash, anti-fuses, poly- or e-fuses into a particular state. Computations on these secrets are performed by driving electrical signals from one register to the next and transforming them using combinatorial circuits consisting of digital gates. Side-channel attacks pick up physically leaked key-dependent information from internal computations, e.g., by observing consumed power [29] or emitted radiation [1], making many straightforward algorithms and implementations insecure. It is clear that from an electronic hardware point of view, security is viewed differently, see, e.g., [32, 46, 50, 51].

Even when no computation is performed, stored secret bits may be leaked. For instance, in [45] it was shown that data can be recovered from flash memory even after a number of erasures. By decapsulating the chip and using scanning electron microscopes or transmission electron microscopes the states of anti-fuses and flash can be made visible. Similarly, a typical computer memory is not erased when its power is turned off giving rise to so-called cold-boot attacks [24]. More radical approaches such as opening up an integrated circuit and probing metal wires or

F. Armknecht (✉)
Horst Görtz Institute for IT Security, Ruhr-University Bochum, Bochum, Germany; Technische Universität Darmstadt, Darmstadt, Germany
e-mail: Frederik.Armknecht@trust.rub.de

A.-R. Sadeghi, D. Naccache (eds.), *Towards Hardware-Intrinsic Security,* Information Security and Cryptography, DOI 10.1007/978-3-642-14452-3_6,
© Springer-Verlag Berlin Heidelberg 2010

scanning non-volatile memories with advanced microscopes or lasers generally lead to a security breach of an algorithm, often immediately revealing an internally stored secret [45].

Given this observation, it becomes natural to investigate security models with the basic assumption: *memory leaks information on the secret key.* Consequently, a recently initiated line of research has focused on the use of new cryptographic primitives that are less vulnerable to leakage of key bits [2, 38]. These works establish security by adapting public-key algorithms to remain secure even after leaking a limited number of key bits. However, no security guarantees can be given when the leakage exceeds a certain threshold, e.g., when the whole non-volatile memory is compromised. Furthermore, they do not provide a solution for the traditional settings, e.g., for securing symmetric encryption schemes.

Here we explore an alternative approach: Instead of making another attempt to solve the problem in an algorithmic manner, we base our solution on a new physical primitive. So-called physically unclonable functions (PUFs) provide a new cryptographic primitive able to store secrets in a non-volatile but highly secure manner. When embedded into an integrated circuit, PUFs are able to use the deep submicron physical uniqueness of the device as a source of randomness [12, 15, 22, 49]. Since this randomness stems from the uncontrollable subtleties of the manufacturing process, rather than hard-wired bits, it is practically infeasible to externally measure these values during a physical attack. Moreover, any attempt to open up the PUF in order to observe its internals will with overwhelming probability alter these variables and change or even destroy the PUF [49].

In this chapter, we take advantage of the useful properties of PUFs to build an encryption scheme resilient against memory leakage adversaries as defined in [2]. We construct a block cipher that explicitly makes use of the algorithmic and physical properties of PUFs to protect against physical *and* algorithmic attacks at the same time. Other protection mechanisms against physical attacks require either additional algorithmic effort, e.g., [26, 36, 41, 47], on the schemes or separate (possibly expensive) hardware measures.

Our encryption scheme can especially be useful for applications such as secure storage of data on untrusted storage (e.g., hard disk) where (i) no storage of secrets for encryption/decryption is needed and keys are only re-generated when needed, (ii) copying the token is infeasible (unclonability), (iii) temporary unauthorized access to the token will reveal data to the adversary but not the key (i.e., forward security is preserved), or (iv) no non-volatile memory is available.

Contribution. Our contributions are as follows:

- *A new cryptographic primitive: PUF-PRF.* We place the PUFs at the core of a pseudorandom function (PRF) construction that meets well-defined properties. We provide a formal model for this new primitive that we refer to as PUF-PRF. PRFs [21] are fundamental primitives in cryptography and have many applications, e.g., see [20, 33, 34].
- *PUF-PRF-based provably secure stream and block ciphers.* One problem with our PUF-PRF construction is that it requires some additional helper data that

inevitably leaks some internal information. Hence, PUF-PRFs cannot serve as a direct replacement for PRFs. However, we present a provably secure stream and block ciphers based on PUF-PRFs that remain secure despite the information leakage. Furthermore, no secret key needs to be stored, protecting the scheme against memory leakage attacks. The tight integration of PUF-PRFs into the cryptographic construction improves the tamper resilience of the overall design. Any attempt at accessing the internals of the device will result in change of the PUF-PRF. Hence, no costly error detection networks or alternative anti-tampering technologies are needed. The unclonability and tamper-resilience properties of the underlying PUFs allow for elegant and cost-effective solutions to specific applications such as software protection or device encryption.

- *An improved and practical PUF-PRF construction.* Although the information leakage through helper data is unavoidable in the general case, the concrete case might allow for more efficient and secure constructions. We introduce SRAM-PRFs, based on so-called SRAM PUFs, which are similar to the general PUF-PRFs but where it can be shown that no information is leaked through the helper data if run in an appropriate mode of operation. Hence, SRAM-PRFs are for all practical purposes a physical realization of expanding PRFs.

Organization. This chapter is organized as follows. First, we give an overview of related work in Sect. 2. In Sect. 2, we define and justify the considered attacker model and provide in Sect. 4 the necessary preliminaries. In Sect. 5, we introduce a formal model for physically unclonable functions (PUFs). Based on this, we define in Sect. 6 a new class of cryptographic primitives, called PUF-PRFs. These can be seen as a physical realization of pseudorandom functions (PRFs) with the main difference that for a correct mode of operation, some additional data needs to be stored, called helper data, that leaks some information on the current inputs. Nonetheless, we explain in Sect. 7 how PUF-PRFs can be deployed for memory-leakage resilient encryption. We present both a stream cipher (Sect. 7.2) and a block cipher (Sect. 7.3) based on PUF-PRFs and prove their security despite the information leakage through the helper data. In Sect. 8, we explain for the concrete case of SRAM-PUFs an improved construction that shares the same benefits like general PUF-PRFs but where it can be argued that the helper data does not leak any information. Finally, in Sect. 9 we present the conclusions.

2 Related Work

In recent years numerous results in the field of physical attacks emerged showing that the classical black box model is overly optimistic, see, e.g., [3, 29, 30, 44, 45]. Due to a number of physical leakage channels, the adversary often learns (part of) a stored secret or is able to observe some intermediate results of the private computations. These observations give him a powerful advantage that often breaks the security of the entire scheme. To cope with this reality, a number of new theoretic adversary models were proposed, incorporating possible physical leakage of this kind. Ishai et al. [26] model an adversary which is able to probe, i.e., eavesdrop, a number of lines carrying intermediate results in a private circuit, and show how to create a secure primitive within this computational leakage model. Later, generalizations such as Physically Observable Cryptography proposed by Micali et al. [36] investigate the situation where only computation leaks information while assuming leak-proof secret storages. Recently, Pietrzak [14, 41] and Standaert et al. [47] put forward some new models and constructions taking physical side-channel leakage into account.

Complementary to the computation leakage attacks, another line of work explored memory leakage attacks: an adversary learns a fraction of a stored secret [2, 38]. In [2] Akavia et al. introduced a more realistic model that considers the security against a wide class of side-channel attacks when a function of the secret key bits is leaked. Akavia et al. further showed that Regev's lattice-based scheme [43] is resilient to key leakage. More recently Naor et al. [38] proposed a generic construction for a public-key encryption scheme that is resilient to key leakage. Although all these papers present strong results from a theoretical security point of view, they are often much too expensive to implement on an integrated circuit (IC), e.g., the size of private circuits in [26] blows up with $O(n^2)$ where n denotes the number of probings by the adversary. Moreover, almost all of these proposals make use of public-key crypto primitives, which introduce a significant overhead in systems where symmetric encryption is desired for improved efficiency.

Besides the information leakage attacks mentioned above, another important field of studies are tampering attacks. Numerous countermeasures have been discussed, e.g., use of a protective coating layer [42] or the application of error detection codes (EDCs) [17, 27]. Observe that limitations and benefits of tamper-proof hardware have likewise been theoretically investigated in a series of works [10, 18, 28, 37].

3 Memory Attacks

In this work we consider an extension of memory attacks as introduced by Akavia et al. [2]. In a nutshell, the attacker in [2] can extract a bounded number of bits of a stored secret. The model allows for covering a large variety of different memory attacks, e.g., cold-boot attacks described in [24]. However, this general model might not adequately capture certain concrete scenarios. For example, feature sizes on ICs have shrunk to nanometer levels and probing such fine metal wires is even for high-end IC manufacturers a difficult task. During a cryptographic computation a secret state is typically (temporarily) stored in volatile memory (e.g., in registers and flip-flops). However, in a typical IC, these structures are relatively small compared to the rest of the circuit, making them very hard to locate and scan properly. Thus, applying these attacks typically is usually significantly physically more involved for the case of embedded ICs than for the non-embedded PC setting where additional measures to access the memory exist, e.g., through software and networks.

On the other hand, storing long-term secrets, such as private keys, requires non-volatile memory, i.e., memory that sustains its state while the embedding device is powered off. Implementation details of such memories like ROM, EEPROM, flash, anti-fuses, poly- or e-fuses and recent results on physical attacks such as [45] indicate that physically attacking non-volatile memory is *much* easier than attacking register files or probing internal busses on recent ICs, making non-volatile memory effectively the weak link in many security implementations.

Motivated by these observations, we consider the following attacker model in this work:

Definition 1 (Non-volatile memory attacker) Let $\alpha : \mathbb{N} \to \mathbb{N}$ be a function with $\alpha(n) \leq n$ for all $n \in \mathbb{N}$, and let S be a secret stored in non-volatile memory. A α-*non-volatile memory attacker* can access an oracle $\mathcal{O}$ that takes as input adaptively chosen a polynomial-size circuits h_i and outputs $h_i(S)$ under the condition that the total number of bits that A gets as a result of oracle queries is at most $\alpha(|S|)$.

The attacker is called a *full non-volatile memory attacker* if $\alpha = $ id, that is the attacker can extract the whole content of the non-volatile memory.

Obviously, a cryptographic scheme can only provide protection against full non-volatile memory attackers if it does not use any long-term secrets that are stored within non-volatile memory. One obvious approach is to require a user password before each invocation. However, this reduces usability and is probably subject to password attacks. In this chapter, we use another approach and make use of a physical primitive called physically unclonable function (PUF), which will be modeled in the next section. PUFs allow to *intrinsically* store permanent secrets which are, according to current state of knowledge, not accessible to a non-volatile attacker.

4 Preliminaries

We start with some basic definitions. An oracle is a (usually probabilistic) algorithm that accepts inputs from an input domain and delivers values from an output domain. Let $\mathcal{O}_0$ and $\mathcal{O}_1$ denote two oracles with the same input and output domain. A *distinguisher* $\mathcal{D}$ for $\mathcal{O}_0$ and $\mathcal{O}_1$ is a (possibly probabilistic) algorithm that has either access to $\mathcal{O}_0$ or $\mathcal{O}_1$ and aims for distinguishing between these two cases. More precisely, $\mathcal{D}$ generates (possibly adaptively chosen) values from the input domain of the oracle, hands these to the oracle it has access to, and receives the outputs of the oracle. At the end, $\mathcal{D}$ outputs either 0 or 1. The *advantage* of $\mathcal{D}$ is defined by

$$\mathrm{Adv}(\mathcal{D}) \stackrel{\mathrm{def}}{=} |\Pr[1 \leftarrow \mathcal{D}^{\mathcal{O}_0}] - \Pr[1 \leftarrow \mathcal{D}^{\mathcal{O}_1}]|, \tag{1}$$

where $\mathcal{D}^{\mathcal{O}_b}$ denotes the case that $\mathcal{D}$ has access to $\mathcal{O}_b$ with $b \in \{0, 1\}$. Furthermore, we define the advantage of distinguishing between $\mathcal{O}_0$ and $\mathcal{O}_1$ as

$$\mathrm{Adv}(\mathcal{O}_0, \mathcal{O}_1) \stackrel{\mathrm{def}}{=} \max_{\mathcal{D}} \mathrm{Adv}(\mathcal{D}), \tag{2}$$

where the maximum is taken over all distinguishers for $\mathcal{O}_0$ and $\mathcal{O}_1$.

For a probability distribution $\mathbb{D}$, the expression $x \leftarrow \mathbb{D}$ denotes the event that x has been sampled according to $\mathbb{D}$. For a set S, $x \xleftarrow{*} S$ means that x has been sampled uniformly random from S. For $m \geq 1$, we denote by $\mathbb{U}_m$ the uniform distribution on $\{0, 1\}^m$. The *min-entropy* $H_\infty(\mathbb{D})$ of a distribution $\mathbb{D}$ is defined by $H_\infty(\mathbb{D}) \stackrel{\text{def}}{=} -\log_2(\max_x \Pr[x \leftarrow \mathbb{D}])$. Min-entropy can be viewed as the "worst-case" entropy [11] and specifies how many nearly uniform random bits can be extracted from it.

For a distribution $\mathbb{D}$ we denote by $\mathcal{O}^{\mathbb{D}}$ an oracle that on input x samples an output $y \leftarrow \mathbb{D}$. Here, it takes care of consistency: same inputs are always responded with the same output. At several occasions, we consider the difficulty of distinguishing between two distributions $\mathbb{D}_0$ and $\mathbb{D}_1$. The advantage of a distinguisher $\mathcal{D}$ is defined in this case as

$$\text{Adv}(\mathcal{D}) \stackrel{\text{def}}{=} |\Pr[1 \leftarrow \mathcal{D}^{\mathcal{O}_{\mathbb{D}_0}}] - \Pr[1 \leftarrow \mathcal{D}^{\mathcal{O}_{\mathbb{D}_1}}]|. \tag{3}$$

We will sometimes abbreviate $\mathcal{D}^{\mathcal{O}_{\mathbb{D}}}$ to $\mathcal{D}^{\mathbb{D}}$. The advantage to distinguish between $\mathbb{D}_0$ and $\mathbb{D}_1$ is defined as

$$\text{Adv}(\mathbb{D}_0, \mathbb{D}_1) \stackrel{\text{def}}{=} \text{Adv}(\mathcal{O}_{\mathbb{D}_0}, \mathcal{O}_{\mathbb{D}_0}). \tag{4}$$

5 Physically Unclonable Functions

In this section, we introduce a formal model for physically unclonable functions (PUFs). In a nutshell, PUFs are physical mechanisms that accept challenges and return responses, that is behaving like functions. The main properties of PUFs that are important in the context of cryptographic applications are the following:

Noise: The responses of PUFs are noisy in the sense that giving the same challenge twice can lead to different (but close) responses.
Non-uniform distribution: The distribution of the responses is usually non-uniform.
Independence: Two different PUFs show completely independent behavior, even if they stem from the same manufacturing process.
Unclonability: No efficient process is known that allows for physically cloning PUFs.
Tamper evidence: Physically tampering with a PUF, e.g., trying to read it out, will most likely destroy its physical structure and hence either make it unusable or turn it into a new PUF.

Remark 1 We want to emphasize that the properties above are of physical nature and hence are very hard to prove in the rigorous mathematical sense. However, they are based on experiments conducted worldwide and reflect the current *assumptions* and observations regarding PUFs, e.g., see [49].

We first provide a formal definition for noisy functions before we give a definition for PUFs.

Definition 2 (Noisy functions) For three positive integers $\ell, m, \delta \in \mathbb{N}$ with $0 \leq \delta \leq m$, a (ℓ, m, δ)-*noisy function* f^* is a probabilistic algorithm accepts inputs (challenges) $x \in \{0, 1\}^\ell$ and generates outputs (responses) $y \in \{0, 1\}^m$ such that the Hamming weight between two outputs to the same input is at most δ. In a similar manner, we define a (ℓ, m, δ)-*noisy family of functions* to be a set of (ℓ, m, δ)-noisy functions.

Definition 3 (Physically unclonable functions) A set $\mathcal{P}$ is a $(\ell, m, \delta; q_{\mathrm{puf}}, \varepsilon_{\mathrm{puf}})$-*family of PUFs* if it is a physical realization of a family of probabilistic algorithms that fulfills the following algorithmic and physical properties.

Algorithmic properties:

- *Noise:* $\mathcal{P}$ is a (ℓ, m, δ)-noisy family of functions with $\delta < \frac{m}{2}$
- *Non-uniform output and independence:* There exists a distribution $\mathbb{D}$ on $\{0, 1\}^m$ such that for any input $x \in \{0, 1\}^\ell$, the following two distributions on $(\{0, 1\}^m)^{q_{\mathrm{puf}}}$ can be distinguished with advantage at most $\varepsilon_{\mathrm{puf}}$.

  1. $(\Pi_1(x), \ldots, \Pi_{q_{\mathrm{puf}}}(x))$ for adaptively chosen $\Pi_i \in \mathcal{P}$.
  2. $(y_1, \ldots, y_{q_{\mathrm{puf}}})$ with $y_i \leftarrow \mathbb{D}$.

  In order to have a practically useful PUF, it should be that $q_{\mathrm{puf}} \approx |\mathcal{P}|$, $\varepsilon_{\mathrm{puf}}$ is negligible and $H_\infty(\mathbb{D}) > 0$.

Physical properties:

- *Unclonability*: No efficient technique is known to physically clone any member $\Pi \in \mathcal{P}$.
- *Tamper evidence*: For any PUF $\Pi \in \mathcal{P}$, any attempt to externally obtain its responses or parameters, e.g., by means of a physical attack, will significantly alter its functionality or destroy it.

A number of constructions for PUFs have been implemented and most of them have been experimentally verified to meet the properties of this theoretical definition. For more details we refer to the literature, e.g., [12, 22, 31, 48, 49]. One important observation we make is that a number of PUF implementations can be efficiently implemented on an integrated circuit, e.g., SRAM PUFs (cf. Sect. 8.2). Their challenge–response behavior can hence be easily integrated with a chip's digital functionality.

Remark 2 Due to their physical properties, PUFs became an interesting building block for protecting against full non-volatile memory attackers. The basic idea is to use a PUF for *implicitly* storing a secret: instead of putting a secret directly into non-volatile memory, it is derived from the PUF responses during run time. A possible instantiation is to store a challenge x and derive the secret from $\Pi(x)$.

As PUFs are, according to the current state of knowledge, neither (efficiently) readable nor clonable by an attacker, a non-volatile memory attacker is not able to evaluate $\Pi(x)$ by herself, hence protecting this value from her. Furthermore, due to the (assumed) independence of PUFs, making the challenge x public does not reveal anything about $\Pi(x)$ (except of what is known about the output distribution), even if the attacker has access to other PUFs from the same manufacturing process. Together with additional measures, e.g., fuzzy extractors (see Sect. 6), this allows to securely store a uniformly random secret which can be used with other cryptographic schemes, e.g., AES for encryption.

6 Pseudorandom Functions Based on PUFs

In the Sect. 5, we explained how to use PUFs for protecting any cryptographic scheme against full non-volatile memory attackers (see Remark 2). In the remainder of this chapter, we go one step further and explore how to use PUFs for protecting against algorithmic attackers *in addition*. For this purpose, we discuss how to use PUFs as a source of reproducible pseudorandomness. This approach is motivated by the observation that certain PUFs behave to some extent like unpredictable functions. This will allow for constructing (somewhat weaker) physical instantiations of (weak) pseudorandom functions.

The notion of pseudorandom functions (PRFs) [21] is established since long in cryptography and many important cryptographic schemes can be constructed from them (see, e.g., [20, 33, 34]). In the following, we first recall (weak) pseudorandom functions, usually abbreviated to (w)PRFs, and show how PUFs can be used (under certain assumptions) to construct a new cryptographic primitive, termed PUF-(w)PRFs.

As PUF-(w)PRFs are weaker than (w)PRFs, known constructions based on (w)PRFs are not directly applicable to PUF-(w)PRFs. We will show in Sects. 7.2 and 7.3a stream cipher and a block cipher, respectively, that are based on PUF-PRFs. First, we present an extended definition of (weak) PRFs.

Definition 4 ((Weak) pseudorandom functions) Consider a family of functions $\mathcal{F}$ with input domain $\{0, 1\}^{\ell}$ and output domain $\{0, 1\}^m$. We say that $\mathcal{F}$ is $(q_{\mathrm{prf}}, \varepsilon_{\mathrm{prf}})$-pseudorandom with respect to a distribution $\mathbb{D}$ on $\{0, 1\}^m$, if the advantage to distinguish between the following two distributions for adaptively chosen pairwise distinct inputs $x_1, \ldots, x_{q_{\mathrm{prf}}}$ is at most $\varepsilon_{\mathrm{prf}}$:

- $y_i = f(x_i)$ where $f \xleftarrow{*} \mathcal{F}$
- $y_i \leftarrow \mathbb{D}$

$\mathcal{F}$ is called *weakly pseudorandom* if the inputs are not chosen by the distinguisher, but uniformly random sampled from $\{0, 1\}^{\ell}$ (still under the condition of being pairwise distinct).

$\mathcal{F}$ is called $(q_{\mathrm{prf}}, \varepsilon_{\mathrm{prf}})$-(weakly)-pseudorandom if it is $(q_{\mathrm{prf}}, \varepsilon_{\mathrm{prf}})$-(weakly)-pseudorandom with respect to the uniform distribution $\mathbb{D} = \mathbb{U}_m$.

Remark 3 This definition differs in several aspects slightly from the original definition of pseudorandom functions, e.g., [4, 5]. First, specifying the output distribution $\mathbb{D}$ allows for covering families of functions which have a non-uniform output distribution, e.g., PUFs. The original case, as stated in the definition, is $\tilde{\mathbb{D}} = \mathbb{U}_m$.

Second, the requirement of pairwise distinct inputs x_i has been introduced to deal with noisy functions where the same input can lead to different outputs. By disallowing multiple queries on the same input, we do not need to model the noise distribution, which is sometimes hard to characterize in practice. Furthermore, in the case of non-noisy (weak) pseudorandom functions, an attacker gains no advantage by querying the same input more than once. Hence, the requirement does not limit the attacker in the non-noisy case.

Observe that the "non-uniform output and independence" assumption on PUFs (as defined in Definition 3) does not automatically imply (weak) pseudorandomness. The first considers the unpredictability of the response to a specific challenge after making queries to *several different* PUFs while the later considers the unpredictability of the response to a challenge after making queries to the *same* PUF.

In the following, we discuss how a family $\mathcal{P}$ of (noisy) PUFs that are (weakly) pseudorandom with respect to some non-uniform distribution can be turned into (weakly) pseudorandom functions (or, more precisely, a somewhat weaker variant of it). Obviously, the main obstacle is to convert noisy non-uniform inputs into reliably reproducible, uniformly distributed random strings. For this purpose, we employ an established tool in cryptography, i.e., *fuzzy extractors* (FE) [13]:

Definition 5 (Fuzzy extractor) A $(m, n, \delta; \mu_{FE}, \varepsilon_{FE})$-*fuzzy extractor* E is a pair of randomized procedures, "generate" (Gen) and "reproduce" (Rep), with the following properties:

Generation: The generation procedure Gen on input $y \in \{0, 1\}^m$ outputs an extracted string $z \in \{0, 1\}^n$ and a helper string (also called helper data) $\omega \in \{0, 1\}^*$.

Reproduction: The reproduction procedure Rep takes an element $y' \in \{0, 1\}^m$ and a bit string $\omega \in \{0, 1\}^*$ as inputs.

Correctness: The correctness property of fuzzy extractors guarantees that if the Hamming distance $\mathsf{dist}(y, y') \leq \delta$ and z, ω were generated by $(z, \omega) \leftarrow \mathsf{Gen}(y)$, then $\mathsf{Rep}(y', \omega) = z$. If $\mathsf{dist}(y, y') > \delta$, then no guarantee is provided about the output of Rep.

Security: The security property guarantees that for any distribution $\mathbb{D}$ on $\{0, 1\}^m$ of min-entropy μ_{FE}, the string z is nearly uniform even for those who observe ω: if $(z, \omega) \leftarrow \mathsf{Gen}(\mathbb{D})$, then it holds that $\mathsf{SD}((z, \omega), (\mathbb{U}_n, \omega)) \leq \varepsilon_{FE}$.

In [13], several constructions for efficient fuzzy extractors have been presented. PUFs are most commonly used in combination with fuzzy extractor constructions based on error-correcting codes and universal hash functions. In this case, the helper data consists of a code-offset, which is of the same length as the PUF output, and the seed for the hash function, which is in the order of 100 bits and can often be reused for all outputs.

Theorem 1 (Pseudorandomness of PUF-FE-composition) *Consider a set $\mathcal{P}$ that is a $(\ell, m, \delta; q_{puf}, \varepsilon_{puf})$-family of PUFs which are $(q_{prf}, \varepsilon_{prf})$-pseudorandom with respect to some distribution $\mathbb{D}$. Let $E = (Gen, Rep)$ be an $(m, n, \delta; H_\infty(\mathbb{D}), \varepsilon_{FE})$ fuzzy extractor. Then, for any $s \in \{1, \ldots, q_{prf}\}$ the advantage of any distinguisher that adaptively chooses pairwise distinct inputs $x_1, \ldots, x_s$ and receives outputs $(z_1, \omega_1), \ldots, (z_s, \omega_s)$ to distinguish the following two distributions is at most $2\varepsilon_{prf} + s \cdot \varepsilon_{FE}$:*

1. *$(z_i, \omega_i) = Gen(\Pi(x_i))$ where $\Pi \xleftarrow{*} \mathcal{P}$*
2. *(z_i, ω_i) where $z_i \leftarrow \mathbb{U}_n$, $(z'_i, \omega_i) = Gen(\Pi(x_i))$ and $\Pi \xleftarrow{*} \mathcal{P}$*

The analogous result holds if $\mathcal{P}$ is $(q_{prf}, \varepsilon_{prf})$-weak-pseudorandom and if the challenges x_i are sampled uniformly random (instead of being adaptively selected), still under the condition of being pairwise distinct.

Proof Let us name the two distributions by $\mathbb{D}_1$ and $\mathbb{D}_2$, respectively. That is it holds:

Distribution $\mathbb{D}_1$: $(z_i, \omega_i) = Gen(\Pi(x_i))$ where $\Pi \xleftarrow{*} \mathcal{P}$
Distribution $\mathbb{D}_2$: (z_i, ω_i) where $z_i \leftarrow \mathbb{U}_n$, $(z'_i, \omega_i) = Gen(\Pi(x_i))$ and $\Pi \xleftarrow{*} \mathcal{P}$

Consider a distinguisher $\mathcal{D}$ that distinguishes between $\mathbb{D}_1$ and $\mathbb{D}_2$. We have to show that $Adv(\mathcal{D}) \leq \varepsilon_{prf} + s \cdot \varepsilon_{FE}$. For this purpose, we introduce two further distributions $\widetilde{\mathbb{D}}_1$ and $\widetilde{\mathbb{D}}_2$, being defined as follows:

Distribution $\widetilde{\mathbb{D}}_1$: $(z_i, \omega_i) = Gen(y_i)$ where $y_i \leftarrow \mathbb{D}$ and $\Pi \xleftarrow{*} \mathcal{P}$
Distribution $\widetilde{\mathbb{D}}_2$: (z_i, ω_i) where $z_i \leftarrow \mathbb{U}_n$, $(z'_i, \omega_i) = Gen(y_i)$, $y_i \leftarrow \mathbb{D}$, and
$\Pi \xleftarrow{*} \mathcal{P}$

By the usual triangular in equation, it holds that

$$Adv(\mathcal{D}) \overset{def}{=} |\Pr[1 \leftarrow \mathcal{D}^{\mathbb{D}_1}] - \Pr[1 \leftarrow \mathcal{D}^{\mathbb{D}_2}]| \tag{5}$$

$$\leq |\Pr[1 \leftarrow \mathcal{D}^{\mathbb{D}_1}] - \Pr[1 \leftarrow \mathcal{D}^{\widetilde{\mathbb{D}}_1}]| + \tag{6}$$

$$|\Pr[1 \leftarrow \mathcal{D}^{\widetilde{\mathbb{D}}_1}] - \Pr[1 \leftarrow \mathcal{D}^{\widetilde{\mathbb{D}}_2}]| + \tag{7}$$

$$|\Pr[1 \leftarrow \mathcal{D}^{\widetilde{\mathbb{D}}_2}] - \Pr[1 \leftarrow \mathcal{D}^{\mathbb{D}_2}]|. \tag{8}$$

Observe that the values in (6), (7), and (8) are upper bounded by $Adv(\mathbb{D}_1, \widetilde{\mathbb{D}}_1)$, $Adv(\widetilde{\mathbb{D}}_1, \widetilde{\mathbb{D}}_2)$, and $Adv(\widetilde{\mathbb{D}}_2, \mathbb{D}_2)$, respectively. We will prove that $Adv(\mathbb{D}_1, \widetilde{\mathbb{D}}_1) \leq \varepsilon_{prf}$, $Adv(\mathbb{D}_2, \widetilde{\mathbb{D}}_2) \leq \varepsilon_{prf}$, and $Adv(\widetilde{\mathbb{D}}_1, \widetilde{\mathbb{D}}_2) \leq s \cdot \varepsilon_{FE}$ what finally shows the claim. To keep the description simple, we consider only the case of non-weak pseudorandom functions. The other case can be shown analogously.

Consider a distinguisher $\mathcal{D}_1$ between $\mathbb{D}_1$ and $\widetilde{\mathbb{D}}_1$. We construct from $\mathcal{D}_1$ another distinguisher $\mathcal{D}'_1$ between $\mathbb{D}$ and the distribution given by $y_i = \Pi(x_i)$ where $\Pi \overset{*}{\leftarrow} \mathcal{P}$ (see also Definition 1). Let $\mathcal{O}_{\mathsf{PRF}}$ denote an oracle that generates the latter distribution. Summing up, we consider two distinguishers:

- Distinguisher $\mathcal{D}_1$ for the two oracles $\mathcal{O}_{\mathbb{D}_1}$ and $\mathcal{O}_{\widetilde{\mathbb{D}}_1}$.
- Distinguisher $\mathcal{D}'_1$ for the two oracles $\mathcal{O}_{\mathbb{D}}$ and $\mathcal{O}_{\mathsf{PRF}}$.

Let $\mathcal{O}$ be the oracle that $\mathcal{D}'_1$ has access to. That is $\mathcal{O} = \mathcal{O}_{\mathbb{D}}$ or $\mathcal{O} = \mathcal{O}_{\mathsf{PRF}}$. $\mathcal{D}'_1$ simulates the oracle for $\mathcal{D}_1$ as follows. Whenever $\mathcal{D}'_1$ receives (adaptively chosen) inputs x_i from $\mathcal{D}_1$, it forwards them to $\mathcal{O}$ and gets back a value y_i. Then, $\mathcal{D}'_1$ computes $(z_i, \omega_i) := \mathsf{Gen}(y_i)$ and hands this tuple to $\mathcal{D}_1$. When $\mathcal{D}_1$ finally produces an output (0 or 1), then $\mathcal{D}'_1$ uses this value as its own output. It is easy to see that if $\mathcal{O} = \mathcal{O}_{\mathbb{D}}$, then the values y_i are sampled according to $\mathbb{D}$ and $\mathcal{D}'_1$ perfectly simulated $\mathcal{O}_{\widetilde{\mathbb{D}}_1}$. In the other case, i.e., $\mathcal{O} = \mathcal{O}_{\mathsf{PRF}}$, $\mathcal{D}'_1$ perfectly simulated $\mathcal{O}_{\mathbb{D}_1}$. Thus, it follows that $\mathsf{Adv}(\mathcal{D}'_1) = \mathsf{Adv}(\mathcal{D}_1)$. As by assumption $\mathsf{Adv}(\mathcal{D}'_1) \leq \varepsilon_{\mathsf{prf}}$ and as $\mathcal{D}_1$ is an arbitrary distinguisher, we get that $\mathsf{Adv}(\mathbb{D}_1, \widetilde{\mathbb{D}}_1) \leq \varepsilon_{\mathsf{prf}}$. In a similar manner one can show that $\mathsf{Adv}(\mathbb{D}_2, \widetilde{\mathbb{D}}_2) \leq \varepsilon_{\mathsf{prf}}$.

It remains to show that $\mathsf{Adv}(\widetilde{\mathbb{D}}_1, \widetilde{\mathbb{D}}_2) \leq s \cdot \varepsilon_{\mathsf{FE}}$. We define a sequence of intermediate distributions $\widetilde{\mathbb{D}}[i]$ as follows: the first i tuples are sampled according to $\widetilde{\mathbb{D}}_2$ and the remaining $s - i$ according to $\widetilde{\mathbb{D}}_1$. In particular it holds that $\widetilde{\mathbb{D}}[0] = \widetilde{\mathbb{D}}_1$ and $\widetilde{\mathbb{D}}[s] = \widetilde{\mathbb{D}}_2$. Thus it holds

$$\mathsf{Adv}(\widetilde{\mathbb{D}}_1, \widetilde{\mathbb{D}}_2) \leq \sum_{i=0}^{s-1} \mathsf{Adv}(\widetilde{\mathbb{D}}[i], \widetilde{\mathbb{D}}[i+1]). \tag{9}$$

Observe that the only difference between two subsequent distributions $\widetilde{\mathbb{D}}[i]$ and $\widetilde{\mathbb{D}}[i+1]$ is that $(z_i, \omega_i) = \mathsf{Gen}(y_i)$ in the first case while (z_i, ω_i) where $z_i \leftarrow \mathbb{U}_n$ and in the second. By the security property of fuzzy extractors (Def. 5), the statistical distance between both distributions is $\leq \varepsilon_{\mathsf{FE}}$. It is common knowledge that this implies $\mathsf{Adv}(\widetilde{\mathbb{D}}[i], \widetilde{\mathbb{D}}[i+1]) \leq \varepsilon_{\mathsf{FE}}$ as well. Hence, the claims follows with (9). This concludes the proof. $\quad\square$

Definition 6 (PUF-(w)PRFs) Consider a family of (weakly)-pseudorandom PUFs $\mathcal{P}$ and a fuzzy extractor $\mathsf{E} = (\mathsf{Gen}, \mathsf{Rep})$ (where the parameters are as described in Theorem 1). A family of PUF-(w)PRFs is a set of pair of randomized procedures, called *generation* and *reproduction*. The generation function $\mathsf{Gen} \circ \Pi$ for some PUF $\Pi \in \mathcal{P}$ takes as input $x \in \{0,1\}^\ell$ outputs $(z, \omega_x) \overset{\mathsf{def}}{=} \mathsf{Gen}(\Pi(x)) \in \{0,1\}^n \times \{0,1\}^*$, while the reproduction function $\mathsf{Rep} \circ \Pi$ takes $(x, \omega_x) \in \{0,1\}^\ell \times \{0,1\}^*$ as input and reproduces the value $z = \mathsf{Rep}(\Pi(x), \omega_x)$.

Remark 4 Observe that although in cryptography usually only PRFs with a reasonably large input space $\{0,1\}^\ell$ and a huge number q_{prf} of allowed queries are considered, we do not put this condition here. The reason is to cover a broad range of PUF-PRF constructions with this model. Whether realizations with small values of ℓ or q_{prf} are useful depends on the actual application. For the sake of consistency, we use continuously the term PUF-PRF and a reader should keep in mind that this does not automatically mean large values of ℓ and q_{prf}.

7 Encrypting with PUF-(w)PRFs

7.1 General Thoughts

A straightforward approach for using PUF-wPRFs against full non-volatile memory attackers would be to use them for key derivation where the key is afterward used in some encryption scheme. However, in this construction PUF-wPRFs would ensure security against non-volatile memory attackers only while the security of the encryption scheme would need to be shown separately.

Theorem 1 actually shows that PUF-(w)PRFs and "traditional" (w)PRFs have in common that (part of) the output cannot be distinguished from uniformly random values. As many cryptographic designs are known where the security can be reduced to the indistinguishability of the deployed PRFs, it is a promising thought to put PUF-(w)PRFs into the core of cryptographic algorithms to achieve protection against non-volatile memory attackers *and* algorithmic attackers at the same time. One might be tempted to plug in PUF-(w)PRFs wherever (w)PRFs are required. Unfortunately, things are not that simple since the information saved in the helper data is also needed for correct execution. It is a known fact that the helper data of a fuzzy extractor *always* leaks some information about the input, e.g., see [25]. Hence, extra attention must be paid when deploying PUF-PRFs in cryptographic schemes.

In the following, we present a stream ciphers and a block cipher-based PUF-(w)PRFs that are secure *although* the helper data is made public. These constructions effectively describe *keyless* symmetric encryption systems which are provably secure. Since there is no digital secret stored in non-volatile memory, even a full non-volatile memory attacker has no advantage in breaking this scheme, which is a clear advantage over conventional key-based ciphers. On the other hand, the absence of a secret key limits the application of this scheme. Since no key can be shared between parties, encrypting digital communication with a PUF-PRF-based cipher is not possible. In other use cases this limitation does not pose a problem, e.g., for encrypting data stored in untrusted or public storage. Also, the scheme does not need to be bound to a certain physical location since the PUF-PRFs can be embedded in a portable device, e.g., a USB dongle or a cell phone.

In this context an interesting observation is that making the helper data public represents a special case of *computational leakage*. In that sense one could say that using PUFs allows to transform memory leakage into computational leakage. We leave for further work the question whether existing schemes that are secure against computational leakage, e.g., [41], can be used in combination with PUF-(w)PRFs.

In the best case, this may yield schemes which are secure against memory leakage and computational leakage at the same time.

7.2 A Stream Cipher Based on PUF-PRFs

A stream cipher is an encryption mechanism that allows for efficient encryption of data streams of arbitrary length. A common approach is to deploy a keystream generator. The basic idea behind keystream generators is to mimic one-time pad encryption. That is, given a short seed, the goal is to stretch the seed to a keystream that "looks as random as possible." Encryption is realized by simply adding the keystream to the plaintext stream, giving the ciphertext stream. A recipient who knows the seed (and who has access to the same keystream generator) can then easily decrypt by first computing the same keystream and then subtracting it from the ciphertext stream.

It is common knowledge that PRFs can be used for constructing pseudorandom sequence generators under certain conditions. Probably the simplest approach is based on PRFs where the output size exceeds the input size, that is $f : \{0, 1\}^\ell \to \{0, 1\}^n$ with $n > \ell$. Given an initial seed $S_0 \in \{0, 1\}^\ell$, one iteratively computes $(S_i \| K_i) \stackrel{\text{def}}{=} f(S_{i-1})$ with $S_i \in \{0, 1\}^\ell$ and $K_i \in \{0, 1\}^{n-\ell}$. S_i represents the next internal state while K_i can be used for the keystream. A more comprehensive treatment of stream ciphers can be found in [19]. In the following, we will show that similar constructions are possible with PUF-(w)PRFs. An interesting fact is that only an authorized user has access to the PUF. This allows the use of the full internal state as output.

Definition 7 (A PUF-PRF based stream cipher) Let $\mathcal{F}$ denote a family of PUF-PRFs as specified in Definition 6 with $n = \ell$. Let $f = (f_{\text{Gen}}, f_{\text{Rep}}) \in \mathcal{F}$, that is $f_{\text{Gen}} = \text{Gen} \circ \Pi$ and $f_{\text{Rep}} = \text{Rep} \circ \Pi$, respectively, for some PUF Π. We define a keystream generator $\mathcal{C}_f$ as follows:

> Input: The input to $\mathcal{C}_f$ is a value $S_0 \in \{0, 1\}^\ell$ which can be public.
> State: The state before the ith round is $S_i \in \{0, 1\}^\ell$.
> Encryption: Given an initial state S_0 and a stream of plaintext blocks $P_1, P_2, \ldots \in \{0, 1\}^\ell$, $\mathcal{C}_f$ computes iteratively $(S_{i+1}, \omega_{i+1}) \stackrel{\text{def}}{=} f_{\text{Gen}}(S_i) \in \{0, 1\}^n \times \{0, 1\}^*$ and outputs $(C_i \stackrel{\text{def}}{=} S_i \oplus P_i, \omega_i)$ for $i \geq 1$.
> Decryption: Given S_0 and a sequence $\{(C_i, \omega_i)\}_{i \geq 1}$, $\mathcal{C}_f$ reconstructs the values $S_{i+1} = f_{\text{Rep}}(S_i, \omega_i)$ and decrypts $P_i = C_i \oplus S_i$.

The correctness of the scheme can be easily derived from the correctness of the fuzzy extractor. The following theorem gives an upper bound on the advantage of distinguishing the keystream from a random bit stream if the keystream is not too long.

Theorem 2 *Let $\mathcal{F}$ denote a family of PUF-PRFs as specified in Definition 7 with $n \geq \ell$. Consider the following two oracles that, on input S_0, generate a sequence $\{(S_i, \omega_i)\}_{i=1}^{s}$ with $s \leq q_{\text{prf}}$. Oracle $\mathcal{O}_1$ generates these values as described for the stream cipher C_f (see Definition 7) for $f \overset{*}{\leftarrow} \mathcal{F}$. Oracle $\mathcal{O}_2$ behaves in principle in the same way but in addition randomizes the values $\{S_i\}_{i \geq 1}$. More precisely whenever a value S_i is computed, it is first replaced by some value $\widetilde{S}_i \overset{*}{\leftarrow} \mathbb{U}_\ell$ before the next values are computed. It holds that*

$$\text{Adv}(\mathcal{O}_1, \mathcal{O}_2) \leq 2\varepsilon_{\text{prf}} + s \cdot \varepsilon_{\text{FE}}. \tag{10}$$

Proof The proof follows directly from Theorem 1. If we set $x_i \overset{\text{def}}{=} S_{i-1}$ and $z_i \overset{\text{def}}{=} S_i$, then in principle we have the same situation here as in Theorem 1 with the only difference being that the choices of the inputs x_i are fixed. That is, the class of considered distinguishers is only a (weaker) subclass of the distinguisher covered by Theorem 1. Hence, $\text{Adv}(\mathcal{O}_1, \mathcal{O}_2)$ can be at most as big as the upper bound derived there which is $2\varepsilon_{\text{prf}} + s \cdot \varepsilon_{\text{FE}}$. This shows the claim. $\square$

Observe that the proof does not translate directly to the case of PUF-wPRFs as the distinguisher can partly choose the inputs, namely x_1, and as the other inputs are computed by following a strict rule instead of being randomly chosen. However, we will show that already a small change is sufficient to make the encryption scheme secure even if the only PUF-wPRFs are used (instead of PUF-PRFs).

Definition 8 (A PUF-wPRF based stream cipher) Let $\mathcal{F}$ denote a family of PUF-wPRFs as specified in Definition 6 with $n \geq \ell$. Let $f = (f_{\text{Gen}}, f_{\text{Rep}}) \in \mathcal{F}$, that is $f_{\text{Gen}} = \text{Gen} \circ \Pi$ and $f_{\text{Rep}} = \text{Rep} \circ \Pi$, respectively, for some PUF Π. We define a keystream generator C'_f as follows:

> Input: The input to C'_f is a value $S'_0 \in \{0,1\}^\ell$ which can be public.
> Encryption: Given an initial state S'_0 and a stream of plaintext blocks $P_1, P_2, \ldots \in \{0,1\}^\ell$, C'_f first choose $\rho \overset{*}{\leftarrow} \mathbb{U}_\ell$ and computes $S_0 \overset{\text{def}}{=} S'_0 \oplus \rho$. Then, it generates iteratively $(S_{i+1}, \omega_{i+1}) \overset{\text{def}}{=} f_{\text{Gen}}(S_i) \in \{0,1\}^n \times \{0,1\}^*$ and outputs $(C_i \overset{\text{def}}{=} S_i \oplus P_i, \omega_i)$ for $i \geq 1$. In addition it reveals the value ρ.
> Decryption: Given S'_0, ρ, and a sequence $\{(C_i, \omega_i)\}_{i \geq 1}$, C'_f sets $S_0 \overset{\text{def}}{=} S'_0 \oplus \rho$, reconstructs the values $S_{i+1} = f_{\text{Rep}}(S_i, \omega_i)$ and decrypts $P_i = C_i \oplus S_i$.

Observe that the only difference between the stream cipher C_f from Definition 7 and C'_f from Definition 8 is that the latter first randomizes the initial state. The reason for this is that it prevents an attacker to adaptively choose the internal states, making C'_f secure for wPRFs as well (instead for PRFs only as in the case of C_f). More precisely we have

Theorem 3 *Let $\mathcal{F}$ denote a family of PUF-PRFs as specified in Definition 7 with $n \geq \ell$. Consider the following two oracles that, on input S_0, generate a sequence*

$\{(S_i, \omega_i)\}_{i=1}^{s}$ *with* $s \le q_{\mathrm{prf}}$. *Oracle* $\mathcal{O}_1$ *generates these values as described for the stream cipher* $\mathcal{C}_f$ *(see Definition 7). Oracle* $\mathcal{O}_2$ *behaves in principle in the same way but in addition randomizes the values* $\{S_i\}_{i \ge 1}$. *It holds that*

$$\mathrm{Adv}(\mathcal{O}_1, \mathcal{O}_2) \le s \cdot (2\varepsilon_{\mathrm{prf}} + \varepsilon_{\mathrm{FE}}). \tag{11}$$

Proof Consider a sequence of oracles $\mathcal{O}[i]$, $i = 1, \dots, s$. $\mathcal{O}[i]$ generates like $\mathcal{O}_1$ the outputs but randomizes in addition the first i values S_i. As S_0 gets randomized via the value ρ, it holds that $\mathcal{O}[0] = \mathcal{O}_1$. Furthermore, one may easily see that $\mathcal{O}[s] = \mathcal{O}_2$.

Now let $i \in \{1, \dots, s-1\}$ and consider two subsequent oracles $\mathcal{O}[i]$ and $\mathcal{O}[i+1]$. By definition, the only difference is that $\mathcal{O}[i+1]$ randomizes the value S_{i+1}. Observe that S_i is uniformly at random sampled in both cases. Hence, we can apply Theorem 1 for the PUF-wPRF case and get

$$\mathrm{Adv}(\mathcal{O}[i], \mathcal{O}[i + 1]) \le 2\varepsilon_{\mathrm{prf}} + \varepsilon_{\mathrm{FE}}. \tag{12}$$

The claim follows from the usual hybrid argument. □

7.3 A Block Cipher Based on PUF-PRFs

One of the most important results with respect to PRFs was developed by Luby and Rackoff in [34]. They showed how to construct pseudorandom permutations from PRFs. Briefly summarized, a pseudorandom permutation (PRP) is a PRF that is a permutation as well. PRPs can be seen as an idealization of block ciphers. Consequently, the Luby–Rackoff construction is often termed as Luby–Rackoff cipher.

Unfortunately, the Luby–Rackoff result does not automatically apply to the case of PUF-PRFs. As explained previously, PUF-(w)PRFs differ from (w)PRFs as they additionally need some helper data for correct execution. First, it is unclear if and how the existence and necessity of helper data would fit into the established concept of PRPs. Second, an attacker might adaptively choose plaintexts to force internal collisions and use the information leakage of the helper data for checking for these events.

Nonetheless, we can show that a Luby–Rackoff cipher based on PUF-wPRFs also yields a secure block cipher. For this purpose, we consider the set of concrete security notions for symmetric encryption schemes that have been presented and discussed in [4]. More precisely, we prove that a randomized version of a three-round Luby–Rackoff cipher based on PUF-PRFs fulfills real-or-random indistinguishability against a chosen-plaintext attacker.[1] In a nutshell, a real-or-random attacker adaptively chooses plaintexts and hands them to an encryption oracle. This oracle

---

[1] Due to the lack of space, we consider here the simplest case, being a three rounds Luby–Rackoff cipher and a chosen-plaintext attackers.

encrypts either received plaintexts (real case) or some random plaintexts (random case). The encryptions are given back to the attacker. Her task is to distinguish between both cases. Thus, eventually she outputs a guess (a bit). The scheme is real-or-random indistinguishable if the advantage of winning the game is negligible (in some security parameter). We recall the formal definition:

Definition 9 (Real-or-random security) [4] An Encryption scheme with encryption mechanism $\mathcal{E}$, decryption mechanism $\mathcal{D}$, and keyspace $\mathcal{K}$ is said to be $(q; \varepsilon)$-secure, in the real-or-random sense, if for any adversary $\mathcal{A}$ which makes at most q oracle queries,

$$\text{Adv}_{\mathcal{A}}^{rr} \stackrel{\text{def}}{=} |Pr[k \leftarrow \mathcal{K} : \mathcal{A}^{\mathcal{O}_{\mathcal{E}_k(\cdot)}} = 1] - Pr[k \leftarrow \mathcal{K} : \mathcal{A}^{\mathcal{O}_{\mathcal{E}_k(\$)}} = 1]|. \quad (13)$$

The notation $\mathcal{O}_{\mathcal{E}_k(\cdot)}$ indicates an oracle which, in response to a query x, returns $y \stackrel{\text{def}}{=} \mathcal{E}_k(x)$, while $\mathcal{O}_{\mathcal{E}_k(\$)}$ is an oracle which, in response to a query x, chooses $x' \stackrel{*}{\leftarrow} \{0, 1\}^{|x|}$ and then returns $y \stackrel{\text{def}}{=} \mathcal{E}_k(x')$.

Next, we first define the considered block cipher, a three-round PUF-PRF-based Luby–Rackoff cipher and prove its security afterward. The working principle is very similar to the original Luby–Rackoff cipher and is displayed in Fig. 1. The main differences are twofold. First, at the beginning some uniformly random value $\rho \in \{0, 1\}^{\ell}$ is chosen to randomize the right part R of the plaintext. Second, the round functions are PUF-wPRFs instead of PRFs.

Definition 10 (3-round PUF-wPRF-based Luby–Rackoff cipher) Let $\mathcal{F}$ denote a family of PUF-wPRFs with input and output length n.[2] The three-round PUF-PRF-

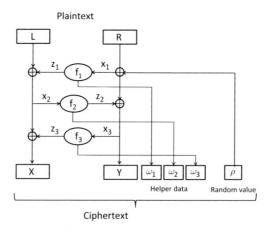

Fig. 1 A randomized three-round Luby–Rackoff cipher based on PUF-PRFs

---

[2] Although the fuzzy extractor usually reduces the output length, such situation can exist if the output length of the PUF is bigger than the input length.

based Luby–Rackoff cipher $\mathcal{C}_{\mathcal{F}}$ uses three different PUF-wPRFs $f_1, f_2, f_3 \in \mathcal{F}$ as round functions. That is each f_i is composed of two functions $\mathsf{Gen} \circ \Pi_i$ and $\mathsf{Rep} \circ \Pi_i$ for an appropriate fuzzy extractor $\mathsf{E} = (\mathsf{Gen}, \mathsf{Rep})$.

Given a plaintext $(L, R) \in \{0, 1\}^n \times \{0, 1\}^n$, first a random value $\rho \xleftarrow{*} \{0, 1\}^n$ is sampled. Then, the following values are computed:

$$x_1 \stackrel{\text{def}}{=} R \oplus \rho, \qquad (z_1, \omega_1) \stackrel{\text{def}}{=} (\mathsf{Gen} \circ \Pi_1)(x_1) \tag{14}$$

$$x_2 \stackrel{\text{def}}{=} L \oplus z_1, \qquad (z_2, \omega_2) \stackrel{\text{def}}{=} (\mathsf{Gen} \circ \Pi_2)(x_2) \tag{15}$$

$$x_3 \stackrel{\text{def}}{=} x_1 \oplus z_2, \qquad (z_3, \omega_3) \stackrel{\text{def}}{=} (\mathsf{Gen} \circ \Pi_3)(x_3) \tag{16}$$

$$X \stackrel{\text{def}}{=} x_2 \oplus z_3, \qquad Y \stackrel{\text{def}}{=} x_3 \tag{17}$$

The ciphertext is $(X, Y, \omega_1, \omega_2, \omega_3, \rho)$.

Decryption works similar to the case of the "traditional" Luby–Rackoff cipher where the helper data ω_i is used together with the Rep procedure for reconstructing the output z_i of the PUF-PRF f_i and the value ρ to "derandomize" the input to the first round function f_1. More precisely the following computations are performed:

$$x_3' \stackrel{\text{def}}{=} Y, \qquad z_3' \stackrel{\text{def}}{=} (\mathsf{Rep} \circ \Pi_3)(x_3', \omega_3), \tag{18}$$

$$x_2' \stackrel{\text{def}}{=} X \oplus z_3', \qquad z_2' \stackrel{\text{def}}{=} (\mathsf{Rep} \circ \Pi_2)(x_2', \omega_2), \tag{19}$$

$$x_1' \stackrel{\text{def}}{=} x_3' \oplus z_2', \qquad z_1' \stackrel{\text{def}}{=} (\mathsf{Rep} \circ \Pi_1)(x_1', \omega_1), \tag{20}$$

$$L' \stackrel{\text{def}}{=} x_2' \oplus z_1', \qquad R' \stackrel{\text{def}}{=} x_1' \oplus \rho. \tag{21}$$

Due to the correctness properties of the fuzzy extractor, one can deduce that $z_i = z_i'$ and $x_i = x_i'$ for $i = 1, 2, 3$. In particular it follows that $L' = L$ and $R' = R$.

Theorem 4 *Let $\mathcal{E}^{\mathcal{F}}$ be the encryption scheme defined in Definition 10 using a family $\mathcal{F}$ of PUF-wPRFs (with parameters as specified in Theorem 1). Then, the advantage of a real-or-random attacker making up to q_{prf} queries is at most $5\varepsilon_{\mathrm{prf}} + 2q_{\mathrm{prf}} \cdot \varepsilon_{\mathrm{FE}} + 3 \cdot \frac{q_{\mathrm{prf}}^2}{2^n}$.*

Proof Let $\{(L^{(i)}, R^{(i)})\}_{i=1,\ldots,q_{\mathrm{prf}}}$ denote the sequence of the adaptively chosen plaintexts, $x_j^{(i)}, z_j^{(i)}$ be the values as specified in Eqs. (14), (15) and (16), $\rho^{(i)}$ the randomly chosen values.

Let $\mathcal{O}_{\mathcal{E}(\cdot)}$ denote the oracle that honestly encrypts given plaintexts while $\mathcal{O}_{\mathcal{E}(\$)}$ encrypts randomly chosen plaintexts. We have to show that $\mathrm{Adv}(\mathcal{O}_{\mathcal{E}(\cdot)}, \mathcal{O}_{\mathcal{E}(\$)}) \leq 5\varepsilon_{\mathrm{prf}} + 2q_{\mathrm{prf}} \cdot \varepsilon_{\mathrm{FE}} + 3 \cdot \frac{q_{\mathrm{prf}}^2}{2^n}$. We prove the claim by defining a sequence of oracles and estimating the advantages of distinguishing between them. The differences between the oracles are that *parts* of the involved values are replaced by some uniform random values. To allow an easier tracking of the differences, we introduce the following notation. Let $\mathcal{V} := \{L, R, x_1, z_1, \omega_1, \ldots, x_3, z_3, \omega_3, \rho, X, Y\}$ be the set of values that occur during one encryption (see Eqs. (14), (15) and (16) in

Definition 10). For $V \subseteq \mathcal{V}$, oracle $\mathcal{O}[V]$ encrypts given plaintexts but during *each* encryption process, the values indicated in V are randomized.[3] For example, it holds that $\mathcal{O}[\emptyset] = \mathcal{O}_{\mathcal{E}(\cdot)}$ (nothing is randomized) and $\mathcal{O}[\{L, R\}] = \mathcal{O}_{\mathcal{E}(\$)}$ (the plaintexts are randomized). Let $\mathcal{D}$ be an arbitrary distinguisher between $\mathcal{O}[\emptyset]$ and $\mathcal{O}[\{L, R\}]$. We will consider the following in equations:

$$\text{Adv}(\mathcal{D}) \overset{\text{def}}{=} |Pr[1 \leftarrow \mathcal{D}^{\mathcal{O}[\emptyset]}] - Pr[1 \leftarrow \mathcal{D}^{\mathcal{O}[\{L,R\}]}]| \tag{22}$$
$$\leq |Pr[1 \leftarrow \mathcal{D}^{\mathcal{O}[\emptyset]}] - Pr[1 \leftarrow \mathcal{D}^{\mathcal{O}[\{x_2\}]}]| + \tag{23}$$
$$|Pr[1 \leftarrow \mathcal{D}^{\mathcal{O}[\{x_2\}]}] - Pr[1 \leftarrow \mathcal{D}^{\mathcal{O}[\{x_1,x_2\}]}]| + \tag{24}$$
$$|Pr[1 \leftarrow \mathcal{D}^{\mathcal{O}[\{x_1,x_2\}]}] - Pr[1 \leftarrow \mathcal{D}^{\mathcal{O}[\{L,R\}]}]|. \tag{25}$$

Similar to the proof of Theorem 1, we will give upper bounds for each expression (23), (24), and (25). Let us start with (23): $|Pr[1 \leftarrow \mathcal{D}^{\mathcal{O}[\emptyset]}] - Pr[1 \leftarrow \mathcal{D}^{\mathcal{O}[\{x_2\}]}]|$. Recall that $x_2 = L \oplus z_1$ (Eq. 15). Thus, randomizing x_2 is equivalent to randomizing z_1 and hence $\mathcal{O}[\{x_2\}] = \mathcal{O}[\{z_1\}]$. By definition, the only difference between $\mathcal{O}[\emptyset]$ and $\mathcal{O}[\{z_1\}]$ is the tuple $(z_1, \omega_1) = \text{Gen}(\Pi_1(x_1))$, namely:

- $(z_1, \omega_1) = \text{Gen}(\Pi_1(x_1))$ in the case of $\mathcal{O}[\emptyset]$ and
- $z_1 \leftarrow \mathbb{U}_n$ and $(z_1', \omega_1) = \text{Gen}(\Pi(x_1))$ in the case of $\mathcal{O}[\{z_1\}]$.

Under the assumption that the values $x_1^{(i)}$ are pairwise distinct, the advantage to distinguish between both cases is at most $2\varepsilon_{\text{prf}} + q_{\text{prf}} \cdot \varepsilon_{\text{FE}}$ according to Theorem 1. Furthermore, as the values $\rho^{(i)}$ are uniformly random, the probability of a collision in the values $x_1^{(i)}$ is at most $\frac{q_{\text{prf}}^2}{2^n}$. As a consequence, we have

$$\text{Adv}(\mathcal{O}[\emptyset], \mathcal{O}[\{x_2\}]) \leq 2\varepsilon_{\text{prf}} + q_{\text{prf}} \cdot \varepsilon_{\text{FE}} + \frac{q_{\text{prf}}^2}{2^n}. \tag{26}$$

Next, we consider the differences between $\mathcal{O}[\{x_2\}]$ and $\mathcal{O}[\{x_1, x_2\}]$. As explained above, randomizing x_2 is equivalent to randomizing z_1. Thus, we can consider the differences between $\mathcal{O}[\{z_1\}]$ and $\mathcal{O}[\{x_1, z_1\}]$, instead. Observe that the value x_1 is involved in the derivation of the following values:

1. $z_1 \leftarrow \mathbb{U}_n$ and $(z_1', \omega_1) = \text{Gen}(\Pi(x_1))$ (Eq. (14)).
2. $x_3 = x_1 \oplus z_2$ (Eq. (16)).

As z_2 is independent of (z_1', ω_1), these two features are independent and can be examined separately. Regarding the first feature, the difference is

1. $z_1 \leftarrow \mathbb{U}_n$ and $(z_1', \omega_1) = \text{Gen}(y_1)$ with $y_1 = \Pi_1(x_1)$ in the case of $\mathcal{O}[\{z_1\}]$.
2. $z_1 \leftarrow \mathbb{U}_n$ and $(z_1', \omega_1) = \text{Gen}(y_1)$ with $y_1 \leftarrow \mathbb{D}$ in the case of $\mathcal{O}[\{z_1, x_1\}]$.

---

[3] As the randomization is done for every encryption, we omit for simplicity the superscripts (i) at the values.

As the PUFs are assumed to be wPRFs, the advantage to distinguish these two cases is at most ε_{prf} if the values $x_1^{(i)}$ are pairwise different and uniformly random. As explained above the probability of a collision in the values $x_1^{(i)}$ is at most $\frac{q_{\text{prf}}^2}{2^n}$. Furthermore, the values $x_1^{(i)}$ are derived by an XOR with a uniformly random value. Taking together, the advantage to distinguish between $\mathcal{O}[\{z_1\}]$ and $\mathcal{O}[\{z_1, x_1\}]$ based on the first feature is at most $\varepsilon_{\text{prf}} + \frac{q_{\text{prf}}^2}{2^n}$.

Now we turn our attention to the second feature: $x_3 = x_1 \oplus z_2$. With the same arguments as above, randomizing x_1 here is equivalent to randomizing z_2. We make here use of the fact that z_2 is only involved in the definition of x_3 (or, equivalently, of Y). Analogously to the case of randomizing z_1 (see above) it follows that the advantage of distinguishing between $\mathcal{O}[\{z_1\}]$ and $\mathcal{O}[\{z_1, x_1\}]$ based on the second feature is at most $2\varepsilon_{\text{prf}} + q_{\text{prf}} \cdot \varepsilon_{\text{FE}} + \frac{q_{\text{prf}}^2}{2^n}$.

Altogether, it follows that

$$\text{Adv}(\mathcal{O}[\{x_2\}], \mathcal{O}[\{x_1, x_2\}]) \leq 3\varepsilon_{\text{prf}} + q_{\text{prf}} \cdot \varepsilon_{\text{FE}} + 2\frac{q_{\text{prf}}^2}{2^n}. \qquad (27)$$

Finally, we have to investigate $\text{Adv}(\mathcal{O}[\{x_1, x_2\}], \mathcal{O}[\{L, R\}])$. Recall that $x_1 \overset{\text{def}}{=} R \oplus \rho$ (Eq. (14)). Thus, it is indistinguishable whether x_1 is randomized or R. Likewise $x_2 \overset{\text{def}}{=} L \oplus z_1$ (Eq. (15)) implies that it indistinguishable whether x_2 or L is realized. This implies that $\mathcal{O}[\{x_1, x_2\}] = \mathcal{O}[\{L, R\}]$ and in particular

$$\text{Adv}(\mathcal{O}[\{x_1, x_2\}], \mathcal{O}[\{L, R\}]) = 0. \qquad (28)$$

Summing up, the advantage of a real-or-random attacker is at most $5\varepsilon_{\text{prf}} + 2q_{\text{prf}} \cdot \varepsilon_{\text{FE}} + 3 \cdot \frac{q_{\text{prf}}^2}{2^n}$ what concludes the proof. $\square$

8 SRAM PRFs

In Sect. 6, we described a general approach for mimicking (w)PRFs by certain PUFs, based on fuzzy extractors. In this section, we show that in the concrete case, other approaches might exist that possibly lead to stronger results. First, we describe the implementation and properties of SRAM PUFs, which are practical instantiations of PUFs that are covered by our PUF model (Definition 3) and are $(2^\ell, 0)$-pseudorandom. Then, we introduce a new mode of operation for these SRAM PUFs which likewise allows for generating uniformly random, reproducible values. As opposed to the approach using fuzzy extractors, we can argue that the helper data does not leak any information, neither on the PUF output itself nor if two outputs are related. This leads to a new cryptographic primitive, termed SRAM-PRFs, that can be viewed as a physical realization of an expanding PRF. Observe that one drawback of SRAM-PRFs is that the hardware size grows exponentially with the input length.

Thus, SRAM-PRFs are rather hardwired long random strings than PRFs in the usual cryptographic sense. As explained in Remark 4, we keep the terminology PRF for consistency. However, SRAM-PRFs cannot be used as a concrete instantiation of PUF-PRFs for our constructions from Sect. 7. This section rather shows up an alternative approach for constructing cryptographic mechanisms based on PUFs despite of the noise problem.

8.1 Physical Implementation Details of Static Random Access Memory (SRAM)

An *SRAM cell* is a physical realization of the logical scheme shown in Fig. 2a. The important part of this construction consists of the two *cross-coupled* inverters.[4] This particular construction causes the circuit to assume two distinct logically stable states, commonly denoted as "0" and "1." By residing in one of both states, one binary digit is effectively stored in the memory cell.

Definition 11 (SRAM) A (ℓ, m)-Static Random Access Memory (SRAM) is defined as a $2^\ell \times m$ matrix of physical SRAM cells, each storing an element from $\{0, 1\}$.

8.2 The SRAM PUF Construction

In order to understand the operation of an SRAM PUF, we pay special interest to the physical behavior of an SRAM cell during device power-up. Let $\widetilde{\mathbf{R}} \in \{0, 1\}^{2^\ell \times m}$ denote the state of the SRAM matrix immediately after a particular power-up of the

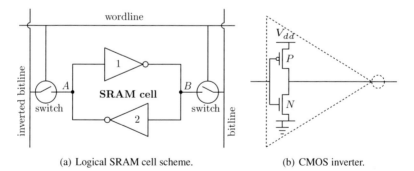

(a) Logical SRAM cell scheme.   (b) CMOS inverter.

Fig. 2 (**a**) Logical scheme of an SRAM cell. In its most common form, it consists of two logical inverters (labeled 1 and 2) taking each others output as input. The cell assumes two logically stable states: $(AB = 01)$ and $(AB = 10)$. (**b**) Typical two-transistor CMOS implementation of an inverter, consisting of a p-channel (P) and an n-channel (N) transistor

---

[4] The output of the first inverter is connected to the input of the second one and vice versa.

memory. A particular row of $\widetilde{\mathbf{R}}$ is addressed by $x \in \{0, 1\}^\ell$ and denoted as $\widetilde{r_x}$. The ith bit of this row is denoted as $\widetilde{r_{x,i}}$.

We introduce the operation of an SRAM PUF by means of the following theorem:

Theorem 5 *Let $\widetilde{R}$ be the noisy power-up state bit matrix that arises after a specific power-up of a particular physical (ℓ, m)-SRAM instantiation as described in Sect. 8.1. The procedure that accepts as input a* challenge $x \in \{0, 1\}^\ell$ *and thereupon returns the row vector* $y = \widetilde{r_x} \in \{0, 1\}^m$ *as a* response *is a realization of a $(\ell, m, \delta; q_{\mathrm{puf}}, \varepsilon_{\mathrm{puf}})$-PUF that is $(2^\ell, 0)$-pseudorandom.*

This PUF construction was originally introduced and experimentally verified in [22] and is commonly called an *SRAM PUF*. A proof sketch for this theorem follows from a number of physical assumptions regarding the SRAM power-up state which are based on physical observations and model-based simulations of SRAM cells:

Definition 12 (SRAM Assumptions) Let $r_{x,i}$ denote the power-up state of a particular SRAM cell in the absence of any stochastical noise component at the time of power-up:

1. This *noise-free* power-up state $r_{x,i}$ of an SRAM cell is static for every power-up of a particular cell, but independently and uniformly distributed over $\{0, 1\}$ for a randomly selected cell.
2. The actual power-up state of an SRAM cell is *noisy* by nature and given by $\widetilde{r_{x,i}} = r_{x,i} \oplus e_{x,i}$, with $e_{x,i}$ a Bernoulli distributed random variable with $\Pr[e_{x,i} = 1] \stackrel{\text{def}}{=} p_{x,i} < \frac{1}{2}$. The noise bit $e_{x,i}$ is drawn independently at every power-up of every SRAM cell.

The first part of the SRAM Assumptions follows from applying basic electrical laws to the implementation of the SRAM cell as shown in Fig. 2a. The implementation scheme of one logical inverter in CMOS technology is shown in Fig. 2b and consists of one p-channel and one n-channel transistor. It is beyond the scope of this work to explain the full transistor-level operation of the cell, which leads to the determination of the power-up state. It suffices to know that the electrical behavior of a transistor is mainly controlled by a physical parameter which is determined at manufacturing time and is called the *threshold voltage* or V_{Th}. The noise-free power-up state $r_{x,i}$ of an SRAM cell is consequently determined by the sign of $V_{\mathrm{Th}}^{P_1} - V_{\mathrm{Th}}^{P_2}$, with $V_{\mathrm{Th}}^{P_1}$ and $V_{\mathrm{Th}}^{P_2}$ the respective threshold voltages of the p-channel transistors in inverters 1 and 2 of the cell. In theory the threshold voltages of identically designed transistors (such as P_1 and P_2) are equal, and hence their difference should be zero. However, since V_{Th} is a parameter determined by a physical manufacturing process, it is never entirely deterministic, but it has a stochastic term due to the manufacturing uncertainties. In fact, the threshold voltages $V_{\mathrm{Th}}^{P_1}$ and $V_{\mathrm{Th}}^{P_2}$ can be assumed to be independently normally distributed random variables with the same mean value, and the sign of their difference is by consequence independently uniformly distributed. The first assumption follows directly from this physical observation.

So far, any noise during power-up has been ignored. However, at temperatures above absolute 0, there will always be some amount of random thermal noise affecting the voltages in the SRAM cells, with the noise power increasing with temperature. The *noisy* power-up state of an SRAM cell is determined by the sign of $V_{\text{Th}}^{P_1} - V_{\text{Th}}^{P_2} - V_{\text{noise}}$, with V_{noise} the amplitude of the noise voltage in the cell at the time of power-up. V_{noise} is drawn independently for every cell from the normal distribution $N(0, \sigma_{\text{noise}}^2)$ and drawn again every time the cell is powered up. The second part of the SRAM assumptions follows from the first part and the fact that the additional V_{noise} term can change the sign of $V_{\text{Th}}^{P_1} - V_{\text{Th}}^{P_2}$, causing a bit error ($e_{x,i} = 1$). Because V_{noise} is assumed to be drawn independently for every SRAM cell, possible bit errors occur independently in distinct cells, with probability $p_{x,i}$. It is clear that this bit error probability increases with rising temperature.

8.3 SRAM PUF Parameters and Experimental Validation

To briefly demonstrate that an SRAM PUF meets the PUF definition (Definition 3), we have to show that **R** contains independent random values, and that the Hamming weight of a random error vector $e_x = (e_{x,0}, \ldots, e_{x,m-1})$ is bounded:

- The values ℓ and m are design parameters of the SRAM memory. Mind that the size of the memory (and hence the implementation cost) rises linearly in m, but exponentially in ℓ.
- From the SRAM assumptions it follows that the theoretical distribution of the SRAM PUF responses has a min-entropy of $\mu = m$ *bit/response* ($\varepsilon_{\text{prf}} = 0$), even if the whole SRAM is queried ($q_{\text{prf}} = 2^\ell$).
- If for an average bit error probability $< \frac{1}{2}$, the value for δ is chosen large enough, then Pr[more than δ bit errors in m bits] can be made negligible.[5]

In [22] an SRAM PUF was constructed on an FPGA and the theoretical values for the min-entropy and the average bit error probability were experimentally verified. The performed experiments indicate that the average bit error probability of the response bits is bounded by 4% when the temperature is kept constant at 20°C, and by 12% at large temperature variations between −20 and 80°C. The probability of more than δ bit errors occurring decreases exponentially with increasing δ according to the Chernoff bound. δ is chosen high enough such that in practice, more than δ bit errors will never be observed. Accurately determining the min-entropy from a limited amount of PUF instances and responses is unattainable. In [35] it was shown that the mean smooth min-entropy of a stationary ergodic process is equal to the mean Shannon entropy of that process. Since the SRAM PUF responses are distributed according to such a stationary distribution (as they result from a physical phenomenon) it was estimated in [23] that its Shannon entropy equals 0.95 bit/cell.

---

[5] For example, choosing δ large enough such that Pr[more than δ bit errors in m bits] $\leq 10^{-9}$ will generally assure that more than δ bit errors will never occur in practice in a single response.

Because the mean smooth min entropy converges to the mean Shannon entropy, $H_\infty^\varepsilon(r_x)$ will be close to $H(r_x)$. Given this motivation, it is safe to put that $H_\infty^\varepsilon(r_x) \approx 0.95 \cdot m$.

8.4 From SRAM PUF to SRAM PRF

Next, we will introduce a new mode of operation that, similarly to the fuzzy extractor approach in the previous section, allows for extracting uniform values from SRAM PUFs in a reproducible way. This approach likewise stores some additional helper data but, as opposed to the case of fuzzy extractors, the helper data does not leak any information on the input. Hence, this construction might be of independent interest for SRAM PUF-based applications. The proposed construction is based on two techniques: Temporal Majority Voting and Excluding Dark Bits.

In general, we denote the individual bits of a PUF response as $y = (y_0, \ldots, y_{m-1})$, with $y_i \in \{0, 1\}$. When performing a response measurement on a PUF Π, every bit y_i of the response is determined by a Bernoulli trial. Every y_i has a most likely value $y_i^{(ML)} \in \{0, 1\}$, and a certain probability $p_i < 1/2$ of differing from this value which we define as its *bit-error probability*. We denote $y_i^{(k)}$ as the kth measurement or sample of bit y_i in a number of repeated measurements.

Definition 13 (Temporal majority voting (TMV)) Consider a Bernoulli distributed random bit y_i over $\{0, 1\}$. We define temporal majority voting of y_i over N votes, with N an odd positive integer, as a function $\text{TMV}_N : \{0, 1\}^N \rightarrow \{0, 1\}$, that takes as input N different samples of $y_i : y_i^{(0)}, \ldots, y_i^{N-1}$ and outputs the most often occurring value in these samples.

We can calculate the error probability $p_{N,i}$ of bit y_i after TMV with N votes as:

$$p_{N,i} \overset{\text{def}}{=} \Pr\left[\text{TMV}_N\left(y_i^{(0)}, \ldots, y_i^{(N-1)}\right) \neq y_i^{(ML)} \right] = 1 - \text{Bin}_{N,p_i}\left(\frac{N-1}{2} \right) \leq p_i, \tag{29}$$

with Bin_{N,p_i} the cumulative distribution function of the binomial distribution. From Eq. (29) it follows that applying TMV to a bit of a PUF response effectively reduces the error probability from p_i to $p_{N,i}$, with $p_{N,i}$ becoming smaller as N increases. Figure 3 shows a plot demonstrating the decreasing average bit-error probability after TMV as a function of the number of votes N. Given an initial error probability p_i, we can determine the number of votes N we need to reach a certain threshold p_T such that $p_{N,i} \leq p_T$. It turns out that N rises exponentially as p_i gets close to $1/2$. In practice, we also have to put a limit N_T on the number of votes we can perform, since each vote involves a physical PUF response measurement. We call the pair (N_T, p_T) a *TMV-threshold*.

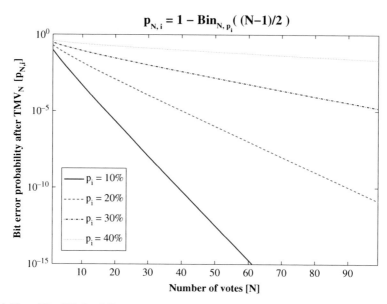

$$P_{N,i} = 1 - Bin_{N,p_i}((N-1)/2)$$

Fig. 3 Plot of Eq. (29) for different p_i

Definition 14 (Dark Bit (DB)) Let (N_T, p_T) be a TMV-threshold. We define a bit y_i to be *dark* with respect to this threshold if $p_{N_T,i} > p_T$, i.e., if TMV with N_T votes cannot reduce the error probability below p_T.

TMV alone cannot decrease the average bit-error probability to acceptable levels (e.g., $\leq 10^{-9}$) because of the non-negligible occurrence of dark bits. We use a *bit mask* γ to identify these dark bits in the generation phase and exclude them during reproduction. Similar to fuzzy extractors, TMV$_{N_T}$ and the consequent identifying of Dark Bits with respect to the TMV-threshold (N_T, p_T), can be used for generating and reproducing uniform values from SRAM PUFs:

Generation: The Gen-procedure takes sufficient measurements of every response bit y_i to make an accurate estimate of its most likely value $y_i^{(ML)}$ and of its error probability p_i. If y_i is dark with respect to (N_T, p_T), then the corresponding bit γ_i in the bit mask $\gamma \in \{0, 1\}^m$ is set to 0 and y_i is discarded, otherwise γ_i is set to 1 and y_i is appended to the bit string s. The procedure Gen outputs a helper string $\omega = (\gamma, \sigma)$ and an extracted string $z = \text{Extract}_\sigma(s)$, with Extract_σ a classical strong extractor[6] with seed σ.

Reproduction: The Rep-procedure takes N_T measurements of a response y' and the corresponding helper string $\omega = (\gamma, \sigma)$, with $\gamma \in \{0, 1\}^m$ as input. If γ_i contains a 1, then the result of TMV$_{N_T}\left(y_i'^{(0)}, \ldots, y_i'^{(N_T-1)}\right)$ is appended

---

[6] See, e.g., [13, 39] for a definition of a strong extractor. Typical seed lengths of strong extractors are in the order of 100 bits, and in most cases the same seed can be reused for all outputs.

to a bit string s', otherwise, y_i' is discarded. Rep outputs an extracted string $z' = \text{Extract}_\sigma(s')$.

A strong extractor [39] is a function that is able to generate nearly uniform outputs from inputs coming from a distribution with limited min-entropy. It ensures that the statistical distance of the extracted output to the uniform distribution is negligible. The required compression rate of Extract_σ depends on the remaining min-entropy μ of the PUF response y after the helper data is observed. We call the above construction a TMV-DB-SRAM-PUF.

Using analogous arguments as in Theorem 1, one can show that the output of a TMV-DB-SRAM-PUF is indistinguishable from random except with negligible advantage. Additionally, in an SRAM PUF as defined by Theorem 5, the most likely value of a bit is independent of whether or not the bit is a dark bit, hence no min-entropy on the PUF output is leaked by the bit mask.[7]

Lemma 1 *For the SRAM PUF as described in Theorem 5, it holds that $\tilde{H}_\infty(y_i|\gamma_i) = H_\infty(y_i)$, i.e. revealing the bit mask bit γ_i does not decrease the min-entropy of the SRAM PUF response bit y_i.*

Proof Starting from the SRAM assumptions as given by Definition 12 and their physical background, it follows that for each SRAM cell, y_i and γ_i are independently distributed random variables, i.e., $\forall a, b \in \{0, 1\} : \Pr[a \leftarrow y_i \cap b \leftarrow \gamma_i] = \Pr[a \leftarrow y_i] \cdot \Pr[b \leftarrow \gamma_i]$. We then start from the definition of conditional min-entropy as introduced by Dodis et al. [13]:

$$
\begin{aligned}
\tilde{H}_\infty(y_i|\gamma_i) &\stackrel{\text{def}}{=} -\log_2 \mathbb{E}_b 2^{-H_\infty(y_i|b \leftarrow \gamma_i)} \\
&= -\log_2 \mathbb{E}_b \max\{\Pr[0 \leftarrow y_i|b \leftarrow \gamma_i], 1 - \Pr[0 \leftarrow y_i|b \leftarrow \gamma_i]\} \\
&= -\log_2 \mathbb{E}_b \max\{\Pr[0 \leftarrow y_i], 1 - \Pr[0 \leftarrow y_i]\} \\
&= -\log_2 \max\{\Pr[0 \leftarrow y_i], 1 - \Pr[0 \leftarrow y_i]\} \stackrel{\text{def}}{=} H_\infty(y_i).
\end{aligned}
$$

$\square$

However, by searching for matching helper strings, an adversary might still be able to find colliding TMV-DB-SRAM-PUF inputs (especially as the input size is small), which can impose a possible security leak. In order to overcome this issue, we present the following way of using a TMV-DB-SRAM-PUF:

Definition 15 (All-at-once mode) Consider a TMV-DB-SRAM-PUF as described above. We define the *all-at-once* mode of operation to be the pair of procedures (Enroll, Eval):

Enrollment: The enrollment procedure Enroll outputs a helper table $\Omega \in \{0, 1\}^{2^\ell \times *}$ when executed. The helper table is constructed by running $\forall x \in \{0, 1\}^\ell$ the generation function $(\text{Gen} \circ \Pi)(x)$, and storing the obtained helper data ω_x as the x-th element in Ω, i.e. $\Omega[x] := \omega_x$.

---

[7] By consequence, also no min-entropy on the PUF *input* is leaked.

Evaluation: The evaluation function $\mathsf{Eval} : \{0, 1\}^\ell \times \{0, 1\}^{2^\ell \times *} \to \{0, 1\}^n$ takes an element $x \in \{0, 1\}^\ell$ and a helper table $\Omega \in \{0, 1\}^{2^\ell \times *}$ as inputs and (after internal computation) outputs a value $\mathsf{Eval}(x, \Omega) = z \in \{0, 1\}^n$, with $z = (\mathsf{Rep} \circ \Pi)(x, \Omega[x])$.

The Enroll-procedure has to be executed before the Eval-procedure, but it has to be run only once for every PUF. Every invocation of Eval can take the same (public) helper table Ω as one of its inputs. However, in order to conceal exactly which helper string is used, it is important that the Eval-procedure takes Ω *as a whole* as input and does not just do a look-up of $\Omega[x]$ in a public table Ω. The all-at-once mode prevents an adversary from learning which particular helper string is used during the *internal* computation.

Definition 16 (SRAM-PRF) An *SRAM-PRF* is a TMV-DB-SRAM-PUF that runs in the all-at-once mode.

Using the arguments given above we argue that SRAM-PRFs are in all practical views a physical realization of PRFs. Next, we demonstrate how to use them as components in an unbalanced Feistel scheme with expanding round functions and discuss some concrete parameters.

8.5 A Concrete Block Cipher Realization Based on SRAM-PRFs

In Sect. 7.3 it was shown how a secure block cipher can be constructed using n-to-n bit PUF-wPRFs. However, when using SRAM PUFs, constructing a PUF-wPRF with a reasonably large input size n is infeasible. Here, we construct a block cipher with the practically feasible expanding SRAM-PRF. As an example, we discuss an expanding Luby–Rackoff cipher. The security of such schemes was studied in [40] and lower bounds on the number of rounds were given. We present details of a SRAM-PRF construction taking an 8-bit challenge as input and producing a 120-bit extracted output. As an instantiation for the PUF, we take an SRAM PUF with an assumed average bit-error probability of 15% and an estimated min-entropy content of 0.95 bit/cell, which are realistic values according to the experimental observations in [22]. We use a TMV-threshold of ($N_T = 99$, $p_T = 10^{-9}$), yielding a safe very low average error probability using a practically feasible number of votes. Simulations and experiments on the SRAM PUF show that about 30% of the SRAM cells produce a dark bit with respect to this TMV-threshold. The strong extractor only has to compress by a factor of $\frac{1}{0.95}$, accounting for the limited min-entropy in the PUF response. An SRAM PUF containing at least $\frac{1}{0.95} \cdot \frac{2^8 \cdot 120}{70\%}$ bits $= 5.6$ KB of SRAM cells is needed to build this PUF-wPRF.

We determine the parameters for a proof-of-concept construction using the *lower* bound of rounds derived in [40]. Observe that this does not automatically imply the security of the scheme but shows a lower bound on the implementation size.

- We construct a block cipher with block length 128.
- In every round, an expanding SRAM-PRF: $\{0, 1\}^8 \rightarrow \{0, 1\}^{120}$ is used. Such a SRAM-PRF uses about 5.6 KB of SRAM cells, as shown earlier. The PUF-wPRF is operated in all-at-once mode.
- The results from [40] suggest to use at least 48 rounds, each with a different PUF-wPRF.
- The entire block cipher uses $48 \cdot 5.6$ KB ≈ 271 KB of SRAM cells. The helper tables also require 5.6 KB each.

Implementing 48 SRAM PUFs using a total of 271 KB of SRAM cells is feasible on recent ICs, and 48 rounds can be evaluated relatively fast. Storing and loading 48 helper tables of 5.6 KB each is also achievable in practice. Observe that the size depends linearly on the number of rounds. The according parameters for more rounds can be easily derived. Reducing the input size of the SRAM-PRF will yield an even smaller amount of needed SRAM cells and smaller helper tables, but the number of rounds will increase. A time-area tradeoff is hence possible.

9 Conclusions

In this chapter we propose a leakage-resilient encryption scheme that makes use of physically unclonable functions (PUFs). The core component is a new PUF-based cryptographic primitive, termed PUF-PRF, that is similar to a pseudorandom function (PRF). We showed that PUF-PRFs possess cryptographically useful algorithmic and physical properties that come from the random character of their physical structures.

Of course, any physical model can only approximately describe real life. Although experiments support our model for the considered PUF implementations, more analysis is necessary. In this context it would be interesting to consider other types of PUFs which fit into our model or might be used for other cryptographic applications. Furthermore, a natural continuation of this works would be to explore other cryptographic schemes based on PUF-PRFs, e.g., hash functions or public-key encryption.

Acknowledgments We thank Stefan Lucks for useful comments and discussions. The work of Berk Sunar was supported by the National Science Foundation Cybertrust grant No. CNS-0831416. The work of Roel Maes is funded by IWT-Flanders grant No. 71369 and is in part supported by the IAP Program P6/26 BCRYPT of the Belgian State and K.U.Leuven BOF funding (OT/06/04).

References

1. D. Agrawal, B. Archambeault, J.R. Rao, P. Rohatgi, in *The Em Side-Channel(s)*, ed. by B.S. Kaliski Jr., C.K. Koç, C. Paar. CHES. Lecture Notes in Computer Science, vol. 2523 (Springer, Berlin, 2002), pp. 29–45

2. A. Akavia, S. Goldwasser, V. Vaikuntanathan, in *Simultaneous Hardcore Bits and Cryptography Against Memory Attacks*, ed. by O. Reingold. TCC. Lecture Notes in Computer Science, vol. 5444 (Springer, Berlin, Heidelberg, 2009), pp. 474–495

3. R.J. Anderson, M.G. Kuhn, in *Low Cost Attacks on Tamper Resistant Devices*. Proceedings of the 5th International Workshop on Security Protocols (Springer, London), pp. 125–136

4. M. Bellare, A. Desai, E. Jokipii, P. Rogaway, in *A Concrete Security Treatment of Symmetric Encryption*. FOCS '97: Proceedings of the 38th Annual Symposium on Foundations of Computer Science (FOCS '97) (IEEE Computer Society, Washington, DC, 1997), p. 394

5. M. Bellare, J. Kilian, P. Rogaway, in *The Security of Cipher Block Chaining*. CRYPTO '94: Proceedings of the 14th Annual International Cryptology Conference on Advances in Cryptology (Springer, London, 1994), pp. 341–358

6. M. Bellare, D. Pointcheval, P. Rogaway, in *Authenticated Key Exchange Secure Against Dictionary Attacks*. EUROCRYPT (Springer, Berlin, 2000), pp. 139–155

7. M. Bellare, P, Rogaway, in *Entity Authentication and Key Distribution*. CRYPTO '93: Proceedings of the 13th Annual International Cryptology Conference on Advances in Cryptology (Springer, London, 1994), pp. 232–249

8. M. Bellare, P. Rogaway, in *Provably Secure Session Key Distribution: The Three Party Case*. STOC '95: Proceedings of the Twenty-seventh Annual ACM Symposium on Theory of Computing (ACM, New York, NY, 1995), pp. 57–66

9. R. Canetti, H. Krawczyk, in *Analysis of Key-Exchange Protocols and Their Use for Building Secure Channels*. EUROCRYPT '01: Proceedings of the International Conference on the Theory and Application of Cryptographic Techniques (Springer, London, 2001), pp. 453–474

10. N. Chandran, V. Goyal, A. Sahai, in *New Constructions for UC Secure Computation Using Tamper-Proof Hardware*. Advances in Cryptology – EUROCRYPT 2008 (Springer, Berlin, Heidelberg, 2008), pp. 545–562

11. B. Chor, O. Goldreich, Unbiased bits from sources of weak randomness and probabilistic communication complexity. SIAM J. Comput. **17**(2), 230–261 (1988)

12. L. Daihyun, J.W. Lee, B. Gassend, G.E. Suh, M. van Dijk, S. Devadash, Extracting secret keys from integrated circuits. IEEE Trans. VLSI Syst. **13**(10), 1200–1205 (Oct 2005)

13. Y. Dodis, R. Ostrovsky, L. Reyzin, A. Smith, Fuzzy extractors: How to generate strong keys from biometrics and other noisy data. SIAM J. Comput. **38**(1), 97–139 (2008)

14. S. Dziembowski, K. Pietrzak, in *Leakage-Resilient Cryptography*. FOCS '08: Proceedings of the 2008 49th Annual IEEE Symposium on Foundations of Computer Science (IEEE Computer Society, Washington, DC, 2008), pp. 293–302

15. B. Gassend, D. Clarke, M. van Dijk, S. Devadas, in *Controlled Physical Random Functions*. Annual Computer Security Applications Conference — ACSAC 2002 (IEEE Computer Society, Washington, DC, 2002), pp. 149

12. B. Gassend, D. Clarke, M. van Dijk, S. Devadas, in *Silicon Physical Unknown Functions*. ed. by V. Atluri. ACM Conference on Computer and Communications Security — CCS 2002 (ACM, New York, NY), pp. 148–160

17. G. Gaubatz, B. Sunar, M. Karpovsky, in *Non-linear Residue Codes for Robust Public-Key Arithmetic*. Workshop on Fault Diagnosis and Tolerance in Cryptography (FDTC '06), Yokohama, Japan, 2006

18. R. Gennaro, A. Lysyanskaya, T. Malkin, S. Micali, T. Rabin, in *Algorithmic Tamper-Proof (ATP) Security: Theoretical Foundations for Security Against Hardware Tampering*. Theory of Cryptography Conference (TCC'04). Lecture Notes in Computer Science, vol. 2951 (Springer, Heidelberg, 2004), pp. 258–277

19. O. Goldreich, *Foundations of Cryptography: Volume 1, Basic Tools* (Cambridge University Press, New York, NY, 2001)

20. O. Goldreich, S. Goldwasser, S. Micali, in *On the Cryptographic Applications of Random Functions*. Proceedings of CRYPTO 84 on Advances in Cryptology (Springer-Verlag New York, Inc., New York, NY, 1985), pp. 276–288

21. O. Goldreich, S. Goldwasser, S. Micali, How to construct random functions. J. ACM **33**(4), 792–807 (1986)

22. J. Guajardo, S. Kumar, G.-J. Schrijen, P. Tuyls, in *FPGA Intrinsic PUFs and Their Use for IP Protection.* ed. by P. Paillier, I. Verbauwhede. Cryptographic Hardware and Embedded Systems — CHES 2007, 10–13 Sept. Lecture Notes in Computer Science, vol. 4727 (Springer, Berlin, Heidelberg, 2007), pp. 63–80
23. J. Guajardo, S. Kumar, P. Tuyls, R. Maes, D. Schellekens, Reconfigurable Trusted Computing with Physical Unclonable Functions. June 2008
24. J.A. Halderman, S.D. Schoen, N. Heninger, W. Clarkson, W. Paul, J.A. Calandrino, A.J. Feldman, J. Appelbaum, E.W. Felten, in *Lest We Remember: Cold Boot Attacks on Encryption Keys.* ed. by P.C. van Oorschot. USENIX Security Symposium. (USENIX Association, Berkeley, CA, 2008), pp. 45–60
25. T. Ignatenko, F. Willems, in *On the Security of the XOR-Method in Biometric Authentication Systems.* Twenty-seventh Symposium on Information Theory in the Benelux, Noordwijk, The Netherlands, 2006, pp. 197–204
26. Y. Ishai, A. Sahai, D. Wagner, in *Private Circuits: Securing Hardware Against Probing Attacks.* CRYPTO, Santa Barbara, CA, 2003, pp. 463–481
27. M. Karpovsky, K. Kulikowski, A. Taubin, in *Robust Protection Against Fault-Injection Attacks on Smart Cards Implementing the Advanced Encryption Standard..* Proceedings of the International Conference on Dependable Systems and Networks (DNS 2004), Florence, Italy, 28 June–1 July 2004
28. J. Katz, in *Universally Composable Multi-Party Computation Using Tamper-Proof Hardware.* Advances in Cryptology – EUROCRYPT, Barcelona, Spain, 20–24 May 2007 (Springer, 2007), pp. 115–128
29. P. Kocher, J. Jaffe, B. Jun, in *Differential Power Analysis.* Proc. of CYRPTO '99 Santa Barbara, CA, USA, 15–19 Aug. Lecture Notes in Computer Science, vol. 1666 (Springer, 1999) pp. 388–397
30. P.C. Kocher, in *Timing Attacks on Implementations of Diffie-Hellman, RSA, DSS, and Other Systems.* ed. by N. Koblitz. CRYPTO. Lecture Notes in Computer Science, vol. 1109 (Springer, London, 1996), pp. 104–113
31. S.S. Kumar, J. Guajardo, R. Maes, G.-J. Schrijen, P. Tuyls, in *The Butterfly PUF: Protecting IP on Every FPGA.* IEEE International Workshop on Hardware-Oriented Security and Trust – HOST 2008, Anaheim, CA, USA, 9 Jun (IEEE, Piscataway, NJ, 2008)
32. K. Lemke, in *Embedded Security: Physical Protection Against Tampering Attacks.* ed. by C. Paar, K. Lemke, M. Wolf. Embedded Security in Cars, Chapter 2 (Springer, Berlin, Heidelberg, 2006), pp. 207–217
33. M. Luby, *Pseudo-Randomness and Applications* (Princeton University Press, Princeton, NJ, 1996)
34. M. Luby, C. Rackoff, How to construct pseudorandom permutations from pseudorandom functions. SIAM J. Comput. **17**(2), 373–386 (1988)
35. U. Maurer, R. Renner, S. Wolf, in *Security with Noisy Data, Part I* (Springer, 2007), pp. 21–44
36. S. Micali, L. Reyzin, in *Physically Observable Cryptography (Extended Abstract).* ed by M. Naor. Theory of Cryptography, First Theory of Cryptography Conference, (TCC 2004), Cambridge, MA, USA, 19–21 Feb, Proceedings. Lecture Notes in Computer Science, vol. 2951 (Springer, 2004), pp. 278–296
37. T. Moran, G. Segev, in *David and Goliath Commitments: UC Computation for Asymmetric Parties Using Tamper-Proof Hardware.* Advances in Cryptology – EUROCRYPT, Istanbul, Turkey, 13-17 April 2008 (Springer, 2008), pp. 527–544
38. M. Naor, G. Segev, in *Public-Key Cryptosystems Resilient to Key Leakage.* Proceedings of the 29th Annual International Cryptology Conference on Advances in Cryptology, Santa Barbara, CA, USA, 16–20 Aug 2009, Lecture Notes in Computer Science, vol. 5677 (Springer Verlag, Berlin, Heidelberg, New York, NY, 2009), pp. 18–35
39. N. Nisan, D. Zuckerman, in *More Deterministic Simulation in Logspace.* STOC '93: Proceedings of the 25th Annual ACM Symposium on Theory of Computing (ACM, New York, NY, 1993), pp. 235–244

40. J. Patarin, V. Nachef, C. Berbain, in *Generic Attacks on Unbalanced Feistel Schemes with Expanding Functions.* ed by K. Kurosawa. Advances in Cryptology - ASIACRYPT 2007, 13th International Conference on the Theory and Application of Cryptology and Information Security, Kuching, Malaysia, 2–6, Dec 2007, Proceedings. Lecture Notes in Computer Science, vol. 4833 (Springer, 2007), pp. 325–341
41. K. Pietrzak, in *A Leakage-Resilient Mode of Operation.* ed. by A. Joux. EUROCRYPT, Cologne, Germany. Lecture Notes in Computer Science, vol. 5479 (Springer, Berlin, Heidelberg, 2009), pp. 462–482
42. R. Posch, Protecting devices by active coating. J. Univers. Comput. Sci. **4**, 652–668 (1998)
43. O. Regev, in *On Lattices, Learning with Errors, Random LOinear Codes, and Cryptography.* Proceedings of the 37th Annual ACM Symposium on Theory of Computing, Baltimore, MD, USA, 22–24 May 2005, pp. 84–93
44. D. Samyde, S. Skorobogatov, R. Anderson, J.-J. Quisquater, in *On a New Way to Read Data from Memory.* SISW '02: Proceedings of the First International IEEE Security in Storage Workshop (IEEE Computer Society, Washington, DC, 2002), p. 65
45. S.P. Skorobogatov, in *Data Remanence in Flash Memory Devices.* ed. by J.R. Rao, B, Sunar. CHES. Lecture Notes in Computer Science, vol. 3659 (Springer, Heidelberg, 2005), pp. 339–353
46. S.W. Smith, Fairy dust, secrets, and the real world. IEEE Secur. Priv. **1**(1), 89–93 (2003)
47. F.-X. Standaert, T. Malkin, M. Yung, in *A Unified Framework for the Analysis of Side-Channel Key Recovery Attacks.* EUROCRYPT, Cologne, Germany, 2009, pp. 443–461
48. G.E. Suh, S, Devadas, in *Physical Unclonable Functions for Device Authentication and Secret Key Generation.* Proceedings of the 44th Design Automation Conference, DAC 2007, San Diego, CA, USA, 4–8 June 2007 (ACM, New York, NY, 2007), pp. 9–14
49. P. Tuyls, G.-J. Schrijen, B. Škorić, J. van Geloven, N. Verhaegh, R. Wolters, in *Read-Proof Hardware from Protective Coatings.* ed. by L. Goubin, M. Matsui. Cryptographic Hardware and Embedded Systems — CHES 2006. Lecture Notes in Computer Science, vol. 4249 (Springer, Heidelberg, 10–13 Oct 2006), pp. 369–383
50. I. Verbauwhede, P. Schaumont, in *Design Methods for Security and Trust.* Proceedings of Design Automation and Test in Europe (DATE 2008), Nice, France, 2007, p. 6
51. S.H. Weingart, in *Physical Security Devices for Computer Subsystems: A Survey of Attacks and Defences.* CHES '00: Proceedings of the Second International Workshop on Cryptographic Hardware and Embedded Systems (Springer, London, 2000), pp. 302–317

Part III
Hardware Attacks

Hardware Trojan Horses

Mohammad Tehranipoor and Berk Sunar

1 What Is the Untrusted Manufacturer Problem?

Over the last two decades we have become dependent on a network of electronic devices that supports a plethora of services, ranging from delivery of entertainment and news to maintenance of business records to filing of legal forms. This network provides a robust platform to handle all kinds of sensitive information at the personal, corporate, or government levels. Furthermore, many physical systems, e.g., the power grid, are currently being connected and to some extent controlled by commands relayed over the very same network. In essence the network permeates and blends into the physical infrastructure.

When we take a closer look at the network we see that it is composed of all kinds of electronic devices put together from integrated circuits (ICs) such as switching devices, analog acquisition boards, and microprocessors that run complex firmware and software. Due to cost-cutting pressures the design and manufacture of the majority of these ICs and other components are outsourced to third parties overseas. In particular, it is expected that by the end of this decade that the majority of the ICs will be fabricated in cheap foundries in China. This particular problem has recently received significant attention in press and in defense-related circles [16, 23].

As for IP cores,[1] many of them are already developed in Eastern European countries by engineers with low wages to keep the overall production cost low. We see a similar trend in software. Many of the low-level device drivers and even higher level software are already outsourced to Indian and Chinese companies. While the economic benefits are clear and many of the manufacturers are honest, outsourcing gives rise to a significant security threat. It takes only one malicious employee to compromise the security of a component that may end up in numerous products.

M. Tehranipoor (✉)
University of Connecticut, Storrs, CT 06269, USA
e-mail: tehrani@engr.uconn.edu

[1] IP-core stands for intellectual property-core and represents third-party hardware modules that achieve a useful isolated task such as audio-video decoders, error correction circuits etc.

A.-R. Sadeghi, D. Naccache (eds.), *Towards Hardware-Intrinsic Security,* Information   167
Security and Cryptography, DOI 10.1007/978-3-642-14452-3_7,
© Springer-Verlag Berlin Heidelberg 2010

Simply stated the untrusted manufacturer problem occurs in two situations:

- an IP-core designer inserts a secret functionality that deviates from the declared specification of the module
- a manufacturer modifies the functionality of the design during fabrication.

Clearly, besides tampering the goal of the malicious manufacturer is to operate covertly. The ever increasing complexity of hardware and software designs makes it much easier for untrusted manufacturers to achieve this goal.

Malicious tampering in manufacturing may manifest itself in a number of ways. During the development of an IC design the entire industry relies heavily on CAD tools. Either the synthesis tools themselves or cell libraries may have been compromised to produce designs with additional hidden functionality. Once the design is formed into an IC mask and shipped off to a foundry, another vulnerability emerges. At the foundry the IC mask may be altered to include new functionality or the ICs may simply be manufactured in such a way as to fail when certain conditions are met or leak information through covert channels. Circuits inserted during the manufacturing process are commonly now referred to as Trojan Circuits.[2] Once the IC is manufactured, more vulnerabilities are in the assembly of the ICs and in the installation of the firmware and higher level software. For example, in a recent news article it was reported that numerous Seagate Hard disks were shipped with a Trojan software that purportedly sent users' information to China [14].

It is clear that manufacturers have plenty of opportunities to inject Trojan functionality at various levels of the implementation in both hardware and software. Moreover, some of these designs may also be portrayed to have innocent bugs which, even when detected, still allow a manufacturer to repudiate *malicious* tampering. Indeed, even innocent manufacturing bugs may be exploited to attack cryptographic schemes [10].

Manufacturing bugs are quite common in the industry, and the common practice is to reduce them to an acceptable level via standard fault coverage tests. However, as it stands now the entire testing process focuses on ensuring the *reliability* of the product and not on the detection of tampered devices. For coverage against malicious faults, ICs could be reverse engineered and any Trojan circuitry could be detected. However, due to the destructive nature of the reverse-engineering process and the expense associated this approach does not appear to be practical.

Hardware Trojan detection is still a fairly new research area that has gained significant attention in the past few years. This chapter presents the current state of knowledge on existing detection schemes as well as design methodologies for improving Trojan detection techniques.

---

[2] To address this problem for ICs to be used in the military, some governments have been subsidizing the operations of a few domestic, high-cost trusted fabrication plants.

2 Hardware Trojans

The IC fabrication process contains three major steps, namely (i) *design* (including IP, models, tools and designer), (ii) *fabrication* (including mask generation, fab, and packaging), and (iii) *manufacturing test*. In today's horizontal design processes, the IPs are designed by IP vendors distributed across the five continents. The IPs are employed by a system-on-chip (SOC) integrator. A SOC design can potentially contain tens of IPs each coming from a different IP vendor. The IPs, models, and standard cells used during the design process by the designer, SOC integrator and post-design processes at the foundry are considered untrusted. The fabrication step may also be considered untrusted since the attacker could substitute Trojan ICs for genuine ICs during transit or subvert the fabrication process itself by implanting a Trojan into the IC mask. The manufacturing test, if only done in the client's (semiconductor company or government agency) production test center, would be considered as trusted.

In general, there are two main options to ensure that a chip used by the client is "authentic," i.e., it performs only those functions originally intended and nothing more. The first option is to make the entire fabrication process trusted. This option is prohibitively expensive with the current trends in the global distribution of the steps in IC design and fabrication. The second option is to verify the trustworthiness of the manufactured chips upon returning to the clients. To achieve this, it would be necessary to define a post-manufacturing step to validate conformance of the chip with the original functional and performance specifications. This new step is called Silicon Design Authentication.

The countermeasures proposed for hardware-based attacks usually modify hardware to prevent carrying a successful attack and to protect IPs or secret keys. However, the types of attacks we are concerned about are fundamentally different. Here, the attacker is assumed to maliciously alter the design before or during fabrication, i.e., the fabricated IC is untrusted. Detection of such alterations is extremely difficult for several reasons [27]:

1. Given the large number of soft, firm, and hard IP cores used in system-on-chips and given the high complexity of today's IPs, detecting a small malicious alteration is extremely difficult.
2. Nanometer IC feature sizes make detection via physical inspection and destructive reverse engineering extremely difficult and costly. Moreover, destructive reverse engineering does not guarantee that the remaining ICs are Trojan free especially when Trojans are selectively inserted into a portion of the chip population.
3. Trojan circuits, by design, are typically activated under very specific conditions (e.g., connected to low-transition probability nets or sense a specific design signal such as power or temperature), which makes them very unlikely to be activated and detected using either random or functional stimuli during test and authentication.

4. Tests used to detect manufacturing faults such as stuck-at and delay faults cannot guarantee detection of Trojans. Such tests target nets in a netlist of a Trojan-free circuit; they do not target multiple nets simultaneously and therefore cannot activate and detect Trojans. Even when 100% fault coverage for all types of manufacturing faults can be obtained, it does not provide any guarantees as far as Trojans are concerned.

5. As physical feature sizes decrease due to improvements in lithography, the process and environmental variations impact the integrity of the circuit parametrics. Detection of Trojans using simple analysis of these parametric signals is ineffective in the presence of variations in addition to measurement noises.

3 A Taxonomy of Hardware Trojans

The first detailed taxonomy for hardware Trojans was developed in [29] to provide the researchers with the opportunity to examine their methods against the different Trojan types. Currently, there is a lack of metrics to evaluate the effectiveness of methods in detecting Trojans thus, developing comprehensive taxonomy can help analyze Trojan detection techniques and measure their effectiveness. As malicious alterations to the structure and function of a chip can take many forms, the authors decompose the Trojan taxonomy into three principle categories as shown in Fig. 1, i.e., according to their *physical, activation,* and *action* characteristics. Although it is possible for Trojans to be hybrids of this classification, e.g., have more than one activation characteristic, this taxonomy captures the elemental characteristics of Trojans and is effective for defining and evaluating the capabilities of various detection strategies. This taxonomy details the vulnerabilities during the

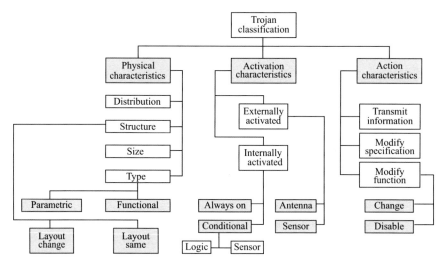

Fig. 1 Detailed Trojan taxonomy

IC fabrication process (from GDSII to fabricated IC); a similar taxonomy can be developed when detecting Trojans in IP cores.

The Physical characteristics category describes the various hardware manifestations of Trojans. The Type category partitions Trojans into functional and parametric classes. The Functional class includes Trojans that are physically realized through the addition or deletion of transistors or gates, while parametric refers to Trojans that are realized through modifications of existing wires and logic. The Size category accounts for the number of components in the chip that have been added, deleted, or compromised while the Distribution category describes the location of the Trojan in the physical layout of the chip. The Structure category refers to the case where the adversary is forced to regenerate the layout to be able to insert the Trojan, then the chip physical form factor may change. Such changes could result in different placement for some or all the design components. Any malicious changes in physical layout can change the delay and power characteristics of the chip would facilitate Trojan detection. In [29], the authors have presented the adversary's abilities for minimizing the probability of detection.

Activation characteristics refer to the criteria that cause a Trojan to become active and carry out its disruptive function. Trojan activation characteristics is partitioned into two categories: labeled externally activated (e.g., by an antenna or a sensor that can interact with the outside world) or internally activated which is divided by Always-on and Condition-based as shown in Fig. 1. Always-on, as the name implies, indicates that the Trojan is always active and can disrupt the function of the chip at any time. This class covers Trojans that are implemented by modifying the geometries of the chip such that certain nodes or paths in the chip have a higher susceptibility to failure. The adversary can insert the Trojans on nodes or paths that are rarely exercised. The Condition-based subclass includes Trojans that are "inactive" until a specific condition is met. The activation condition can be based on the output of a sensor that monitors temperature, voltage, or any type of external environmental condition, e.g., electromagnetic interference, humidity, altitude, temperature. Or it can be based on an internal logic state, a particular input pattern or an internal countervalue. The Trojan in these cases is implemented by adding logic gates and/or flip-flops to the chip, and therefore is represented as a combinational or sequential circuit.

Action characteristics identify the types of disruptive behavior introduced by the Trojan. The classification scheme shown in Fig. 1 partitions Trojan actions into three categories: Modify function, Modify specification, and Transmit information. As the name implies, the Modify function class refers to Trojans that change the chip's function through additional logic or by removing or bypassing existing logic. The Modify specification class refers to Trojans that focus their attack on changing the chip's parametric properties, such as delay in cases where the adversary modifies existing wire and transistor geometries. Finally, Transmit information class would transmit key information to adversary.

Many types of Trojans have been designed by the researchers to evaluate their detection techniques by targeting them in an IC [5, 19, 25, 28, 30]. To imitate adversaries' Trojan insertion, the authors in [5] classified the components needed

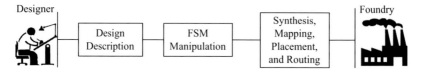

Fig. 2 Insertion of HTH during design process of an IP core

for a hardware Trojan horse (HTH) into three categories: trigger, storage, and driver. Trigger incites the planned HTH. After a trigger occurs, the action to be taken can be stored in memory or sequential circuit. Driver implements the action prompted by the trigger. Based on the above classification, [5] presents a systematic approach to insert Hardware Trojans into the IC by pre-synthesis manipulation of the circuit's structure. Such model addresses the issue of trust in intellectual property cores designed by either a third party vendor or system integrator where a large number of IP cores developed by many vendors are used.

Figure 2 shows an abstracted view of the design process. The Trojan designer composes the high-level design description to find the computation model of the circuit that can be represented by a finite state machine (FSM). HTH can be inserted to the circuit by altering the FSM and embedding states into it. The modified FSM should have trigger as an input and a driver hidden in the structure of the FSM. It can be systematically hidden in the design by merging its states within the states of the original design's FSM. Thus, the HTH would be inseparable (unremovable) from the original design's functionality. A stealth communication, which uses the medium for legitimate communications, can be used as covert channels to transfer confidential data from the working chips. This Trojan embedding approach provides a low-level mechanism for bypassing the higher level authentication techniques.

4 A High-Level Attack: Shadow Circuits

The authors in [19] investigate different types of attacks on a design at the register transfer level (RTL). Specifically, the authors examine the possibility of designing hardware Trojans that are able to evade state-of-the-art detection methodologies, as well as passing functional test. In [21], the authors consider the malicious circuit design space and introduce hardware components that can enable a number of attacks. In particular, they design and implement the Illinois Malicious Processor with a modified CPU. The malicious modifications allow memory access and shadow mode mechanisms. The former permits the attacker to violate the operating system (OS) isolation expectations while the latter admits stealthy execution of a malevolent firmware. The attacks were evaluated on an FPGA development board by modifying the VHDL code of the Leon processor that is an open source SPARC v8 processor including the memory management unit. The overhead in logic is less than 1% for both modifications, while the timing overhead was about 12%.

The authors further design and implement three potential attacks: (i) a privilege escalation attack, which gives the intruder access to the root without checking the credentials or generating log entries, (ii) a login backdoor in the shadow mode letting the intruder to log in as a root without using a password, and (iii) a service for stealing the passwords and sending them to the attacker. This chapter concludes that hardware tampering is practical and could support a variety of attacks while it is also hard to detect. Note that mechanisms for actively controlling the IC can also be used for designer malicious circuit insertion. For example, manipulation of the states in the finite state machine which cannot be reversed engineered can be used for embedding the Trojan circuitry by providing mechanisms for remotely activating, controlling, and disabling [4].

5 Trojan Detection Methodologies

A number of Trojan detection methodologies have been developed over the past few years. Without loss of generality, the methods are categorized as side-channel analysis and Trojan full activation which are mainly chip-level solutions in addition to architectural-level Trojan detection solutions. The detection methods in addition to design-for-hardware-trust techniques are presented in this section.

5.1 Trojan Detection Using Side-Channel Signal Analysis

Side-channel signals, including timing and power, can be used for Trojan detection. Trojans typically change the parametric characteristics of the design, e.g., degrade performance, change power characteristics, and/or introduce reliability problems in the chip regardless of the Trojan type. As a result, power/delay characteristics of wires and gates in the affected circuit are influenced. Power-based side-channel signals provide visibility of the internal structure and activities within IC, which enables Trojans to be detected without fully activating them. Timing-based side channel can be used to detect the presence of the Trojan if the chip is tested using efficient delay tests that are sensitive to small changes in the circuit delay along the affected paths and are able to effectively differentiate Trojans from process variations.

5.1.1 Transient Power-Based Analysis

The work in [3] was first to propose the use of side-channel profiles such as power consumption and electromagnetic emanation for Trojan circuit detection. The authors first generate the power signature profiles of a small number of ICs randomly selected from a batch of manufactured ICs. These ICs serve as golden masters (Trojan-free ICs). Once profiled the golden masters undergo a rigorous destructive reverse-engineering phase where they are compared piece by piece against the original design. If Trojan free, the ICs are then accepted as genuine ICs and their profiles will serve as power templates. The remaining ICs now can be tested efficiently and in a nondestructive manner by simply applying the same stimuli and building their power profiles. The profiles are compared using statistical techniques such as principle component analysis against the templates obtained from the golden masters.

To understand why this technique works and where it breaks down we need to consider the makeup of the power measurements:

1. power consumption of the original circuit;
2. measurement noise which can be drastically reduced via repeated measurements and averaging;
3. noise introduced by process variations which is unique to each IC and cannot be removed; and
4. the power consumption of the Trojan circuit.

The authors simulated their technique on two representative circuits with a 16-bit counter, 8-bit sequential comparator, and a 3-bit combinational comparator Trojan circuit. In all three cases the Trojan circuits were identified in the eigenvalue spectrum of the power profiles.[3] Figure 3 shows the eigenvalue spectrum of the genuine circuit before and after the Trojan circuit is introduced. Note that the Trojan circuit stand out for eigenvalues beyond 12.

With the side-channel-based technique Trojans of various sizes and functionalities may be detected as long as their power consumption contribution statistically stands out from that of process noise. Note that for detection the Trojan need not be active nor draw a significant power. As long as the Trojan changes the parasitic loading of the genuine circuit, it might change the statistics of the circuit and become

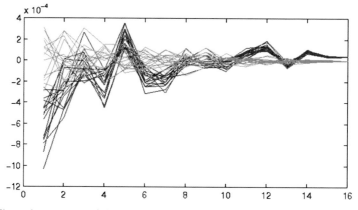

Fig. 3 Eigenvalue spectrum of circuit with and without Trojan

---

[3] The smallest Trojan became only noticeable when the power profiles were sampled at the low-noise parts of the signal.

noticeable. Alternatively, the technique will fail in the presence of excessive process noise.

The authors of [28] argue that most Trojans inserted into a chip require power supply and ground to operate. The Trojans can be of different types and sizes and their impact on the circuit power characteristics could be very large or very small. The chapter develops a multiple-supply transient-current integration methodology to detect the hardware Trojan. Then a Trojan isolation method based on localized current analysis is introduced. The current is assumed to be measured from various power ports or controlled collapse chip connections (C4s) on the die. Random patterns are applied to increase the switching in the circuit in a test-per-clock fashion.

The amount of current a Trojan can draw could be so small that it can be submerged into envelop of noise and process variations effects, and therefore it may be undetectable by conventional measurement equipments. However, Trojan detection capability can be greatly enhanced by measuring currents locally and from multiple power ports/pads. Figure 4 shows the current (charge) integration methodology for detecting hardware Trojans presented in [28]. There are four power ports on the die. The golden die can be identified using an exhaustive test for a number of randomly selected dies. It can also be identified using the pattern set that will be used in the current integration method by comparing the results against each other for all the patterns in an exhaustive fashion. If the same results (within the range of variations) are obtained for all the selected dies, they can be identified as Trojan free. The paper assumes that adversary will insert the Trojans randomly in a selected number of chips [28]. After identifying the golden dies, the worst-case charge is obtained (dashed line in the figure) in response to the pattern set. The worst-case charge is based on the worst-case process variations in one of the genuine ICs. Next, the pattern set is applied to each chip and the current would be measured for each pattern locally via the power ports or C4 bumps.

The figure shows the current waveform of n number of patterns applied to the chips. The figure also illustrates the charge variations with time for all the current waveforms obtained after applying the patterns. The charge corresponds to the area produced by each current waveform. $Q_n(t)$ denotes the accumulative charge after

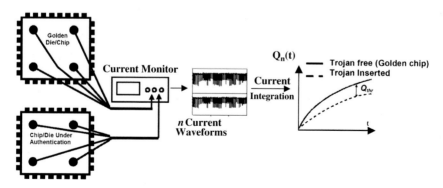

Fig. 4 Current integration (charge) method

applying n patterns. Q_{thr} is the charge threshold to detect a Trojan which is in fact the resolution measurement defined by the instrumentation. When applying the patterns, the charge increases and is compared continuously against the worst-case charge calculated for golden dies. Once the difference between the two curves ΔQ is greater than Q_{thr} then a Trojan is detected. The number of patterns, n, is expected to be very small for large Trojans and large for very small Trojans and application time is expected to be low since the patterns are applied in a test-per-clock fashion.

A region-based transient power signal analysis method to reduce the impact of increasing levels of process variations and leakage currents was proposed in [25]. A region is defined as a portion of the layout that receives the majority of its power from a set of surrounding power ports or C4 bumps. Figure 5a shows a six metal layer power grid with nine power ports (PPs). Measurements are done through each power port individually by applying patterns. The transient-current detection technique (known as I_{DDT}) is based on a statistical analysis of the I_{DDT} waveform areas generated at the nine power ports as a test sequence is simulated on design. For each orthogonal pairing of power ports, a scatter plot is constructed. In this chapter several different process models are used for Trojan-free and Trojan-inserted designs. Figure 5b shows that using a prediction ellipse derived from a Trojan-free design with different process models, it is possible to distinguish between Trojan-inserted and Trojan-free designs. The dispersion in the Trojan-free data points is a result of uncalibrated process and test environment (PE) variations.

However, regional analysis by itself is not sufficient for dealing with the adverse effects of PE variations on detection resolution. Signal calibration techniques are used to attenuate and remove PE signal variation effects to fully leverage the resolution enhancements available in a region-based approach. Calibration is performed on each power port and for each chip separately, and it measures the response of each power port to an impulse. The response of each power port X (PP_X) is normalized by the sum of current drawn from power ports in the same row as PP_X. The normalized values of power ports make calibration matrix. After applying each test pattern, the response is calibrated using calibration matrix. The results presented in

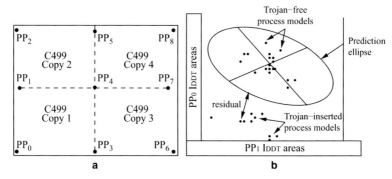

Fig. 5 (a) Architecture of simulation model and (b) Example scatterplot for PP01 for one of the experiments

this chapter shows that calibration can increase distance between Trojan-free and Trojan-inserted (residual in Fig. 5b) designs under different process parameters.

Recently, a number of approaches for gate-level timing/power characterization by nondestructive measurements have been proposed [6]. Each measurement forms one equation. After a linear number of measurements are taken, a system of equations for mapping the measured characteristics to the gate-level will be formed. The authors in [24] exploit the formulation of gate-level characterization by linear programing and by singular value decomposition to detect Trojans. Both timing and static power measurements are used. The Trojan detection task is performed by constraint (equation) manipulation. The method attempts at finding the measurement matrix with the highest rank and derives a number of heuristics for detecting the gates with inconsistent characteristics compared to their original specified characteristics. Learn and test and re-substitution statistical validation techniques are used for estimating the bounds for the normal (nonmalicious) characteristics. The experiments consider the error in noninvasive measurements but do not consider the process variations. The evaluation results are promising as the gate-level characterization can be done with a high accuracy. The gate-level characterization methods can find the characteristics of controllable (sensitizable) gates. This controllability is known to be high for static power measurements and testing (IDDQ testing). The authors in [6] use statistical convergence of the gate-level estimation and signal integrity for Trojan detection. They find efficient robust approximations for the gate power consumptions and identify the malicious insertions by multiple consistency checking.

5.1.2 Timing-Based Analysis

In [22], the authors propose a delay-based physical unclonable function (PUF) for Hardware Trojan detection. The method utilizes sweeping clock delay measurement technique to measure selected register-to-register path delays. Trojans can be detected when one or a group of path delays are lengthened above the threshold determined by process variations level. The path delay measurement architecture is shown in Fig. 6. The main circuit is a register-to-register combinational path that is to be characterized and the registers on this path are triggered by the main system clock (CLK1). The components outside the box are part of the testing circuitry. The *shadow register* takes the same input as the destination register in the main circuit, but it is triggered by the shadow clock (CLK2), which runs at the same frequency as CLK1 but at a controlled phase offset. The results that are latched by the destination register and the shadow register are compared during every clock period. If the comparison result is unequal, the path delay is characterized with a precision of the skew step size. To overcome the problem that temperature will affect path delay, this method incorporates on-die temperature monitor, which uses a ring oscillator as the clock input of a counter to measure operating temperature. Since the oscillator is embedded within the main circuitry and its switching frequency is temperature dependent, then the authenticator can calculate the effective response from the reported temperature and delay signature. Although effective, this

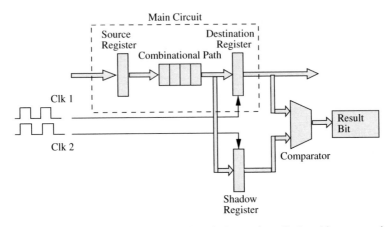

Fig. 6 Path delay measurement architecture using shadow register. Such architecture can be used for IC authentication and Trojan detection

technique suffers from the very large area overhead when targeting today's large designs with millions of paths.

In [20], the authors propose a new fingerprint generating method using path delay information of the whole chip. There are many delay paths in a chip and each one represents one part of the characteristic of the entire chip. The timing features can be used to generate a series of path delay fingerprints. No matter how small the Trojan is compared to the whole size of the chip, it can be significant in the path view and may be detected. The entire testing procedure includes three steps:

1. Path delay gathering of nominal chips: In this step, many chips are selected from a fabricated design. High-coverage input patterns are then run on the sample chips and high-dimensional path delay information is collected. Then, the sample chips are checked under reverse engineering to ensure they are genuine circuits.
2. Fingerprint generation: According to the path delays, a series of delay fingerprints are generated and mapped to a lower dimension space.
3. Trojan detection: All other chips are checked under the same test patterns. Their delay information is reduced to low dimension and compared to the delay fingerprints.

The method uses statistical analysis to deal with process variations. Since there may be millions of paths in today's circuits, measuring all paths and especially the short ones is practically infeasible.

5.2 Trojan Activation Methods

Trojan activation strategies can accelerate Trojan detection process and have, in some cases, been combined with power analysis during implementation. If a portion of the Trojan circuitry is activated, more dynamic power will be consumed by the

Trojan circuit that will further help differentiate the power traces of Trojan-inserted and Trojan-free circuits. The existing Trojan activation schemes can be categorized as follows.

5.2.1 Region-Free Trojan Activation

Region-free Trojan activation refers to methods that do not rely on the region and depend on accidental or systematic activation of Trojans. For instance, [18] presents a randomization-based probabilistic approach to detect Trojans. The authors show that it is possible to construct a unique probabilistic signature of a circuit using specific probability on its inputs. Input patterns are applied based on a specific probability to IUA and outputs are compared with original circuit. In case there is a difference between the outputs, existence of a Trojan is reported. For Trojan detection in a manufactured IC, we can only rely on applying patterns based on such probability to obtain a confidence level whether original design and fabricated chip are the same.

Analyzing rare nets combinations in a design is found in [30]. These rare activated nets are used as Trojan trigger. At the same time, nets with low observability are used as payload as shown in Fig. 7. The authors generate a set of vectors to activate such nets and suggest combining it with traditional ATPG test vectors to activate the Trojan and propagate its impact if the Trojan was connected to these nets.

5.2.2 Region-Aware Trojan Activation

In [7], the authors develop a two-stage test generation technique that targets magnifying the difference between the IUA and the genuine design power waveforms. In the first stage (circuit partitioning), region-aware pattern helps identify the potential

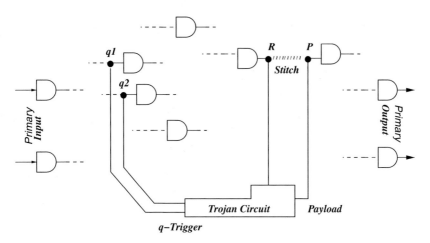

Fig. 7 Trojan circuit model with rare triggering condition

Trojan insertion regions. Next, new test patterns concentrating on the identified regions are applied to magnify the disparity between original and Trojan-inserted circuits (activity magnification). In the circuit partitioning stage, to detect a Trojan circuit, the activity within a portion of the circuit needs to increase while the activity for the rest of the circuit must be minimized simultaneously. The flip-flops in a circuit are classified into different groups depending on structural connectivity. For activity magnification stage, based on the comparison of the relative difference in the power profiles for the genuine and Trojan circuits using the vector sequence generated in Stage 1, the regions (set of flip-flops) that exhibit increased relative activity are identified. In this stage, more vectors for the specific region(s), marked as possible Trojan regions, are generated using the same test generation approach as the circuit partitioning stage.

Magnifying Trojan contribution by minimizing circuit activity is discussed in [8]. This is done by keeping input pins unchanged for a number of clock cycles. Thus, circuit activity comes from the state elements of design. The overall switching activity is therefore reduced and can be limited to specific portions of the design helping Trojan localization. Different portions of the design can be explored by changing input vectors to localize a Trojan. At the same time, each gate is supplied with two counters, namely *TrojanCount* and *NonTrojanCount*. With each vector, if the number of transitions at the output of a gate is more than a specific threshold, its *TrojanCount* would increase and vice versa. The *TrojanCount/NonTrojanCount* ratio, called gateweight, indicates activity of a gate. High gateweight ratio means the gate is considerably impacted by Trojan since there is relatively high power difference corresponding to the gate activation.

Given that the type and size of Trojan are unknown to the test engineer, both region-free and region-aware methods must be applied. If the inputs of Trojan circuit come from part of the circuit where they are functionally dependent (i.e., part of the same logic cone), then the region-aware method can be effective. However, if the Trojan inputs are randomly selected from various parts of the circuit, then region-free methods could increase the probability of detection.

6 Design-for-Hardware-Trust Techniques

Current design practices do not support the effective side-channel signal analysis as well as pattern generation for Trojan detection. The CAD and test community has long benefited from design-for-testability and design-for-manufacturability. In this section, we look closely into some of the methods proposed by hardware security and trust community to improve Trojan detection and isolation by changing/modifying the design flow. We call these methods design-for-hardware-trust. The methods help in (i) preventing the insertion of Trojans, (ii) easier detection of Trojans, and (iii) effective IC authentication.

In [26], the authors developed a methodology to increase the probability of generating a transition in functional Trojan circuits and analyze the transition generation time. Transition probability is modeled using geometric distribution and is estimated

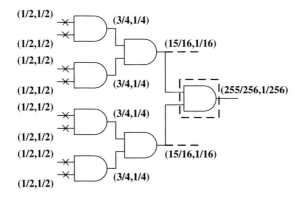

Transition probability at Trojan output = 255/65536

Average clock cycle per transition by GD = 255.6

Average clock cycle per transition by simulation = 255.6

Fig. 8 Analyzing transition probability in the original circuit

based on the number of clock cycles to generate a transition on a net. To increase transition probability of nets whose transition probability is lower than a specific probability threshold, an efficient dummy flip-flop insertion procedure is presented. The procedure identifies nets with low-transition probability and inserts dummy flip-flops such that it increases the transition probability. Figure 8 shows a circuit with a T_{gj} as a Trojan gate. The transition probability at the output of the gate is extremely low. However, after adding dummy scan flip-flop as shown in Fig. 9 to a

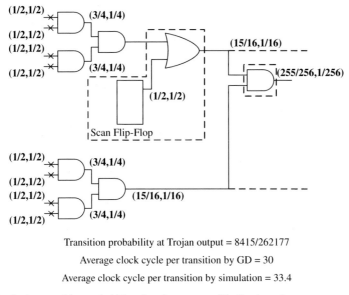

Transition probability at Trojan output = 8415/262177

Average clock cycle per transition by GD = 30

Average clock cycle per transition by simulation = 33.4

Fig. 9 Analyzing transition probability after dummy scan flip-flop insertion

net with low-transition probability, the transition probability at the output of Trojan increased considerably; similarly the average number of clock cycles per transition has decreased.

It should be noted that dummy flip-flops are inserted in a way not to alter the design's functionality and performance. The effectiveness of dummy flip-flop insertion is examined by evaluating different transition probability thresholds for various Trojan circuits. The relationships among authentication time, the number of required transitions in Trojan circuit, and the tester clock were studied in detail. These parameters would help determine the transition probability threshold of a design. The transition probability threshold, in turn, provides an estimation of area overhead induced by inserted dummy flip-flops. This method help detecting Trojans in two ways: (i) it can improve power-based side channel signal analysis methods by increasing the number of Trojan switchings and (ii) it provides an opportunity for fully activating a Trojan circuit and observing the erroneous response at the output of the circuit.

An inverted voltage scheme is proposed in [9] to magnify Trojan activity. As Trojan is assumed to be activated under rare condition, IC inputs could be changed so that rare combinations are created to activate Trojan. For example, for an AND gate with four inputs a rare condition would be when all its inputs are "1," with probability 1/16. The goal of the scheme is to change the functionality of Trojan to remove the rare condition. Exchanging power (V_{DD}) and ground (GND) of a gate changes its function and reduces the noise margin as the output swings between $V_{DD} - V_{TH}$ and V_{TH} (V_{TH} is the voltage threshold of the transistor). Thus, AND changes to NAND and "1" at the output of a NAND Trojan is not a rare value anymore and its probability is 15/16 as shown in Fig. 10. This method would face

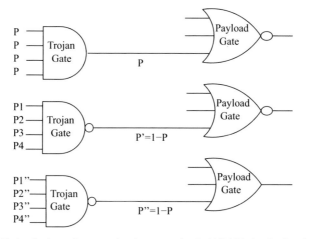

Fig. 10 (*top*) Trojan logic under normal voltage supply, (*middle*) Trojan logic where only Trojan gate is affected by inverted voltage supply and (*bottom*) Trojan logic where both Trojan and Payload gate are affected by inverted voltage supply

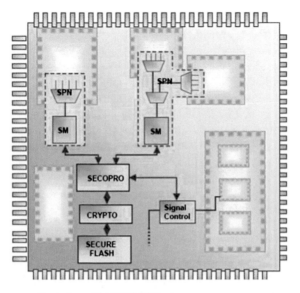

Fig. 11 A system-on-chip design with DEFENSE logic

the difficulty of switching between power and ground for each gate on the circuit as current power distribution networks are not design to support it.

Employing reconfigurable Design-For-Enabling-Security (DEFENSE) [1] logic to the functional design is proposed to monitor the system operation of IC and detect unexpected or illegal behavior. When an attack is detected, the first step is to deploy countermeasures such as disabling a suspect block or forcing a safe operational mode. Figure 11 shows the architecture of a system-on-chip with inserted DEFENSE. Signal Probe Networks (SPNs) are configured to select a subset of the monitored signals and transport them to Security Monitors. A Security Monitor is a programable transaction engine configured to implement an FSM to check user-specified behavior properties of the current signals. The Security and Control Processor (SECOPRO) reconfigures SPNs to select the groups of signals to be checked by SMs and reconfigures Security Monitors to perform the required checks. All the configurations are encrypted and stored in the Secure Flash memory. The security checks are application dependent and circuit dependent. This approach can detect attacks and prevent the damage caused by attacks those are inserted during every phase of the design flow. It is concealed from attackers trying to reverse engineer the device as the reconfigurable logic is blank (un-programed) in a powered-off chip. On the other hand, the security checkers are not accessible from either the functional logic or from the embedded software. Similarly, SECOPRO is invisible for the other on-chip application processors. However, this approach cannot detect the Trojan unless the chip is fabricated and will not be able locate it. To increase the efficiency of the method, a large number of nets in the circuit must be selected for observation which could increase the area overhead.

The authors in [12] introduce a design methodology called On-Demand Transparency. The objective is to facilitate Trojan detection based on logic testing. They define a special mode of operation (called transparent mode) for a multi module system. In this operation mode, a signature is generated upon application of a specific input. The hope is that Trojan tampering would affect the signature and would reveal the Trojan. The selected nodes are those that are assumed to be most susceptible to Trojan attacks, guided by the controllability and observability values. Note that this type of signature generation for circuit watermarking and finding watermark tampering was proposed and used earlier in literature. The drawback of the method proposed in [12] is that it is well known that the states of a finite state machine contain many do not cares and there are exponentially many states. Thus, generating a signature that would be affected by each kind of tampering is an extremely challenging problem for larger circuits.

7 Circuit Obfuscation as a Countermeasure

A completely different approach in tackling the Trojan circuit problem is to camouflage or obfuscate the genuine circuit. For instance, some critical IC designs are requested to be manufactured along with several non-critical ICs or the true function of an IC is buried in complex logic. In fact, this technique has its roots in software obfuscation where the functionality of a piece of code is scrambled into a functionality equivalent code to prevent unauthorized parties from making modifications typically to bypass some authentication function. In the IC industry the obfuscation approach has been used for many years to protect the intellectual property contained in the IC design. Hence, Trojan detection seems like a natural application for circuit obfuscation. Since the technique requires modification to the actual circuit, it presents an example of a Trojan prevention technique rather than a detection technique.

Obfuscation against Trojan circuits was briefly mentioned in [15]. A concrete technique was proposed in [11]. In [11] the authors focus on changing the functionality and structure of the IP core (by modifying the gate-level netlist) to obfuscate the design and to embed authentication features. After the design is manufactured the device may be checked by verifying the authentication response for a particular input. In [13] a key-based obfuscation technique is proposed to achieve security against hardware Trojans. The obfuscation scheme works by modifying the state transition function of a given circuit. The technique is based on the premise that an intelligent adversary would first locate a rare event in the circuit to insert the Trojan circuit in order to make it stealthy. The authors propose to add fictitious states to the circuit to prevent an adversary to locate a rare event state and inject stealthy Trojans. Moreover, if the Trojans are inserted to activate at a fictitious state value (that will never occur during normal mode operation) the Trojan will become benign. The authors implemented the proposed obfuscation technique on the s1196

ISCAS-89 benchmark circuit with 20,000 random instances of four-input Trojans with a 4-bit obfuscation key. The simulation shows that the obfuscation hides the true signal probabilities of a large number of nodes and makes a significant fraction of the Trojans benign.

Despite its elegance and dual purpose (Trojan prevention & IP protection) there are several disadvantages of the obfuscation approach:

- The original circuit needs to be modified. This adds more burden to the designer and makes testing more difficult.
- Regular netlists are highly optimized and do not contain enough redundancy to implement obfuscation, unless the circuit is augmented with artificial redundancy. Thus the technique bears significant overhead compared to testing methods which bear little overhead.
- It is difficult to define metrics to quantify the attained security level by obfuscation. Most metrics such as identification rates for randomly injected Trojans implicitly assume a fairly primitive adversarial model. For instance, obfuscation will not prevent an attacker who only intends to disrupt the operation of the IC, i.e., implement a Denial of Service attack.

Acknowledgments The authors would like to thank Farinaz Koushanfar of Rice University for her contribution to the Trojan detection part of the chapter.

References

1. M. Abramovici, P. Bradley, in *Integrated Circuit Security - New Threats and Solutions*. Proceedings of Cyber Security and Information Infrastructure Research Workshop (CSIIRW), (Oak Ridge National Laboratory, Oak Ridge, TN, USA, 2009)
2. S. Adee, The hunt for the kill switch. IEEE Spectrum **45**(5), 34–39 (May 2008)
3. D. Agrawal, S. Baktir, D. Karakoyunlu, P. Rohatgi, B. Sunar, in *Trojan Detection Using IC Fingerprinting*. Proceedings of the 2007 IEEE Symposium on Security and Privacy, Oakland, CA, USA, 20–23 May 2007, pp. 296–310
4. Y. Alkabani, F. Koushanfar, in *Active Hardware Metering for Intellectual Property Protection and Security*. USENIX Security Symposium, Boston, MA, USA, 6–10 August 2007, pp. 291–306
5. Y. Alkabani, F. Koushanfar, in *Extended Abstract: Designer's Hardware Trojan Horse*. Proceedings of Hardware-Oriented Security and Trust 2008 (HOST), Anaheim, CA, USA, 13–14 June 2008, pp. 82–83
6. Y. Alkabani, F. Koushanfar, in *Efficient Approximations for IC Trojan Detection*. Proceedings of the International Conference on Computer-Aided Design (ICCAD), San Jose, CA, USA, 2–5 Nov 2009
7. M. Banga, M. Hsiao, in *A Region Based Approach for the Identification of Hardware Trojans*. Proceedings of Workshop on Hardware-Oriented Security and Trust (HOST), Anaheim, CA, USA, 13–14 June 2008, pp. 40–47
8. M. Banga, M. Hsiao, in *Novel Sustained Vector Technique for the Detection of Hardware Trojans*. Proceedings of the International Conference on VLSI Design, New Delhi, India, 5–9 Jan 2009 , pp. 327–332

9. M. Banga, M. Hsiao, in *VITAMIN: Voltage Inversion Technique to Ascertain Malicious Insertion in ICs*. Proceedings of Workshop on Hardware-Oriented Security and Trust (HOST), San Francisco, CA, USA, 27 July 2009, pp. 104–107

10. E. Biham, Y. Carmeli, A. Shamir, in *Bug Attacks*. Advances in Cryptology. Crypto. Lecture Notes in Computer Science, vol. 5157 (Springer, 2008), Santa Barbara, CA, USA, 17–21 Aug 2008, pp. 221–240

11. R.S. Chakraborty, S. Bhunia, in *Hardware Protection and Authentication Through Netlist Level Obfuscation*. Proceedings of the International Conference on Computer-Aided Design (ICCAD), San Jose, CA, USA, 10–13 Nov 2008, pp. 674–677

12. R.S. Chakraborty, S. Paul, S. Bhunia, in *On-Demand Transparency for Improving Hardware Trojan Detectability*. Workshop on Hardware-Oriented Security and Trust (HOST), Anaheim, CA, USA, 13–14 June 2008, pp. 48–50

13. R. Chakraborty, S. Bhunia, in *Security Against Hardware Trojan Through a Novel Application of Design Obfuscation*. Proceedings of the International Conference on Computer-Aided Design (ICCAD), San Jose, CA, USA, 2–5 Nov 2009, pp. 113–116

14. China virus found in Seagate drives in Taiwan: report Reuters News Agency, Nov 12, 2007. http://www.reuters.com/article/idUSTP20376020071112

15. DARPA TRUST Program Information, http://www.acq.osd.mil/dsb/reports/2005-02-HPMS_Report_Final.pdf

16. Defense Science Board Task Force, High Performance Microchip Supply, February 2005. http://www.acq.osd.mil/dsb/reports/2005-02-HPMS_Report_Final.pdf

17. Innovation at risk: Intellectual Property Challenges and Opportunities, Semiconductor Equipment and Materials Industry (SEMI), 2008.

18. S. Jha, S. K. Jha, in *Randomization Based Probabilistic Approach to Detect Trojan Circuits*. Proceedings of the IEEE High Assurance Systems Engineering Symposium, Nanjing, China, 3–5 Dec 2008, pp. 117–124

19. Y. Jin, N. Kupp, Y. Makris, in *Experiences in Hardware Trojan Design and Implementation*. Workshop on Hardware-Oriented Security and Trust (HOST), San Francisco, CA, USA, 27 June 2009, pp. 50–57

20. Y. Jin, Y. Makris, in *Hardware Trojan Detection Using Path Delay Fingerprint*. Proceedings of Hardware-Oriented Security and Trust (HOST), Anaheim, CA, USA, 13–14 June 2008

21. S. King, J. Tucek, A. Cozzieand C. Grier, W. Jiang, Y Zhou, in *Designing and Implementing Malicious Hardware*. Usenix Workshop on Large-Scale Exploits and Emergent Threats (LEET), San Francisco, CA, USA, 14 Apr 2008, pp. 1–8

22. J. Li, J. Lach, in *At-Speed Delay Characterization for IC Authentication and Trojan Horse Detection*. Proceedings of Workshop on Hardware-Oriented Security and Trust (HOST), Anaheim, CA, USA, 13–14 June 2008, pp. 8–14

23. J. Lieberman, Whitepaper on National Security Aspects of the Global Migration of the US Semiconductor Industry. http://lieberman.senate.gov/documents/whitepapers/semicon-ductor.pdf; June 2003

24. M. Potkonjak, A. Nahapetian, M. Nelson, T. Massey, in *Hardware Trojan Horse Detection Using Gate-Level Characterization*. Design Automation Conference (DAC), San Francisco, CA, USA, 26–31 July 2009

25. R. Rad, X. Wang, J. Plusquellic, M. Tehranipoor, in *Taxonomy of Trojans and Methods of Detection for IC Trust*. Proceedings of the International Conference on Computer-Aided Design (ICCAD), San Jose, CA, USA, 10–13 Nov 2008

26. H. Salmani, M. Tehranipoor, J. Plusquellic, in *New Design Strategy for Improving Hardware Trojan Detection and Reducing Trojan Activation Time*. Workshop on Hardware-Oriented Security and Trust (HOST), San Francisco, CA, USA, 27 July 2009

27. M. Tehranipoor, F. Koushanfar, A Survey of Hardware Trojan Taxonomy and Detection. IEEE Design and Test of Computers, Jan/Feb 2010, pp. 10–25

28. X. Wang, H. Salmani, M. Tehranipoor, J. Plusquellic, in *Hardware Trojan Detection and Isolation Using Current Integration and Localized Current Analysis*. Proceedings of the International Symposium on Fault and Defect Tolerance in VLSI Systems (DFT), 2008

29. X. Wang, M. Tehranipoor, J. Plusquellic, in *Detecting Malicious Inclusions in Secure Hardware: Challenges and Solutions*. Proceedings of IEEE Intlernational Workshop on Hardware-Oriented Security and Trust (HOST), Anaheim, CA, USA, 13–14 June 2008, pp. 15–19
30. F. Wolff, C. Papachristou, S. Bhunia, R. Chakraborty, *Towards Trojan Free Trusted ICs: Problem Analysis and Detection Scheme*. Proceedings of Design, Automation and Test in Europe (DATE), Munich, Germany, 10–14 Mar 2008, pp. 1362–1365

Extracting Unknown Keys from Unknown Algorithms Encrypting Unknown Fixed Messages and Returning No Results

Yoo-Jin Baek, Vanessa Gratzer, Sung-Hyun Kim, and David Naccache

1 Introduction

In addition to its usual complexity postulates, cryptography silently assumes that secrets can be physically protected in tamper-proof locations.

All cryptographic operations are physical processes where data elements must be represented by physical quantities in physical structures. These physical quantities must be stored, sensed, and combined by the elementary devices (*gates*) of any technology out of which we build tamper-resistant machinery. At any given point in the evolution of a technology, the smallest logic devices must have a definite *physical extent*, require a *minimum time* to perform their function, and dissipate a minimal *switching energy* when transiting from one state to another.

The rapid development of sophisticated digital communication systems has created new academic interest in physical secret information leakage (cryptophthora) [1]. According to our estimates more than 500 papers and 20 Ph.D. mémoires were published as we write these lines.

While most aspects of side-channel leakage are now well understood, no attacks on totally unknown algorithms are known to date.

By *totally unknown* we mean that no information on the algorithm's mathematical description (including the plaintext size), the microprocessor, or the chip's power consumption model is available to the attacker.

This chapter describes such a side channel that we experimented successfully on a test chip.

The precise assumptions we make sure that the attacker is given a device $\mathbf{H}_{\hat{\mathfrak{K}}}(\bullet)$ keyed with an unknown key $\hat{\mathfrak{K}}$ and a physically similar blank device $\mathbf{H}_{\bullet}(\bullet)$ that he can re-key at wish. The attacker only knows the plaintext size and the key size (in other words the target is not even assumed to return the decrypted plaintext to the attacker).

Y.-J. Baek (✉)
System LSI Division, Samsung Electronics Co., Ltd., Suwon, Korea
e-mail: yoojin.baek@samsung.com

A.-R. Sadeghi, D. Naccache (eds.), *Towards Hardware-Intrinsic Security,* Information
Security and Cryptography, DOI 10.1007/978-3-642-14452-3_8,
© Springer-Verlag Berlin Heidelberg 2010

Given a key k and a ciphertext c, $\mathbf{E}_k(c)$ will denote the power consumption curve (Emanation) obtained when decrypting c under k. Although variants of our attack with variable ciphertexts exist, throughout this chapter we will work with a constant c (e.g.,the all-zero ciphertext). We will hence abridge notations by writing $d_k = \mathbf{E}_k(c)$.

Knowledge of the microprocessor's word size w (8 bits, 16 bits, or 32 bits) is not mandatory but may accelerate the attack. Usual countermeasures might slow down the attack or thwart it.

2 The Intuition

The intuition behind our idea is the following: we start by collecting the power traces of the target and average these to get a cleaner representation of $d_{\mathfrak{K}}$.

Then, assuming that the target is an 8-bit machine, we know that the device's power consumption when decrypting c under an arbitrary candidate key k should in principle coincide with $d_{\mathfrak{K}}$ on all samples until the moment where k and $\mathfrak{K}$ start being manipulated (sensed) by the devices.

Consequently, the sample-wise subtraction $\Delta_k = d_{\mathfrak{K}} - d_k$ produces a differential curve whose beginning is flat up to a point where Δ_k suddenly becomes noisy.

Now, being an 8-bit machine, the microprocessor cannot manipulate more than one byte at a time. Hence as soon as we guess correctly $\mathfrak{K}$'s first byte, we expect to see the flat part of Δ_k extended. We can therefore progressively guess key bytes ahead, until we get a completely flat Δ_k. A point at which $\mathfrak{K} = k$.

Note that since the microprocessor does not necessarily begin to work with the key's first byte,[1] the experiment needs to be restarted for each potential byte position, until the key is entirely discovered.

Hence, if $\mathfrak{K}$ is n-word long, recovering $\mathfrak{K}$ will require $2^{w-1}(n^2+n) < 2^w|\mathfrak{K}|^2/w^2$ differential experiments.

Substituting a few practical values ($w = 8, 16, 32$ and $\mathfrak{K} = 64, 96, 128, 160, 256$) into this formula we get the results given in Table 1. However, as we will soon see, the attack requires a very limited number of interactions with the target as the big bulk of experimentations is done *with the engineering sample*. Hence, the above work-factors can be divided by any factor representing the number of blank engineering samples on which the attacker will run the attack in parallel.

Table 1 Attack complexity. Entries express 2^x experiments

	$\mathfrak{K} = 64$	$\mathfrak{K} = 96$	$\mathfrak{K} = 128$	$\mathfrak{K} = 160$	$\mathfrak{K} = 256$
$w = 8$	13	14	15	16	17
$w = 16$	19	20	21	22	23
$w = 32$	34	35	35	36	37

---

[1] For example, the algorithm might start mixing the ciphertext's fifth byte with the key's third byte.

Caveat

The present work assumes the following:

- The implementation $\mathbf{H}_k(c)$ allows fixing c to some arbitrary value (c is possibly unknown to the attacker but common to the target and the sample).
- $\mathbf{H}$ is deterministic and stateless. That is, if $\mathbf{H}_k(c)$ is evaluated twice, the device runs exactly the same instructions and manipulates exactly the same data.
- The emanation function $\mathbf{E}$ is such that for the c used in the attack:

$$k \neq k' \quad \Rightarrow \quad \mathbf{E}_k(c) \neq \mathbf{E}_{k'}(c)$$

3 Notations and Statistical Tools

Statistics provide procedures for evaluating likelihood, called *significance tests*. In essence, given two collections of samples, a significance test evaluates the probability that both samples could rise by chance from the same parent group. If the test's answer turns out to be that the observed results could arise by chance from the same parent source with very low probability we are justified in concluding that, as this is very unlikely, the two parent groups are most certainly different. Thus, we judge the parent groups on the basis of our samples and indicate the degree of reliability that our results can be applied as generalizations. If, on the other hand, our calculations indicate that the observed results could be frequently expected to arise from the same parent group, we could have easily encountered one of those occasions, so our conclusion would be that a significant difference between the two samples was not proven (despite the observed difference between the samples). Further testing might, of course, reveal a genuine difference, so it would be wrong to claim that our test *proved* that a significant difference did not exist; rather, we may say that a significant difference was *not demonstrated on the strength of the evidence presented*, which of course, leaves the door open to reconsider the situation if further evidence comes to hand at a later date.

We run the target and the engineering sample ℓ times and collect ℓ physical signals for each device. We denote by i the acquisition's serial number. The device's emanation can be an array $\{d_k[i, 0], d_k[i, 1], \ldots, d_k[i, \tau - 1]\}$ (e.g., power consumption) or a table representing the simultaneous evolution of ℓ quantities (e.g., electromagnetic emanation samples or microprobes) during τ clock cycles. For the sake of simplicity, we will not treat the table case in this chapter and focus on power curves.[2] The attack will hence operate on $d_k[i, t]$ and use (existing) significance tests as a basic building block:

Definition 1 When called with two sufficiently large samples X and Y, a significance test $S(X, Y)$ returns a probability α that an observed difference in some feature of X and Y could rise by chance assuming that X and Y were drawn

---

[2] Generalizing the attack to spatial signals such as electromagnetic radiations is straightforward.

Table 2 Hypothesis tests

Test S	Notation	Description
Distance of means	DoM-test	[3], (pp. 240–242) 7.9
Goodness of fit	GoF-test	[3], (pp. 294–295) 9.6
Sum of ranks	SoR-test	[3], (pp. 306–308) 10.3

from the same parent group. The minimal sample size required to run the test is denoted size(S).

While we arbitrarily restrict our choice in this work to the three most popular hypothesis tests (see Table 2): the *distance of means, goodness of fit,* and *sum of ranks,* a variety of other hypothesis tests can be used for implementing our attack. The reader may find the description of these procedures in most undergraduate textbooks (e.g., [2, 3]) or replace them by any custom procedure compatible with Definition 1.

4 The Attack

We start by selecting a significance test S (e.g., among DoM-test, GoF-test, and SoR-test) and run the target $\ell = $ size(S) times, collecting at each run an emanation curve $d_{\mathfrak{K}}[i, 0 \leq t \leq \tau - 1]$. We will not need the target anymore during the attack.

As the attack starts we reset a global clock cycle counter $c \leftarrow 0$. c will track the attack's progression on the power curve and represent the rightmost curve coincidence point between the target and the engineering sample.

Denoting by $n = |\mathfrak{K}|/w$ the key-size measured in machine words, we assume that we are given a procedure denoted Exhaust(k, i, c). Here k stands for an intermediate key candidate and $0 \leq i < n$ is a new word position to exhaust in k. Exhaust returns a new key value k_i and an associated new latest coincidence clock cycle $c_i \geq c$.

$$(k_i, c_i) \leftarrow \text{Exhaust}(k, i, c)$$

The attack progressively fills an n-bit string $s = \{s_0, \ldots, s_{n-1}\}$, where positions set to one stand for key words that have been already exhausted in k.

In practical implementations, we keep at each **for** iteration a few best scoring j values and backtrack if the algorithm hits a dead-end (due to measurement inaccuracies or noise).

At the end of the process k is output as a hypothesis for $\mathfrak{K}$'s value (or as a confirmed key if the target returns a plaintext).

4.1 The Exhaust *Routine*

Exhaust gets as input a partially filled key k, a word index i, and a latest coincidence point c. The routine exhausts the 2^w possible values at position i, selects

---

Algorithm 1 Attack Algorithm

```
let s ← {0, ..., 0}
let k ← {0₂ʷ, ..., 0₂ʷ}
while s ≠ {1, ..., 1} do
    for i ← 0 to n − 1 do
        if sᵢ = 0 then
            (kᵢ, cᵢ) ← Exhaust(k, i, c)
        end if
    end for
    j ← index for which cⱼ = maxₛᵢ₌₀(cᵢ)
    sⱼ ← 1
    k ← kⱼ
    c ← cⱼ
end while
```

---

the value which optimizes the coincidence with the target, and returns this optimal coincidence point c_i to the caller.

We will denote by $k_{i \leftarrow e}$ the key k where the ith word was set to the value e and by $\{d_{k_{i \leftarrow e}}[u, 0], \ldots, d_{k_{i \leftarrow e}}[u, \tau - 1]\}$ the τ-sample acquisition collected during the uth experiment, where $\mathbf{H}$ was keyed with $k_{i \leftarrow e}$ (we perform ℓ such experiments per $k_{i \leftarrow e}$ value). We compute for $0 \leq e \leq 2^w - 1$ and $0 \leq t \leq \tau - 1$:

$$\alpha_e[t] \leftarrow S(\{d_{k_{i \leftarrow e}}[1, t], d_{k_{i \leftarrow e}}[2, t], \ldots, d_{k_{i \leftarrow e}}[\ell, t]\}, \{d_{\Re}[1, t], d_{\Re}[2, t], \ldots, d_{\Re}[\ell, t]\})$$

and match the $\alpha_e[t]$ to entire clock cycles by computing:

$$\gamma_e[n] = \sum_{t \in \text{cycle } n} \alpha_e[t]$$

Assuming that at each clock cycle $n \geq c$ the random variable $\gamma_X[n]$ follows a normal distribution (with mean μ_n and standard deviation σ_n that we can easily estimate from the 2^w measurements available), we compute for each e the probability λ_e (p value) that $\gamma_e[n]$ would fall beyond the number of σ_n-s that separate $\gamma_e[n]$ from μ_n. The lowest λ_e determines our choice of e but the exploration of the curve will continue until no deviation bigger than $2\sigma_n$ (i.e., a p value of 0.95%) is found, in which case we consider that we have hit the next de-synchronization point and report it by returning to the caller.

To distinguish complete coincidence[3] from complete de-synchronization both of which are characterized by deviations smaller than $2\sigma_n$, we compare the value of σ_n to a threshold σ' set experimentally. Cycles for which no deviation bigger than $2\sigma_n$ was found and $\sigma_n < \sigma'$ are considered are completely coinciding whereas cycles for which no deviation bigger than $2\sigma_n$ was found and $\sigma_n \geq \sigma'$ are considered an indication of complete de-synchronization.

---

[3] For example, operations that do not manipulate the key nor the ciphertext.

5 Practical Experiments

We tried the new attack on an experimental (unprotected) test chip containing the AES key $\mathfrak{K}$ = 0x0f1571c947d9e8590cb7add6af7f6798. Figure 1 shows the stacking of all the Δ_k for k values of the form $(i|\text{random})$ with $0 \leq i \leq 255$.

Δ_{0x0f}, shown in red, is indeed flatter than the other differential curves as the experiment starts. $d_{\mathfrak{K}}$ and $d_{\mathfrak{K}}$ curves were obtained by averaging 1,380 power traces (1 GHz sampling rate).

Closer inspections of this graphic (Figs. 2 and 3) reveal that the first significant difference between the red curve and the other curves appears between samples 5,500 and 6,500.

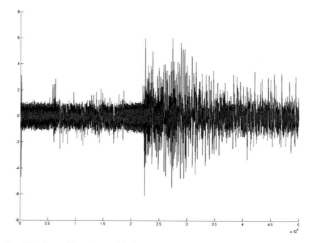

Fig. 1 Δ_k for k of the form $i|\text{random}$ with $0 \leq i \leq 255$. Samples 1–50,000

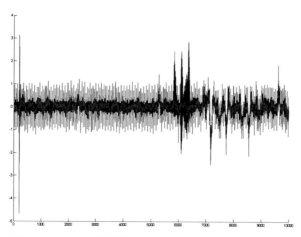

Fig. 2 Excerpt of Fig. 1: samples 1–10,000

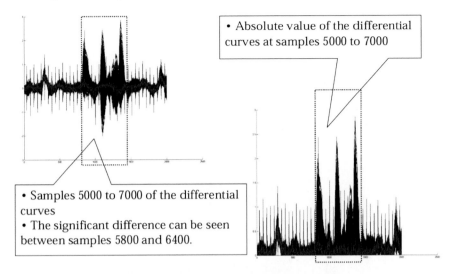

- Absolute value of the differential curves at samples 5000 to 7000

- Samples 5000 to 7000 of the differential curves
- The significant difference can be seen between samples 5800 and 6400.

Fig. 3 Excerpt of Fig. 1: samples 5,000–7,000 and the corresponding $|\Delta_k|$

Before launching the statistical analysis on $|\Delta_k|$, we have noted (Fig. 4) the repeated presence of spikes due to noise. The signal was hence pre-filtered as shown in Fig. 5, we denote the filtered signal by $f(|\Delta_k|)$.

Table 3 Variance test λ_e values for several signals. Samples 5,800–6,400

e	$\lambda_e(\Delta_k)$	$\lambda_e(|\Delta_k|)$	$\lambda_e(f(|\Delta_k|))$
0x0f	0.0183	0.0074	0.0030
0x5e	0.0207	0.0087	0.0063
0x1d	0.0219	0.0093	0.0064
0x5f	0.0222	0.0100	0.0071
0x4e	0.0245	0.0105	0.0075
0x3f	0.0259	0.0114	0.0080
0xce	0.0290	0.0114	0.0081
0x3e	0.0291	0.0130	0.0082
0xee	0.0295	0.0140	0.0089
0x3a	0.0320	0.0141	0.0101

We have tried several non-parametric tests: the sign test for zero median, Wilcoxon's signed rank test for zero median and the DoM-test. All three tests failed to distinguish the red curve. However, as shown in Table 3, the variance test clearly declared the red curve. Indeed, one is reasonably founded in expecting the red curve's variance to be lower than those of the other differential curves.

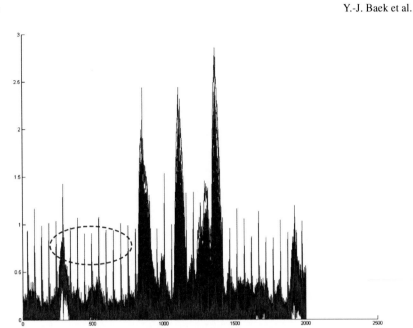

Fig. 4 Circled spikes: noise

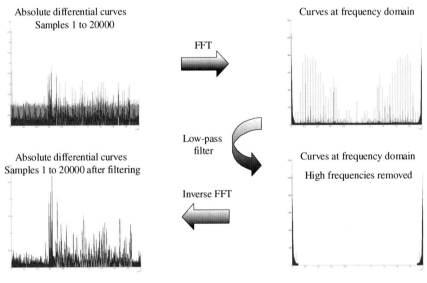

Fig. 5 Filtering the noise

6 Implications and Further Research

The experiments reported in this chapter underline the risk of distributing engineering samples of improperly protected tamper-resistant devices. We showed that this is important even in contexts where the algorithm is unknown to the attacker, a setting frequent in telecommunications and pay-TV applications.

From a technical standpoint, it would be interesting to devise variants of the attack that use other decision criteria (e.g., exploring the correlation coefficient curve between the target's power consumption and the engineering sample) or devise specific countermeasures against this new class of attacks.

As a final note, we observed in several of our experiments that the key was disclosed before the algorithm even started to work. The attack had actually detected key's transfer from non-volatile memory to the target's RAM.

References

1. P. Kocher, J. Jaffe, B. Jun, in *Differential Power Analysis*, Advances in Cryptology - CRYPTO'99, Santa Barbara, CA, USA. Lecture Notes in Computer Science, vol. 1666 (Springer Verlag, Berlin, Heidelberg, 1999), pp. 388–397
2. R. Langley, *Practical Statistics* (Dover Publications, New York, NY, 1968)
3. I. Miller, J. Freund, R. Johnson, *Probability and Statistics for Engineers* (Prentice Hall, Englewood Cliffs, NJ, 1990)
4. F. Wilcoxon, Individual comparisons by ranking methods. Biometrics **1**, 80–83 (1945)

Part IV
Hardware-Based Policy Enforcement

License Distribution Protocols from Optical Media Fingerprints

Ghaith Hammouri, Aykutlu Dana, and Berk Sunar

1 Introduction

According to the Business Software Alliance about 35% of the global software market, worth $141 Billion, is counterfeit. Most of the counterfeit software is distributed in the form of a compact disc (CD) or a digital video disc (DVD) which is easily copied and sold in street corners all around the world but mostly in developing countries. Given the severity of the problem at hand, a comprehensive solution taking into account the manufacturing process, economical implications, ease of enforcement, and the owner's rights needs to be developed. While this is an enormous undertaking requiring new schemes at all levels of implementation, in this work, we focus only on a small part of the problem, i.e., secure fingerprinting techniques for optical media.

To address this problem the SecuRom technology was introduced by Sony DADC. The technology links the identifiers produced to executable files which may only be accessed when the CD is placed in the reader. The main advantage of this technology is that it can be used with existing CD readers and writers. While the specifics of the scheme are not disclosed, in practice, the technology seems to be too fragile, i.e., slightly overused CDs become unidentifiable. Another problem is at the protocol level. The digital rights management (DRM) is enforced too harshly, therefore significantly curtailing the rights of the CD owner.

In this chapter we take advantage of CD manufacturing variability in order to generate unique CD fingerprints. The approach of using manufacturing variability to fingerprint a device or to build cryptographic primitives has been applied in several contexts. A popular example is a new hardware primitives called *physical unclonable functions* (PUFs). These primitives were proposed for tamper-detection at the physical level by exploiting *deep-submicron and nanoscale* physical phenomena

G. Hammouri (✉)
CRIS Lab, Worcester Polytechnic Institute, Worcester, MA 01609-2280, USA
e-mail: hammouri@wpi.edu

A preliminary version of this chapter appeared in [10].
This material is based upon work supported by the National Science Foundation under Grant No. CNS-0831416.

A.-R. Sadeghi, D. Naccache (eds.), *Towards Hardware-Intrinsic Security,* Information Security and Cryptography, DOI 10.1007/978-3-642-14452-3_9,
© Springer-Verlag Berlin Heidelberg 2010

to build low-cost tamper-evident key storage devices [9, 15, 16, 19]. PUFs are based on the subtleties of the operating conditions as well as random variations that are imprinted into an integrated circuit during the manufacturing process. This phenomenon, i.e., manufacturing variability, creates minute differences in circuit parameters, e.g., capacitances, line delays, threshold voltages, in chips which otherwise were manufactured to be logically identical. Therefore, it becomes possible to use manufacturing variability to uniquely fingerprint circuits. These techniques are fully explained in chapter "Physically Unclonable Functions: A Study on the State of the Art and Future Research Directions" by Roel Maes and Ingrid Verbauwhede, this volume. More recently, another circuit fingerprinting technique was introduced. The technique exploits manufacturing variability in integrated chips to detect Trojan circuits inserted during the manufacturing process [1]. This technique is discussed in detail in chapter "Hardware Trojan Horses" by Mohammad Tehranipoor and Berk Sunar, this volume.

Another secure fingerprinting technology named RF-DNA was developed by Microsoft Research [6]. The RF-DNA technology provides unique and unclonable physical fingerprints based on the subtleties of the interaction of devices when subjected to an electromagnetic wave. The fingerprints are used to produce a cryptographic certificate of authenticity (COA) which when associated with a high value good may be used to verify the authenticity of the good and to distinguish it from counterfeit goods. More details about this technique can be found in chapter "Anticounterfeiting: Mixing the Physical and the Digital World" by Darko Kirovski, this volume. Another application of manufacturing variability is fingerprinting paper objects. In [5] the authors propose Laser Surface Authentication which uses a high-resolution laser microscope to capture the image texture from which the fingerprint is developed. In a more recent proposal, a cheap commodity scanner was used to identify paper documents [4]. While most of the results cited above were developed in the last decade, the idea of using physical fingerprints to obtain security primitives is not new at all. According to [6], access cards based on physical unclonable properties of media have been proposed decades ago by Bauder in a Sandia National Labs technical report [2].

Our Contribution: We introduce a method which exploits CD manufacturing variability to generate unique fingerprints from logically identical CDs. The biggest advantage of our approach is that it uses the electrical signal generated by the photodiode of a CD reader. Thus no expensive scanning or imaging equipment of the CD surface is needed. This means that regular CD readers can implement the proposed method with minimal change to their design. We investigate the new approach with a study of over 100 identical CDs. Furthermore, we introduce a new technique, called the threshold scheme, for utilizing fuzzy extractors over the Lee metric without much change to the standard code offset construction [7]. The threshold scheme allows us to use error correcting codes working under the Hamming metric for samples which are close under the Lee metric. The threshold scheme is not restricted to CDs, and therefore can serve in any noisy fingerprinting application where the Lee metric is relevant. With the aid of the proposed fuzzy extractor we give specific parameters and a code construction to convert the derived fingerprints into 128-bit cryptographic keys. In this chapter we also present a family of license distribution protocols which take advantage of the extracted fingerprint. The presented protocols are essential for demonstrating usage scenarios for the fingerprinting techniques. However, we stress that these protocols are an example of a much larger set of possible usage scenarios.

The remainder of the chapter is organized as follows. In Sect. 2, we discuss the physical aspects of CD storage, the sources of manufacturing variability, and the statistical model capturing the CD variability. Section 3 presents experimental data to verify our statistical model. In Sect. 4 we discuss the fingerprint extraction technique and determine the parameters necessary for key generation. We discuss the robustness of the fingerprint in Sect. 5 and present a family of license distribution protocols which utilize the extracted fingerprint in Sect. 6. Finally, we conclude in Sect. 7.

2 Pits and Lands

On a typical CD data are stored as a series of lands and pits formed on the surface of the CD. The pits are bumps separated by the lands to form a spiral track on the surface of the CD. The spiral track starts from the center of the CD and spirals outward. It has a width of about 0.5 μm and a 1.6 μm separation. The length of the land or pit determines the stored data. The encoding length can assume only one of nine lengths with minimum value in the range of 833–972 nm up to a maximum of 3, 054–3, 563 nm with increments ranging from 278 to 324 nm. Note that the range is dependent on the speed used while writing the CD. To read the data on the CD the reader shines a laser on the surface of the CD and collects the reflected beam. When the laser hits the pits it will reflect in a diffused fashion thus appearing relatively dark compared to the lands. Upon the collection of the reflected beam, the reader can deduce the location and length of the lands and pits which results in reading the data on the CD.

CDs are written in two ways, pressing and burning. In pressed CDs a master template is formed with lands and pits corresponding to the data. The master template is then pressed into blank CDs in order to form a large number of copies. In burned CDs, the writing laser heats the dye layer on the CD-R to a point where it turns dark, thus reflecting the reading laser in a manner consistent with physical lands. Note that burned CDs will not have physical lands and pits but will act as if they had these features. Figures 1 and 2 show the lands and pits of a pressed CD. We captured Fig. 1 using an optical microscope and Fig. 2 using a scanning electron microscope.

2.1 Source of Variation

Similar to any physical process, during the writing process CDs will undergo manufacturing variation which will directly affect the length of the lands and pits. For

Fig. 1 Lands and pits image
using an optical microscope

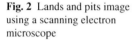

Fig. 2 Lands and pits image
using a scanning electron
microscope

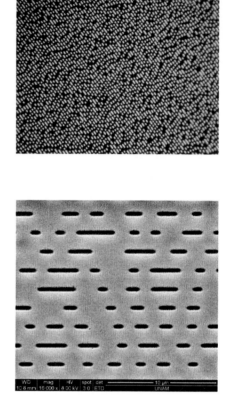

burned CDs this variability will be a direct result of the CD velocity while writing takes place. This velocity is assumed to be at a fixed rate between 1.2 and 1.4 m/s where the velocity variation during writing should be within ±0.01 m/s [8]. Pressed CDs are manufactured by molding thermoplastics from a micro or nanostructured master prepared by lithographic methods. The molding process itself is optimized for replication fidelity and speed with typical replication variations on the order of tens of nanometers [17]. The molding process involves contacting the thermoplastic with the master slightly above the glass transition temperature of the material, with a preset pressure for a brief amount of time, cooling the master and the thermo-plastic to below the glass transition temperature, and demolding. Local variations of polymer material's mechanical and thermal properties, local variations of the tem-perature and pressure all potentially lead to variations in the imprinted structures. The thermal stresses induced during cooling and demolding also potentially lead to variations. In this chapter we aim at using the small variation in the length of lands and pits to form a unique fingerprint for each CD. In the next section we characterize the length features of lands and pits.

2.2 Single Location Characterization

Together lands and pits form the full spiral track. Therefore, it makes sense to fingerprint only lands or pits. The length of both lands and pits will follow similar distributions which is why we will simply use the term *location* to refer to either of them. We label the lengths of n consecutive locations by starting from a reference point on the track as $L_1, L_2, \ldots, L_n$. In the ideal setting $L_i = c_i \cdot L$ for a small constant integer $c_i \in [3, 4, \ldots, 11]$ and $L \approx 300$ nm. However, due to the subtle variations we discussed in the previous section we expect $L_i = c_i \cdot L + \ell_i$. The variable ℓ_i is expected to be quite small compared to L_i, and therefore difficult to measure precisely. Still our measurements should be centered around the ideal length. Hence, quite naturally across all identical CDs we model L_i as a random variable drawn from a Gaussian distribution $\mathcal{H}_i = N(M_i, \Sigma)$ where $M_i = c_i \cdot L$ and Σ denote the mean and the standard deviation, respectively.[1]

Here we are assuming that regardless of the location, the standard deviation Σ will be the same. This is a quite a realistic assumption since Σ essentially captures the manufacturing variability which should affect all locations similarly. The more precise the manufacturing process is, the less of a standard deviation we would expect $\mathcal{H}_i$ to have. A perfect manufacturing process would yield $\Sigma = 0$ and would therefore give all CDs the same exact length of a specific location across all identical CDs. On the other hand, for better identification of CDs we would like $\mathcal{H}_i$ to have a relatively large Σ.

In a typical CD reader, the reading laser is reflected from the CD surface back into a photodiode which generates an electrical signal that depends on the intensity of the reflected laser. Therefore, the electrical signal is expected to depict the shape of the CD surface. If these electrical signals are used to measure the length of any given location, we expect these measurements to have a certain level of noise following a Gaussian distribution. So for location i on CD_j we denote this distribution by $\mathcal{D}_{ij} = N(\mu_{ij}, \sigma)$. The noise in the length measurements is captured through the standard deviation σ. Since this quantity mainly depends on the reader's noise, we assume that its same for all CDs and all CD locations. Contrary to Σ, to identify different CDs using the length information of CD locations we would like to see a relatively small σ.

3 Experimental Validation

To validate the statistical model outlined in the previous section, we conducted extensive experiments on a number of CDs. We directly probed into the electrical signal coming out of the photodiode constellation inside the CD reader. The intensity of this signal will reflect the CD surface geometry, and therefore can be used to

---

[1] $N(\mu, \sigma)$ is a normal distribution with mean μ and standard deviation σ.

study the length of the CD locations. To sample the waveform we used a 20 GHz oscilloscope. Each CD was read a number of times in order to get an idea of the actual $\mathcal{D}$ distribution. Similarly, we read from the same locations of about 100 identical CDs in order to generate the $\mathcal{H}$ distribution. Each collected trace required about 100 MB of storage space. Moreover, synchronizing the different traces to make sure that the data were captured from the same location of the CD was quite a challenge. We had to assign a master trace which represented the locations we were interested in studying and then ran the other traces through multiple correlation stages with the master to finally extract synchronized signals from the same locations on different CDs. Automating the process in order to accurately capture these massive amount of data was a time-consuming challenge. However, we note that all this work would be almost trivially eliminated if we had access to the internal synchronization signals of the CD reader chip. The captured signals were then further processed using Matlab to extract the location lengths and obtain the distributions. After processing, we extracted the length of 500 locations (lands) on the CDs. We used commercially pressed CDs for all the experiments reported in this chapter.[2]

Figure 3 shows the histogram of lengths extracted from 550 reads for a randomly chosen location on one CD. The mean length of the histogram is about $\mu_{ij} = 958$ nm. This histogram captures the $\mathcal{D}$ distribution. The other locations observe similar distributions with different mean lengths which will depend on the encoded information. When considering data coming from different locations and different CDs we obtain $\sigma = 20$ nm (with an average standard deviation of 2 nm on σ). This will be a good estimate for the noise observed during probing of the electrical signals. These results verify the assumption that the noise in the electrical signal can be approx-

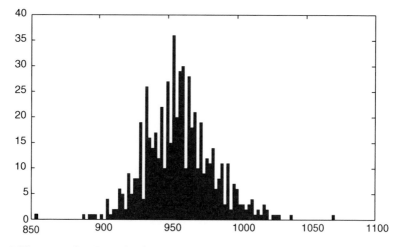

Fig. 3 Histogram of reads coming from the same location on the same CD

---

[2] We have verified a similar behavior for burned CDs. Not surprisingly, data coming from burned CDs had a much larger variation and were easier to analyze.

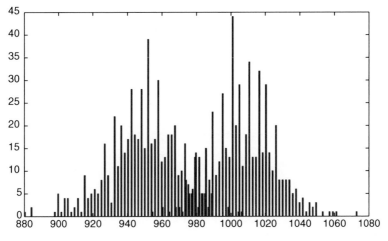

Fig. 4 Histograms of reads coming from the same location on two identical CDs

imated as Gaussian noise. Note that with Gaussian noise simple averaging can be used to substantially reduce the noise level. As we are interested in studying the behavior of the location lengths across different CDs, we next shift our attention to two CDs before we look at a larger batch of CDs. Figure 4 captures a histogram for the length of the same location on two identical CDs. What is important here is the distance between the two Gaussian distributions. The larger this distance becomes the easier it is to identify CDs. Our basic thesis for fingerprinting CDs is that the length of a single location will vary across multiple identical CDs. As pointed out earlier, this behavior can be modeled with the Gaussian distribution $\mathcal{H}_i$. The histogram in Fig. 4 captures this for two CDs. To generalize these results and estimate the $\mathcal{H}_i$ distribution we need a larger sample space. The major problem here is that

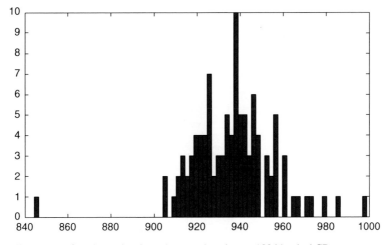

Fig. 5 Histograms of reads coming from the same location on 100 identical CDs

each data point needs to come from a different CD. Therefore, to obtain a histogram which clearly depicts a Gaussian we would need to test on the order of 500 CDs. This was not possible as each CD required substantial time, computing power and storage space in order to produce final data points. However, we were able to carry out this experiment for about 100 CDs. Each CD was read about 16 times to reduce the noise. Finally, we extracted the lengths of 500 locations for each of the CDs. Figure 5 depicts the histogram over 100 CDs for a randomly chosen location out of the 500 extracted locations. The histogram in Fig. 5 has a mean of about 940 nm. Over all locations, Σ had a mean of 21 nm (with an average standard deviation of 1.8 nm on Σ). The histogram in Fig. 5 looks similar to a Gaussian distribution generated from 100 data points. However, it would be interesting to get a confirmation that with more data points this histogram would actually yield a Gaussian. To do so, we normalized the lengths of each location by subtracting the average length for that particular location. Since the distribution for each location had roughly the same Σ the normalization process effectively made all these distributions identical with a mean of 0 and a standard deviation of Σ. We then collected all these data points (on the order of 50,000 points) and plotted the corresponding histogram. This is shown in Fig. 6. The histogram of Fig. 6 strongly supports our thesis of normally distributed location lengths across different CDs. One might observe a slight imbalance on the positive side of the Gaussian. This behavior seems to be a result of the DC offset observed while reading some of the CDs. Fortunately, this will not pose a problem for our fingerprinting technique as we will be normalizing each batch of data to have a mean of zero, thus removing any DC components. We finish this section by showing the histogram in Fig. 7. The main purpose of this histogram is to confirm that what we are studying is in fact the length of data locations written on the CD. We elaborated earlier that on a CD data are stored in discrete lengths ranging from about 900 nm to about 3, 300 nm taking nine steps in increments of about 300 nm. We build the histogram in Fig. 7 using the data collected from 500 locations over the 100 CDs without normalizing each location's length to zero. In Fig. 8 we show a

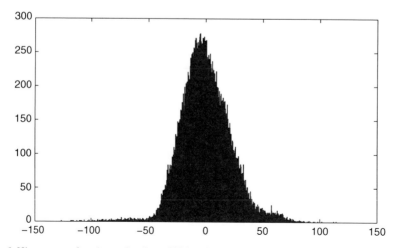

Fig. 6 Histograms of reads coming from 500 locations on 100 identical CDs

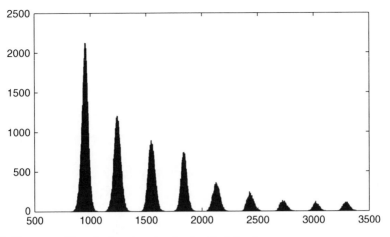

Fig. 7 Histogram of location lengths using the electrical signal

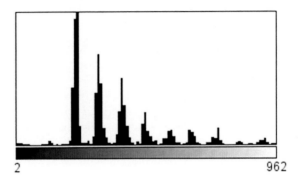

Fig. 8 Histogram of location areas using electron microscope images

similar histogram with data extracted by processing images coming from a scanning electron microscope.

4 CD Fingerprinting

There are many challenges in deriving a robust and secure fingerprint. One important issue is the reading noise. Similar to a human fingerprint, we saw in the previous section that the readings used to extract the CD fingerprint are inherently noisy. The extraction of a deterministic and secure fingerprint from noisy data has been previously studied in the literature [7, 12, 13]. Most relevant to our work is the fuzzy extractor technique proposed by Dodis et al. in [7]. For the remainder of this section we will present a quick review of the fuzzy extractor technique and then discuss how this technique can be modified and applied to the CD setting. Moreover, we will discuss the experimental results and present various bounds needed to achieve high levels of security.

4.1 Fuzzy Extractors

Loosely speaking a fuzzy extractor is a technique to extract an almost uniform random string from a given input such that it is possible to reproduce the same output string from a noisy version of the input. In [7] the authors show how a fuzzy extractor can be built using an error correcting code along with a universal hashing function. Their construction requires that the output of the fingerprint (the biometric data in their language) be represented as an element of $\mathcal{F}^n$ for some field $\mathcal{F}$ and an integer n which represents the size of the fingerprint. Moreover, it is naturally assumed that the noise experienced by the fingerprint is upper bounded by a constant distance from the original fingerprint in order to guarantee identical reproduction of the extracted key. We start by quoting the following theorem introduced in [7], and then give the specific construction which the theorem describes.

Theorem 1 *([7]) Given any* $[n, k, 2t + 1]_{\mathcal{F}}$ *code* C *and any* m, ε, *there exists an average-case* $(M, m, \ell, t, \varepsilon)$-*fuzzy extractor, where* $\ell = m + kf - nf - 2\log(\frac{1}{\varepsilon}) + 2$. *The generation algorithm* GEN *and the recovery algorithm* REP *are efficient if* C *has efficient encoding and decoding.*

We explain the parameters in the theorem by outlining an actual construction. This construction is proposed in [7] and further explained in [9]. As stated in the theorem, C is an error correcting code over the field $\mathcal{F}$, where $f = \log(|\mathcal{F}|)$.[3] For the construction we will also need a family of universal hashing functions **H**.[4] The generation algorithm GEN takes the fingerprint $x \in \mathcal{F}^n$ as input and outputs the triplet (k, w, v). Here, x is drawn from some distribution X over $\mathcal{F}^n$ which has min-entropy m. Note that in our context the parameter m captures the entropy provided by the CD variability. GEN starts by computing $w = x + c$ for a randomly chosen code word $c \in C$ and then computes the key $k = h_v(x) \in \{0, 1\}^\ell$ for some string v chosen uniformly at random such that $h_v \in$ **H**. The recovery algorithm REP takes in the *helper data* (w, v) along with x', a noisy version of the fingerprint x, and returns the key k. REP starts by computing $c' = w - x'$ which is a noisy version of c. If the Hamming distance between x and x' is less than t then so will the Hamming distance between c and c'. Therefore, using the error correcting code C, REP can reproduce c from c'. Next, REP computes $x = w - c$ and consequently compute $k = h_v(x)$ which will conclude the recovery algorithm. All that remains to be defined is the parameter ε which captures the security of the fuzzy extractor. Specifically, if the conditional min-entropy[5] $H_\infty(X|I)$ (meaning X conditioned on I)[6] is larger than m then $\mathbf{SD}((k, (w, v), I), (U_\ell, (w, v), I)) \leq \varepsilon$ where $\mathbf{SD}(A, B) = \frac{1}{2}\sum_v |\Pr(A = v) - \Pr(B = v)|$ is the statistical distance between

---

[3] Note that all logarithms in this chapter are with respect to base 2.

[4] For details on universal hashing the reader is referred to [3].

[5] The definition of min-entropy is $H_\infty(A) = -\log(\max_a \Pr[A = a])$.

[6] Typically we use the | operator to mean concatenation. This will be the only part of the chapter where it will have a different meaning.

two probability distributions A and B. Finally, U_ℓ is the uniform distribution over $\{0, 1\}^\ell$ and I is any auxiliary random variable.

With this construction we will have a clear way to build a fuzzy extractor. However, the key size ℓ and the security parameter ε will both depend on m and the code used. Moreover, the code will depend on the noise rate in the fingerprint. We finish this section by relating the min-entropy and the error rate of the fingerprint. Recall that x is required to have a min-entropy of m and at the same time using the above construction x will have n symbols from $\mathcal{F}$. To merge these two requirements we define the average min-entropy in every symbol $\delta = m/n$. We also define v to be the noise rate in the fingerprint x and $F = |\mathcal{F}|$. With these definitions we can now prove the following simple bound relating the noise rate and the min-entropy rate δ/f.

Proposition 1 *For the fuzzy extractor construction of Theorem 1, and for any meaningful security parameters of $\varepsilon < 1$ and $\ell > 2$ we have $H_F(v) < \frac{\delta}{f}$, where H_F is the F-ary entropy function.*

Proof From Theorem 1 we now have that $\ell = m + kf - nf - 2\log\left(\frac{1}{\varepsilon}\right) + 2$. Let $A = \ell + 2\log\left(\frac{1}{\varepsilon}\right) - 2 = m + kf - nf$. From the conditions above we now have that $A > 0$ and therefore $m + kf - nf > 0$. Let $R = k/n$ which yields $(\delta + Rf - f)n > 0$ and therefore $R > 1 - \delta/f$. Using the sphere packing bound where $R \leq 1 - H_F(v)$ we immediately get $H_F(v) < \frac{\delta}{f}$. $\square$

As it is quite difficult to calculate the min-entropy for a physical source we will estimate this quantity over the symbols of x. The bound given above will give us an idea whether the min-entropy in the symbols of x will be sufficient to handle the measured noise rate. Next we shift our attention to the fingerprint extraction technique. Note here that we still did not address how the data extracted from the CDs will be transformed into the fingerprint x.

4.2 Fingerprint Extraction

In Sect. 3 we described how the empirical data suggest that every CD has unique location lengths. These location lengths as can be seen from Fig. 7 will have different values depending on the encoded information. Moreover, we discussed earlier that the raw data measured from the electrical signal will sometimes have different DC offsets. Therefore, it is important to process the data before the different locations can be combined together in order to produce the final fingerprint x. The first step in processing the data coming from every location on every CD is to remove the signal noise. To achieve this, the length of every location on a CD is averaged over a number of readings. Since we are assuming Gaussian noise, the noise level σ will scale to $\sigma/\sqrt{a}$, where a is the number of readings used for averaging. Next, we normalize the data using the ideal average of each location. As the ideal location lengths are discretized it becomes easy to find the ideal length for every location and subtract it from the measured lengths. This will guarantee that all location lengths have similar distributions as will be seen in Fig. 6. Finally, to remove the DC component we need a second normalizing step. We subtract the mean of the reading coming from different locations of the same CD. Figures 9 and 10 show the variation in the length of 500 locations for two identical CDs after being averaged and normalized. Each figure contains three traces with an added horizontal shift to set the traces apart. The top two traces in each figure are obtained from readings taken at different times using one CD reader. The bottom trace in each figure was obtained 3 months after the first two traces using a second CD reader with a different brand and model. The vertical axis represents the variation in nanometers from the ideal length of that location. These figures clearly support the idea of identical CDs having different fingerprints which are reproducible from different readers. We still need to outline a technique to extract a final fingerprint. Even after the previous averaging and normalization steps we will still have errors in the length readings. Although we will be using a fuzzy extractor to correct the errors, the biggest challenge toward achieving an efficient extraction technique will be the nature of these errors. The noise is Gaussian over the real values of the lengths. This means that even when the data are discretized the error will manifest itself more as a shift error from the ideal length rather than a bit flip error. Unfortunately, the Hamming metric does not naturally accommodate for this kind of error. Moreover, if we assume that every location length of the CD will be a symbol in the extracted fingerprint, then the error rate would be very high as it is very difficult to get the same exact length for the CD locations. A more natural distance metric in this situation would be the Lee metric [14]. However, this will require finding long codes that have good decoding performance under the Lee metric. To solve this problem we propose a *threshold* scheme which uses the Hamming distance while allowing a higher noise tolerance level. The threshold scheme also works naturally with the fuzzy extractor construction of Theorem 1. Table 1 shows a formulation of the threshold scheme applied to the CD setting. The threshold τ solves the error correcting problem with respect to the Lee distance. In particular, τ helps control the error rate which arises when treating the real values as symbols over some field. Without a threshold scheme ($\tau = 0$), the error rate will be very high. On the other hand, if τ grows too large then the error rate will be low. However, the Hamming distance between the extracted fingerprint originating from different CDs will decrease thus decreasing distinguishability between CDs. An important aspect about the threshold scheme is that it is very simple to compute and does not require previous knowledge of the distribution average.

4.3 Entropy Estimation and 128-Bit Security

The previous sections dealt with the theoretical aspects of extracting the CD fingerprint. In this section we take more of an experimental approach where we are interested in computing actual parameters. The most important parameters that we

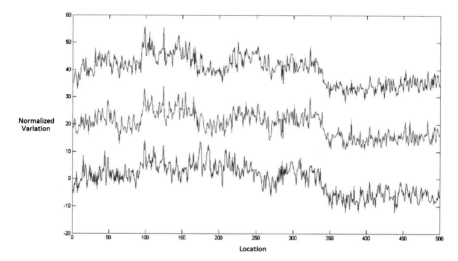

Fig. 9 Length variation over 500 locations from CD1 with the bottom trace taken 3 months after the top two traces

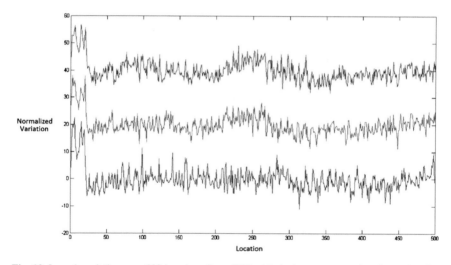

Fig. 10 Length variation over 500 locations from CD2 with the bottom trace taken 3 months after the top two traces

need to estimate are the entropy of the source (the CD variability) and the noise level. With these two parameters the rest of the parameters can be determined. The first and hardest task here will be to decide the amount of entropy generated by the source. In [9, 11] the authors use a universal source coding algorithm in order to estimate the secrecy rate. In particular it was proposed to use the context-tree weighting method (CTW) [21]. What is quite useful about the CTW algorithm is

Table 1 Formulation of the threshold scheme for CD fingerprint extraction

---

Threshold Scheme: (**GEN,REP**) parameterized by $M, m, \ell, t, \varepsilon, l, C, \mathbf{H}, \tau = 2^s$

GEN: $(k, w, v) \leftarrow \text{GEN}(CD j)$

1. Obtain (a) samples for the length of each of the n locations on $CD j$.
2. Generate $z = z_n \ldots z_1$:

   a. Average the lengths over a samples,
   b. Subtract the ideal mean from the averaged reads,
   c. Normalize the sequence to have a zero mean and set that to z.

3. Find u such that $-2^{u-1} \leq z_i \leq 2^{u-1} - 1$ for all i, and shift z_i to $0 \leq z_i \leq 2^u - 1$.
4. Shift the binary representation of z_i left by l bits, round to an integer and set to $\hat{z}_i$.
5. Form $z_{2,i}$, the lowest $s + 1$ bits of $\hat{z}_i$, and $x_i = z_{1,i}$, the remaining bits of $\hat{z}_i$.
6. Set $x = x_n \ldots x_1$ to be the fingerprint template.
7. Choose a random code word $c \in C$, such that $c = c_n \ldots c_1$.
8. Compute $w_i = (x_i | z_{2,i}) + (c | \tau)$ and form $w = w_n \ldots w_1$.
9. Randomly choose v to compute $k = h_v(x)$ where $h_v \in \mathbf{H}$, and output (k, w, v).

REP: $k \leftarrow \text{REP}(CD j, w, v)$

1. Generate $z' = z'_n \ldots z'_1$ as $\hat{z} = \hat{z}_n \ldots \hat{z}_1$ was generated in Steps 1 through 4 of GEN.
2. Set c'_i to be the highest $u + l - s - 1$ bits of $w_i - z'_i$.
3. Use C to correct $c' = c'_n \ldots c'_1$ to $c = c_n \ldots c_1$.
4. Compute $x_i = w_i - c_i$.
5. Form $x = x_n \ldots x_1$ and return $k = h_v(x)$.

---

that in [20] it was shown that for any binary stationary and ergodic source X, the compression rate achieved by CTW is upper bounded by the min-entropy $H_\infty(X)$ as the length of the input sequence approaches infinity. This is a good indication about the entropy produced by the source provided enough bits are fed to the algorithm. To apply this algorithm to our setting we start by using the data coming from the 100 CDs. On each CD we collected data from 500 locations and processed the data with a threshold value of $\tau = 2^2$. The final data came out to be in the range $[0, 2^5 - 1]$ and we did not use any fractional bits so, $l = 0$. With these parameters the size of the symbols was $f = 2$. This means that every CD produced 1,000 bits. The data were fed into the CTW algorithm which resulted in a compression rate of about 0.83 bits of entropy per extracted bit. Recall here that these samples were not averaged over multiple reads. Therefore the error rate is quite high. When we averaged over 16 samples the combined entropy rate became 0.71. This is expected since the noise will add to the entropy. In order to get a more precise estimate for the min-entropy we decided to average over 225 reads. With this many reads we had to restrict our sample to only 14 CDs as the amount of data quickly becomes large. With the new sample the compression rate of the CTW algorithm was about 0.675 which seemed to be a good estimate of our min-entropy. For this sample, the average error rate is $P_e = 0.08$. On the other hand the collision probability P_c, the probability of extracting similar bits between two different CDs, is about 0.46.

Proposition 1 suggests that for a noise rate of 0.08 and $f = 2$ the entropy of the source should be at least 0.40 which translates to $\delta = 0.8 < 1.35$, and therefore we conclude that we have enough entropy in our source. However, with this level of entropy we are placing stringent conditions on R, i.e., the rate of the error correcting code.[7] To relax the restriction on the code rate we took a closer look at our source bits. Ideally the two bits would have the same entropy. However, looking at Figs. 9 and 10 and multiple similar figures, we clearly see that there is a degree of dependency between the adjacent locations. There is a low probability of a sharp change in the length variability from one location to its neighbor. With this observation we would suspect that the most significant bit will have less entropy as it is less likely to change across adjacent locations. To verify this observation, we applied the CTW algorithm to each of the two extracted bits separately. For the most significant bit, the entropy for the cases of no averaging, averaging over 16 reads, and averaging over 225 reads was $1, 0.9$, and 0.6 bits of entropy, respectively. When we repeated this process for the least significant bit we obtained $1, 1$, and 0.98 bits of entropy, respectively. Clearly, we have more entropy in the least significant bit. It seems reasonable to only use the least significant bit to form the fingerprint and the final key. This would effectively increase the entropy of our source while very slightly affecting the error rate and the collision rate. For this least significant bit scheme we obtained $P_e = 0.08$ and $P_c = 0.46$.

We now have $P_e = 0.08$, $\delta = 0.98$, and $f = 1$. With these parameters we can build a fuzzy extractor which can extract secure keys from CD fingerprints. For a 128-bit key we set $\ell = 128$. Similarly, to achieve a fuzzy extractor output which reveals very little information about the fingerprint we set $\varepsilon = 64$. Using the equation of Theorem 1 we require that the error correcting code in the fuzzy extractor should satisfy $k \geq 190 + 0.02n$. Note that although $P_e = 0.08$, this is the expected error rate. For a practical scheme we require the fuzzy extractor to correct around a 0.17 error rate. These parameters can now be satisfied using a binary BCH code of $[255, 45, 88]$. More specifically, we define a code word containing seven code words of this BCH code, which will make $n = 1,785$. With this construction the failure probability[8] P_{fail} will be on the order of 10^{-6}. Note here that treating the seven code words separately to generate separate parts of the key would significantly decrease ε but will decrease the failure probability. Therefore, in our failure probability we treat the seven code words as a single entity. As we noted earlier, our data suffer from higher error rates due to the external connections which we used. With an on-chip process we can expect the error rate to drop significantly.

---

[7] Recall from the proof of Proposition 1 that $R \geq A/nf + (1 - \delta/f)$ for a security level of at least $A = \ell + 2\varepsilon - 2$.

[8] Here, $P_{\text{fail}} = 1 - \left(\sum_{i=0}^{t=43} \binom{255}{i} P_e^i (1 - P_e)^{255-i} \right)^7$.

5 Robustness of the Fingerprint

A CD fingerprint can be used to tie software licenses to individual CDs where the software is stored. Under this use scenario it becomes important to address the robustness of the fingerprint. In all our experiments the data collected came from locations in the same sector of the CD. In a real application readings would typically be collected from different sectors, thus ensuring that a scratch or any physical damage to a specific location will not render the CD fingerprint useless.

Another important concern regarding the robustness of the fingerprint is that of aging. Although no quantitative estimate of fingerprint durability can be given within the scope of this work, mechanisms related to viscoelastic relaxation in optical disc patterns need to be discussed briefly. Optical discs are printed on polymeric substrates, which have glass transition temperatures typically above 150°C. The viscosity of such materials are temperature dependent and governed by an Arrhenius type exponential temperature dependence, described by an activation energy defined by the glass transition temperature. In its simplest form, the Arrhenius model assumes that the rate of change is proportional to $e^{\frac{-E_a}{kT}}$ where E_a is the activation energy, k is the Boltzmann constant (an invariant physical parameter), and T is the absolute temperature (temperature in degrees Kelvin). Even at lower temperatures (natural operating and storage temperature range of the optical disc), viscosity of the polymer remains finite. During the molding process, most of the internal stresses are relieved upon cooling, resulting in fluctuations in the nanoscale structure of the bit patterns. The pressed discs have a thin metal coating, which is typically coated on to the polymer disc by evaporation or sputter coating, that results in the increase of the surface temperature by up to 50°C. This process is also likely to be a source of local thermoelastic stress buildup which relaxes over the lifetime of the CD. In a first-order approximation, the disc material can be thought of as a Kelvin–Voigt material, and creep relaxation can be approximated by a single time-constant exponential behavior. In such a case, most of the viscoelastic relaxation will occur at the early stages of disc production, and latter timescales will have less of an effect. It may be speculated that the fingerprints due to length fluctuations of 25 nm upon 300 nm characteristic bit length will persist within at least 10% of the CD lifetime, which is predicted to be 217 years at 25°C and 40% relative humidity conditions. This gives an estimated 20-year lifetime for the fingerprint [18]. Due to the exponential dependence of the relaxation on time, by recording the signature on a slightly aged optical disc (months old), the persistence of the signature can be increased.

6 License Distribution Protocol

Once extracted the optical media fingerprints may be used in a number of applications. More specifically, in this section we focus our attention on license distribution protocols, i.e., techniques for allowing/restricting access to the content of a CD. The only users that are allowed to access the CD content are the ones with valid licenses handed out at the purchase by the original producer of the CD. This scenario is vital for digital rights management (DRM) in software and digital media distribution.

In the simplest license distribution scheme the license key is delivered in printed form (on the CD or its case). The offline license distribution scheme can be summarized as follows: Identical copies of an installation software are pressed on the CDs. A unique license key is printed on the cover of each CD. The CDs are sold to customers. Customers run the installation software on their machines. During installation each customer enters the license key manually to the installer. The installer checks the validity of the license. If valid, the installer copies the software to the target machine.

This scheme is convenient as no network connection is required to complete the installation process. However, the offline scheme has a number of shortcomings. The licenses are checked for membership to the set of valid licenses. This means that licenses are *not* tied to CDs. In other words, the software on an authentic CD can be trivially copied for a fraction of a dollar to another CD, which will install under any valid license key. One way to curb rampant CD counterfeiting is to require an online license registration step, where a central database is contacted after verification of the license which checks whether the number of valid installations is exceeded. If not, the software is installed and the central database is updated accordingly. While this simple online scheme is effective it is inconvenient as it requires an online connection. Furthermore, the restriction on the number of installations is highly restrictive and unfair to paying owners of the CD. The main difficulty in binding a key to a CD is that CDs are pressed with one template since the production of the template bears significant cost. This means that CDs coming off a production line necessarily share an identical digital image. Here we propose to use the physical fingerprints of the CDs as an identifier. We bind this identifier to the license and thereby achieve unique licenses that are intrinsically tied to CDs without changing the manufacturing process.

In this section we present three protocols. These protocols differ in the assumptions that are made for the validity of the protocol. We assume that the data pressed on the CD are some software equipped with an installer. Moreover, in all the following protocols we assume that the CD reader can extract the noisy CD fingerprint x from publicly known addresses on the CD and then return x to the PC. Before we present the protocols we present the notation. Recall from the extraction procedure presented in Table 1 that (w, v) is the helper data. Also recall that k is the cryptographic fingerprint extracted from the CD. We use $E_K(\cdot)$ to indicate a secure block cipher encryption using private key K. For clarity the reader can assume this encryption algorithm to be AES. For public key cryptography we use $E_{K_e}(\cdot)$ to denote encryption using the public key K_e and $D_{K_d}(\cdot)$ to denote decryption using a private key K_d.

6.1 Simple Distribution Protocol

The simple distribution protocol shown in Table 2 demonstrates the simplest protocol to distribute licenses using CD fingerprints. Our goal here is to minimize the assumptions and the changes made to current CD readers. *Assumptions*:

- The installer keeps a secret key K which cannot be compromised by an attacker.

Table 2 Simple distribution protocol

Pre-distribution:

1. Data are pressed onto CD.
2. CD is processed to extract (w, v, k).
3. The triplet $(w, v, E_K(k))$ is printed on the CD or its package as the license key.

Authentication Protocol:

1. The reader extracts the noisy fingerprint x' from the CD and returns it to the installer.
2. The installer prompts the user to enter the license key.
3. The user enters the triplet $(w, v, E_K(k))$.
4. The installer runs the REP algorithm using (w, v, x') to retrieve k'.
5. If $E_K(k') = E_K(k)$ the installer grants access and installs the software, otherwise it halts.

Strengths and Weaknesses: This protocol is straightforward. If an attacker tries to copy the CD he will be changing the CD fingerprint x. Therefore, the installer will fail to extract the correct key k from the CD and thus $E_K(k') \neq E_K(k)$. That said, the attacker has a number of ways in which he can attack this protocol. One way is to record the fingerprint x obtained from the first CD by directly reading the communication between the CD reader and the PC. To use x the attacker will have to override the driver of the CD reader and force it to feed the installer with the recorded x rather than the actual fingerprint of the copied CD. This attack is feasible but will require a highly skilled attacker. Another attack that can be carried out against this protocol is to completely eliminate the installer. This problem seems fundamental because there will always be a need for some entity which enforces the license.

6.2 Secure Reader Protocol

In the previous protocol the biggest weakness was in the installer. In the secure reader protocol shown in Fig. 3 we shift the license enforcing to the CD reader. This approach requires more changes to the CD reader but will provide a higher level of security.

Assumptions:

- The CD reader can perform the entire extraction process of Table 1.
- The CD reader can carry out public key encryption and decryption using a public key/private key pair (K_e, K_d).
- The CD reader can carry out private key decryption and encryption.

Table 3 Secure reader protocol

Pre-distribution:

1. Data D are encrypted using a randomly chosen key K to produce $E_K(D)$.
2. $E_K(D)$ is pressed onto the CD.
3. CD is processed to extract (w, v, k).
4. The string R is computed as $R = K \oplus k$.
5. The triplet $(w, v, E_{K_e}(R))$ is printed on the CD or its package as the license key.

Authentication protocol:

1. The installer prompts the user to enter the license key.
2. The user enters the triplet $(w, v, E_{K_e}(R))$ which is passed to the CD reader.
3. The reader extracts the noisy fingerprint x' and uses (w, v) to extract k'.
4. Using the private key K_d the reader computes $R = D_{K_d}(E_{K_e}(R))$.
5. The reader computes $K' = R \oplus k'$.
6. Using K' the reader decrypts the data D and sends it back to the PC.

Strengths and Weaknesses: This protocol shifts all the verification steps on to the CD reader. The installer does not have any control over the data. Even if an attacker copies the CD, the fingerprint will change therefore causing $K' \neq K$ which will prevent access to the data content. Note that the encryption key K never leaves the CD reader and therefore the encrypted data cannot be decrypted outside the reader. Even with this level of security, one should realize that once the data are decrypted and sent to the PC any entity can copy the data. This is again a fundamental problem which will exist as long as the data are accessed in an open format. This problem can be addressed by assuming a trusted operating system which guarantees the security of the information even if it is accessed in open format.

6.3 Online Distribution Protocol

In the previous protocols we assumed that the user will have to enter the helper data (w, v). This is realistic in most scenarios but sometimes it is convenient to have an entirely automated system. In the online distribution protocol shown in Fig. 4 we consider a scenario where the software is purchased online. This situation is becoming more common in the real world. It would be convenient to tie the software image sent via the Internet to the physical CD where the image will be written. The important change in this protocol is that the CD writer can write to different locations of the CD at different times. Note that in this protocol we are back to making minimal assumptions on the reader, namely, the ability to extract the noisy fingerprint x.

Table 4 Online distribution protocol

Downloading phase:

1. The user connects to the server and purchases the software D.
2. The server sends back a random string R to be written to sector 1 of a blank CD.
3. The user inserts a blank CD on which the writer burns the image R to sector 1.
4. The CD reader extracts the noisy fingerprint x from Sector 1.
5. The fingerprint x is sent back to the server.
6. The server computes the triplet (w, v, k) using the extraction method of Table 1.
7. The server returns the string $(E_k(D), E_K(w, v))$ to the user.
8. The CD writer burns $(E_k(D), E_K(w, v))$ to sector 2 of the blank CD.

Authentication protocol:

1. The reader extracts the noisy fingerprint x' from sector 1 on the CD.
2. The reader returns x' along with $(E_k(D), E_K(w, v))$ read from sector 2 on the CD.
3. Using K the installer computes (w, v) from which it extracts k'.
4. The installer retrieves D by decrypting $E_k(D)$ using the extracted key k'.

Assumptions:

- The CD writer can write to different sectors of the CD at different times.
- The installer keeps a secret key K which cannot be compromised by an attacker.

Strengths and Weaknesses: This protocol requires minimal changes to the CD reader. The strength of the protocol lies in having all relevant data on the CD. Any attempt to copy the CD will change the fingerprint and will therefore result in $k' \neq k$. Therefore, the installer will not be able to retrieve D since it will be using the wrong key. Similar to the second protocol once the data are in open format an attacker can copy the data circumventing any protection mechanism. The advantage this protocol has over the first protocol is that the installer is essential for the retrieval of the data. As long as the attacker is not granted access to the data in its open format he cannot bypass the installer.

7 Conclusion

In this chapter we showed how to generate unique fingerprints for any CD. The proposed technique works for pressed and burned CDs, and in theory can be used for other optical storage devices. We tested the proposed technique using 100 identical CDs and characterized the variability across the studied CDs. We also gave specific parameters and showed how to extract a 128-bit cryptographic keys. Moreover, we presented three protocols which employ the extracted fingerprint in license distribution protocols. Our work here does not provide final solutions but rather a new door of research in the area of CD IP-protection.

References

1. D. Agrawal, S. Baktir, D. Karakoyunlu, P. Rohatgi, B. Sunar, in *Trojan Detection Using IC Fingerprinting*. Proceedings of S&P 2007, Oakland, CA, USA, 20–23 May 2007 (IEEE Computer Society, Los Alamitos, CA, 2007), pp. 296–310
2. D.W. Bauder, *An Anti-Counterfeiting Concept for Currency Systems*. Research Report PTK-11990, Sandia National Labs, Albuquerque, NM, USA, 1983
3. L. Carter, M. Wegman, Universal hash functions. J. Comput. Syst. Sci. **18**(2), 143–154 (1979)
4. W. Clarkson, T. Weyrich, A. Finkelstein, N. Heninger, J.A. Halderman, E.W. Felten, in *Fingerprinting Blank Paper Using Commodity Scanners*. Proceedings of S&P 2009, Oakland, CA, May 2009 (IEEE Computer Society, Los Alamitos, CA, 2009)
5. R.P. Cowburn, J.D.R. Buchanan, Verification of authenticity. U.S. Patent Application 2007/0028093, 27 July 2006
6. G. DeJean, D. Kirovski, in *RF-DNA: Radio-Frequency Certificates of Authenticity*. Proceedings of CHES 2007. Lecture Notes in Computer Science, vol. 4727 (Springer, Heidelberg, 2007), pp. 346–363
7. Y. Dodis, R. Ostrovsky, L. Reyzin, A. Smith, Fuzzy extractors: how to generate strong keys from biometrics and other noisy data. SIAM J. Comput. **38**(1), 97–139 (2008)
8. European Computer Manufacturers' Association, *Standard ECMA-130: Data Interchange on Read-Only 120mm Optical Data Disks (CD-ROM)*, 2nd edn. (ECMA, Geneva, 1996)
9. J. Guajardo, S.S. Kumar, G.J. Schrijen, P. Tuyls, in *FPGA Intrinsic PUFs and Their Use for IP Protection*. Proceedings of CHES 2007. Lecture Notes in Computer Science, vol. 4727 (Springer, Heidelberg, 2007), pp. 63–80
10. G. Hammouri, A. Dana, B. Sunar, in *CDs Have Fingerprints Too*. ed. by C. Clavier, K. Gaj. Proceedings of the 11th Workshop on Cryptographic Hardware and Embedded Systems (CHES 2009) Lecture Notes in Computer Science, vol. 5747 (Springer-Verlag, Heidelberg, Germany, 2009), pp. 348–362
11. T. Ignatenko, G.J. Schrijen, B. Skoric, P. Tuyls, F. Willems, in *Estimating the Secrecy-Rate of Physical Unclonable Functions with the Context-Tree Weighting Method*. Proceedings of ISIT 2006, Seattle, WA, USA, 9–14 July 2006 (IEEE, Washington, DC, 2006), pp. 499–503
12. A. Juels, M. Sudan, A fuzzy vault scheme. Designs Codes Cryptogr. **38**(2), 237–257 (2006)
13. A. Juels, M. Wattenberg, in *A Fuzzy Commitment Scheme*. Proceedings of CCS 1999 (ACM, New York, NY, 1999), pp. 28–36
14. C. Lee, Some properties of nonbinary error-correcting codes. IRE Trans. Inf. Theory **4**(2), 77–82 (1958)
15. D. Lim, J.W. Lee, B. Gassend, G.E. Suh, M. van Dijk, S. Devadas, Extracting secret keys from integrated circuits. IEEE Trans. VLSI Syst. **13**(10), 1200–1205 (2005)
16. P.S. Ravikanth, *Physical One-Way Functions*. Ph.D. thesis, Department of Media Arts and Science, Massachusetts Institute of Technology, Cambridge, MA, USA, 2001
17. H. Schift, C. David, M. Gabriel, J. Gobrecht, L.J. Heyderman, W. Kaiser, S. Köppel, L. Scandella, Nanoreplication in polymers using hot embossing and injection molding. Microelectronic Eng. **53**(1–4), 171–174 (2000)
18. D. Stinson, F. Ameli, N. Zaino, *Lifetime of Kodak Writable CD and Photo CD Media* (Eastman Kodak Company, Digital & Applied Imaging, Rochester, NY, 1995)
19. P. Tuyls, G.J. Schrijen, B. Skoric, J. van Geloven, N. Verhaegh, R. Wolters, in *Read-Proof Hardware from Protective Coatings*. Proceedings of CHES 2006. Lecture Notes in Computer Science, vol. 4249 (Springer-Verlag, Heidelberg, 2006), pp. 369–383

20. F.M.J. Willems, The context-tree weighting method: extensions. IEEE Trans. Inf. Theory **44**(2), 792–798 (1998)
21. F.M.J. Willems, Y.M. Shtarkov, T.J. Tjalkens, The context-tree weighting method: basic properties. IEEE Trans. Inf. Theory **41**(3), 653–664 (1995)

Anti-counterfeiting: Mixing the Physical and the Digital World

Darko Kirovski

1 Introduction

Counterfeiting is as old as the human desire to create objects of value. For example, historians have identified counterfeit coins just as old as the corresponding originals. Archeological findings have identified examples of counterfeit coins from 500 BC netting a 600+% instant profit to the counterfeiter [2]. Test cuts were likely to be the first counterfeit detection procedure – with an objective to test the purity of the inner structure of the coin. The appearance of counterfeit coins with already engraved fake test cuts initiated the cat-and-mouse game between counterfeiters and original manufacturers that has lasted to date [2].

It is hard to assess and quantify the market for counterfeit objects of value today. With the ease of marketing products online, it seems that selling counterfeit objects has never been easier. Industries under attack include the software and the hardware, the pharmaceutical, the entertainment, and the fashion industry. For example, it is estimated that between 7 and 8% of world trade,[1] 10% of the pharmaceuticals market,[2] and 36% of the software market[3] are counterfeited. Consequently, there exists demand for technologies that can either resolve these problems or significantly reduce the breadth of the search space for origins of counterfeits.

D. Kirovski (✉)
Microsoft Research, Redmond, WA 98052, USA
e-mail: darkok@microsoft.com

[1] According to the Interpol, World Customs Organization and International Chamber of Commerce estimates that roughly 7 and 8% of world trade every year is in counterfeit goods.
[2] In a study with the US Food and Drug Administration, Glaxo-Smith-Kline estimated that counterfeit drugs account for 10% of the global pharmaceuticals market.
[3] The Business Software Alliance estimates that 36% of software sales worldwide are counterfeit.

A.-R. Sadeghi, D. Naccache (eds.), *Towards Hardware-Intrinsic Security,* Information   223
Security and Cryptography, DOI 10.1007/978-3-642-14452-3_10,
© Springer-Verlag Berlin Heidelberg 2010

1.1 Classification

We classify the related illegal trade into two groups:

- *Piracy* – where the buyer is confident that the purchased object is not genuine due to an uncharacteristically low price or some other form of discrepancy with respect to the original product. However, the buyer still willingly executes the trade. Such transactions do not gain substantial revenue to the pirate; hence, it is arguable what percentage of losses due to such events could be accounted as lost revenue for the legal copyright owner. First, buyers of such products are usually unlikely to purchase the original product. Second, one could argue that frequently pirated products, due to their public display and widespread usage, actually establish the pirated brand and consequently raise its value.
- *Counterfeits* – where the seller fools the buyer into believing that the merchandise is authentic and collects the full "legal-market" price on the product. In this case, the adversary collects substantial revenue with profit margins typically higher than that of the original manufacturer due to lack of development and marketing costs.

This classification is important as one could argue that it is perhaps impossible to address the first class using only technological means. On the other hand, we recognize that a suspecting buyer or a supply chain inspector could engage in a test of authenticity to address the latter problem. Clearly, a technology designed to help the buyer in the latter case is of no use in the case of piracy.

To the best of our knowledge there does not exist a study which breaks down illegal trade estimates into the above categories; however, for certain markets such as pharmaceuticals and supply chains for airplane parts nearly all illegal trade can be claimed to be counterfeited. Looking into hundreds of billions of dollars lost to counterfeits each year, we want to establish a set of requirements for a growing class of anti-counterfeiting technologies that construct certificates of authenticity using random hard-to-copy objects whose multi-dimensional features are crypto-graphically signed to ensure reliable and convenient authentication.

2 Desiderata for Anti-counterfeiting Technologies

A certificate of authenticity (COA) is a digitally signed physical object of fixed dimensions that has a random unique structure which satisfies the following requirements:

R1 inexpensive to manufacture – The cost of creating and signing original COAs is small, relative to a desired level of security

R2 expensive to copy – The cost of manufacturing a COA instance is several orders of magnitude lower than the cost of exact or near-exact replication of the unique and random physical structure of this instance

R3 inexpensive to authenticate off-line – The cost of verifying the authenticity of a signed COA off-line is small, again relative to a desired level of security

R4 robust – COA must be robust to the environmental elements such as changes in humidity and temperature, and ordinary wear and tear

The key to the analysis of a COA instance is the extraction of its "fingerprint," i.e., a set of features that reliably represents its multi-dimensional structure. This process is typically based on a specific physical phenomenon and produces a cardinality-N vector of numbers $\mathbf{x} \in \mathbb{R}^N$. This imposes that

R5 physical one-way function – It should be computationally difficult to construct an object of fixed dimensions with a "fingerprint" $\mathbf{y}$ such that $||\mathbf{x} - \mathbf{y}|| < \delta$, where $\mathbf{x}$ is a given "fingerprint" of an unknown COA instance and δ bounds the proximity of $\mathbf{x}$ and $\mathbf{y}$ with respect to a standardized distance metric $|| \cdot ||$.

This requirement establishes COA instances as physical one-way functions. By having access only to the "fingerprint" the adversary should face a difficult task of producing an object that has a near-equivalent "fingerprint." For example, when such a COA is attached to a credit card, it would prevent its physical replication by an adversary who obtains all the digital information stored on the card. Such an attack, often referred to as skimming, according to a Nielsen Report is responsible for about US $2B annual loss relative to a US $16B aggregate profit to credit card companies worldwide (data from 2008) [6].

R6 repetitiveness – The noise that stems from reading the "fingerprint" for a specific COA instance by different readers, in different environments, and/or at different read-out misalignments should be such that the probability of a false negative is smaller than a certain desired constant, $\Pr[||\mathbf{x} - \mathbf{y}|| < \delta] \leq \varepsilon_{FN}$, where $\mathbf{x}$ denotes the "fingerprint" read-out of an issued COA instance and $\mathbf{y}$ denotes a "fingerprint" read-out for the same instance during an arbitrary in-field verification

R7 non-collision – The probability of a false positive should be smaller than a certain desired constant, $\Pr[||\mathbf{x} - \mathbf{y}|| < \delta] \leq \varepsilon_{FP} \ll \varepsilon_{FN}$, where $\mathbf{x}$ denotes the "fingerprint" read-out of an issued COA instance and $\mathbf{y}$ denotes the "fingerprint" read-out for any other distinct instance

R8 "fingerprint" interdependence – "Fingerprint" samples collected over a large topological neighborhood should be mutually dependent. In addition, accurate mathematical modeling of this dependence should be as computationally expensive as possible. This dependence ensures that an adversary cannot forge the "fingerprint" one sample at a time – if such an attack is possible with a high success rate, typically its cost is linearly proportional to the number of "fingerprint" samples [21]

Requirement **R8** is one of the crucial requirements in thwarting attacks that do not aim at manufacturing a near-exact copy of the authentic COA instance. Instead, the adversary aims to launch a simple search process that adjusts the object topography so as to fit an authentic "fingerprint." Each iteration of this process would adjust a group of samples at a time. This attack could be even computational if requirement **R5** is not satisfied.

For example, COAs based on fibers relatively sparsely embedded in paper typically do not satisfy **R8** [21]. Positioning of a single fiber in this case is not dependent upon the remaining fibers; thus, the adversary can orient these fibers on paper one by one. If this process is accurate, the cost of recreating a single COA instance is small.

R9 tamper evidence – A COA instance could be conceived to represent a tamper-evident feature, i.e., a seal. If opening a specific package can be done exclusively via destroying the COA instance, and reproduction and reassembly of the signed seal is not easily attainable, then we could use such an instance as a tamper evidence.

R10 visual inspection of the verification path – The observed randomness is scanned using a hardware device; however, the verification path from the random object to the measurement circuitry/COA scanner must not be obstructed by adversarial hardware and/or software. That is why random features in COA instances should have relatively large minimum geometries so that they can be inspected visually. In addition, contactless (optical, wireless) measurements are preferred as static points of contact between a COA instance and a related scanner which represent a perfect opportunity for the adversary to intercept the verification path.

3 Digitizing the Physical World

COA systems as defined in Sect. 2 enable elegant off-line verification using a trusted device that contains the public-key of the COA issuer. In this section, we review how traditional public-key cryptography can be used to bind a product instance, COA's physical random features, and arbitrary information that the issuer desires to associate with the product. A simple protocol is adopted from [13, 14, 21] and presented in Fig. 1.

When creating a COA instance, the issuer digitally signs its "fingerprint" using traditional public-key cryptography. First, the "fingerprint" is scanned, digitized, and compressed into a fixed-length bit string f. Next, f is concatenated to an arbitrary information t that the issuer wants to associate with the product instance (e.g., product ID, expiration date, MSRP, coupon offers). The combined bit string $w = f \| t$ is then signed. Several signing protocols could be used here, for example,

- the Bellare–Rogaway protocol, PSS-R [5], for signing messages with message recovery where an arbitrary signing mechanism such as RSA [29] could be used; the resulting signature s is then encoded directly onto the COA instance using an existing storage technology such as an RFID or

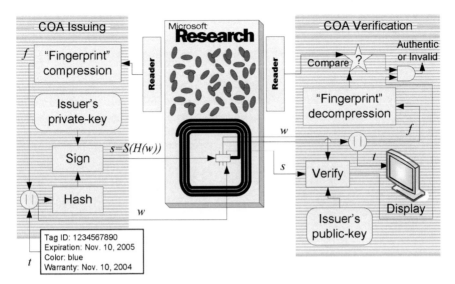

Fig. 1 Block diagram of the key steps involved in issuing and verifying a COA instance

- the issuer could also use a traditional IEEE1363-like cryptographic signature [15, 19], s, attach the plaintext w, and then encode $s||w$ directly onto the COA instance using an existing storage technology such as an RFID.

The resulting tag that contains both the COA and the associated RFID is now attached to a product whose authenticity the issuer wants to vouch. The association of the COA's "fingerprint," the issuer's private key, and the product protected using the COA can be verified in-field off-line using a device that has trusted access to the issuer's public-key. Secure delivery of this key calls for a simple public-key infrastructure where the device is only expected to maintain the public-key of a certifying authority, and the issuer stores a certificate, i.e., its public-key signed by the private key of the trusted party, onto the RFID.

Verification of the tag is done using a simple protocol. First, the corresponding signature, s, from the RFID is verified against the issuer's public-key [5, 19]. In case the integrity test is successful, the original "fingerprint" f (stored in the RFID or extracted from a PSS-R signature) and the associated data, t, are extracted from w. The verifier proceeds to scan in-field the actual "fingerprint," f', of the attached COA instance, i.e., obtain a new reading of the instance's physical properties, and compare them with f. If the level of similarity between f and f' exceeds a predefined and statistically validated threshold δ, the verifier declares the instance to be authentic and displays t. In all other cases, the reader concludes that the COA instance is not authentic.

In order to counterfeit protected objects, the adversary needs to

(i) compute the private key of the issuer – a task which can be made arbitrarily difficult by adjusting the key length of the used public-key crypto-system [19, 29], or

(ii) devise a manufacturing process that can exactly replicate an already signed COA instance – a task which is not infeasible but requires substantial expense by the malicious party – the forging cost dictates the value that a single COA instance can protect [21], or

(iii) misappropriate signed COA instances – a responsibility of the organization that issues COA instances. For example, one possibility is to collect tags from already sold products, attach them to counterfeits, and sell them as authentic merchandise. One way to address this problem is to assign two COAs for each product, one that vouches for product's genuineness, another that vouches that the product is new. Retailer's responsibility is to devalue (i.e., tear apart) the latter COA when the product is sold – an action that is trivial to verify at the back-end of the supply chain (i.e., the retailer would have to send all torn COAs back to the supply chain inspector). The same procedure can be used to signal and/or value product's "nth-owner."

4 Applications

COA instances are generic "objects of value." They have a fully horizontal perspective of possible applications. The value that one COA instance could maximally represent approximately equals the cost to forge this instance [21]. Inexpensive verification makes COAs particularly attractive for several traditional applications as well as for a myriad of new ones. Currency, checks, money orders, credit cards, license and product tags, warranties, receipts, endorsements, ownership documents, proofs of purchase/return, proof of repair, coupons, tickets, seals, tamper-evident hardware can all be produced using COAs.

COAs whose "fingerprints" satisfy requirement **R5**, i.e., they do not reveal their physical structure in a straightforward fashion, could be used against skimming credit cards and falsifying personal identification documents such as passports, visas, driver's licenses, and national ID cards. Then, by accessing full credit card information from a merchant database (e.g., holder's name, card's number and expiration date, PIN code, and COA's "fingerprint"), it would be still difficult for the adversary to create a physical copy of the original credit card produced by the issuing bank. To complete the operation, the adversary would have to gain physical access to the original credit card and accurately scan its 3D structure (e.g., using X-rays or other 3D imaging systems). Finally, the adversary would still have to build the 3D object, a task that requires significant cost due to **R2**.

5 Review of Existing Methodologies

COA instances can be created in numerous ways. For example, when covering a surface with an epoxy substrate, its particles form a lowrise but random 3D landscape which uniquely reflects light directed from a certain angle. COAs based upon this idea were first proposed by Bauder and Simmons from the Sandia National Labs and were used for weapons control during the Cold War [3]. To the best of our knowledge this is the first design of COAs based upon the fact that individual instances are difficult to near-exactly manufacture by a malicious well-financed party.

Fiber-based COA: Bauder and Simmons were also the first to propose COAs created as a collection of fibers randomly positioned in an object using a transparent gluing material which permanently fixes fibers' positioning [3, 4, 31]. Readout of the random structure of a fiber-based COA could be performed in numerous ways using the following fact: if one end of a fiber is illuminated, the other end will also glow. Bauder proposed fiber-based COAs for banknote protection – fibers in that proposal were fixed using a semi-transparent material such as paper [4]. To the best of our knowledge, only few efforts have followed the pioneering work by Bauder and Simmons. Church and Littman have worked on extraction of random optical fiber patterns in the context of currency anti-counterfeiting [8, 10]. The first COA system based upon fiber-infused paper and public-key cryptography was developed by Chen et al. [7, 21]. While efficient and inexpensive, fiber-based COAs do not satisfy **R5** and thus could be vulnerable to malicious attackers who conquer a technology for fiber placement on paper. Although such a technology is not available, we categorize its objective as 2+D manufacturing and speculate that it is substantially easier than manufacturing purely random 3D topologies.

Speckle scattering: Pappu was the first to create a class of physical one-way functions via speckle scattering [27, 28]. A speckle pattern is a random intensity pattern produced by the mutual interference of coherent wavefronts that are subject to phase differences and/or intensity fluctuations. Pappu focused on Gabor wavelets to produce short digests of the natural randomness collected from the optical phenomenon. His Ph.D. thesis has a solid survey of the related but scarce work [27]. Škorić was the first to match experimentation and theoretical bounds on the amount of randomness exhibited by keys formed from speckle [32]. Speckle scattering is a phenomenon sensitive to microscopic changes to the source of scattering, hence, it is difficult to build practical COAs that satisfy **R4**; in addition, it is poorly understood how speckle filtering addresses **R6** and **R8**.

Far-field RF: Finally, COAs in the electromagnetic domain have been proposed by several companies [9, 11, 12, 20, 30], all of them aiming to detect COA's random structure in the far field. The basic idea with these proposals was to identify a certain set of resonant features of dielectric and/or conducting materials in the far field[4] as a complement to RFID communication. As a consequence all proposals suffer from the inability to satisfy **R5** and **R8**, thus presenting relatively easy targets to malicious parties who understand their re-radiation principles. In addition, far field detection is prone to spoofing and jamming by a sophisticated attacker; thus, such schemes often have difficulties satisfying requirement **R10**. Because the detection is taking place in the far-field, these systems operate in the "expensive" 60 GHz frequency

---

[4] We define far field as distance which is multiple wavelengths away from the source of the electromagnetic (re)-radiation.

range making COA verification unnecessarily expensive with current semiconductor technologies.

Physically unclonable functions based upon forced variability in semiconductor manufacturing have been reviewed in Sect. 5.2.

5.1 RF-DNA

The first technology that has focused on identifying radio-frequency "fingerprints" of dielectric and conductive resonators in the near field was developed by DeJean and Kirovski [13, 14]. Their COA proposal, RF-DNA, is based upon the basic re-radiation (including radio-wave resonance, Rayleigh, and Mie scattering) principles described within the generalized extinction theorem [16, 26, 35]. RF-DNA is substantially different from far-field RF schemes, as it aims to capture in its "fingerprint" an accurate image of the variability exerted by the electromagnetic field close[5] to the source of re-radiation, i.e., COA instance. The imaging is done using a dense matrix of patch antennae, each of them capable of transmitting and/or receiving RF waves in the 5–6 GHz RF sub-band.

The technology fares well with respect to the set of desiderata. Each COA instance costs less than a cent, with the COA reader expected to cost less than US $100 and as low as several dollars in mass production. Near-exact replicas would demand true 3D scanning and manufacturing of a specific 3D topology. It is robust to wear and tear as the "fingerprint" read-out is contactless. Its physical one-way function can be well formulated mathematically via the inverse design problem over the Maxwell equations [1], which is an ill-defined problem [33] of exceptional, yet never formally proven,[6] computational difficulty. Even solving the forward simulation problem over the Maxwell equations accurately is an exceptionally challenging task for state-of-the-art electromagnetic field solvers [25]. Typically, the noise that stems from simulations is well over 3 dB with respect to physical measurements [25], whereas the expected noise due to misalignment, environmental factors, and variances in the manufacturing of COA readers should be well within 0.5 dB [14]. DeJean and Kirovski have shown that the detection performance in their system results in negligible rates of false positives and negatives [14]. The COA reader by design enforces that "fingerprint" readouts are dependent across different neighboring transmitter–receiver pairings, thus by minimally altering small part of her design, the adversary would affect the "fingerprint" components of many (or almost all) transmitter-to-receiver responses.

One of the open issues related to RF-DNA is its weak "fingerprint" robustness with respect to noise stemming from re-radiating objects that could be attached to COA instances as products. This is a limitation within requirement **R4** that has not been yet explored.

---

[5] Less than one wavelength from the resonator.

[6] To the best of the author's knowledge.

5.2 Challenge/Response COA Systems

COA systems that rely on small, imperceptible features, such as distinct semiconductor circuits created using manufacturing variability, cannot be verified using a passive protocol because the adversary can always hard-wire the resulting digitized "fingerprint" into an appropriate part of the hardware state and fool the verifier that it is measuring the actual unique manufacturing variability [24]. An active, challenge-based protocol on the other hand, would require that the verifier contains either (i) an accurate but private description of the distinct circuitry in each COA instance so that it can compute the expected response from the COA under test or (ii) a limited set of valid challenge/response pairs.

One major disadvantage of type (i) solutions is the fact that the private circuit description must be kept away from the adversary (otherwise, the adversary can use this description to manufacture the exact same circuit[7]); thus, the verification process must be launched online. To the best of the author's knowledge circuits of type (i) have not been proposed to date.

There exist several proposals for type (ii) circuits [17, 18, 23, 34], often referred to as physically unclonable functions, PUFs. The disadvantages of type (ii) proposals are necessity for online verification and high cost of storage and bandwidth that needs to be allocated to support the overall COA system among others. Table 1 presents a comparison of expenses related to enabling anti-counterfeiting using type (ii) semiconductor-based PUFs and RF-DNA. As one can observe a major cost

Table 1 Comparison table of expenses related to enabling anti-counterfeiting using type (ii) semiconductor-based PUFs and RF-DNA

Property	Semiconductor circuit PUF	RF-DNA
Storage at server	K challenge–response pairs need to be stored, where K equals the anticipated number of verifications per PUF	None
Storage at tag	Yes, on-chip. Needs to store public-key of certification authority	Yes, passive RFID Cost: 2–6 cents
Cost of tag	$\sim$ zero if embedded to protect chip *or* several cents if used to protect-object, i.e., similar to the cost of smartcard	$\sim$ cost of RFID + cost of signing the RF "fingerprint"
Communication w/server during in-field verification	Server-auth TLS handshake with key exchange, then ask for challenge, receive challenge, compute/send response, receive decision	None
Cost of server farm	Linearly proportional to expected peak number of concurrent server–PUF connections	None
Verification requirements	Communication channel to server	RF-DNA scanner

[7] We remind the reader that the variability is enforced, not unavoidable when manufacturing such circuits.

factor is the secure connectivity with the issuing server that needs to be established for a PUF to be verified. Conversely, if PUFs are embedded within an existing semi-conductor product, i.e., integrated chip, in order to protect it, their manufacturing cost is negligible, yet their verification still requires a communication channel to the issuing server.

Apart from probing and brute-force reverse engineering semiconductor PUFs, one specific group of attacks that has not been launched on imperceptive chal-lenge/response PUF systems yet is collusion of correct challenge/response pairs from a single and/or multiple distinct PUF instances with an objective to reverse engineer the underlying random circuit and its points of enforced variability. In addition, power and delay analysis attacks are also possible against such schemes [22]. We note that such attacks are not viable against COA systems outlined using requirements **R1–10**.

6 Conclusion

Randomness is natural to many processes in the physical world. Early ideas by Bauder and Simmons have resulted in a growing list of technologies that aim at using such randomness to create distinct, cryptographically secure, and hard-to-forge certificates of authenticity, i.e., physically unclonable functions. In this chapter, we introduced a set of desiderata for such technologies and showed how state of the art fares with respect to this set. We identified RF-DNA, a proposal by DeJean and Kirovski, as a technology that appears to address well all of the require-ments from this set.

References

1. D.B. Avdeev, Three-dimensional electromagnetic modelling and inversion: from theory to application. Surv. Geophys. **26**, 767–799 (2005)
2. K. Barry, Counterfeits and Counterfeiters: The Ancient World. Available on-line at: http://www.ancient-times.com/newsletters/n13/n13.html.
3. D.W. Bauder, Personal Communication
4. D.W. Bauder, *An Anti-Counterfeiting Concept for Currency Systems*. Research Report PTK-11990, Sandia National Labs, Albuquerque, NM, 1983
5. M. Bellare, P. Rogaway, in *The Exact Security of Digital Signatures How to Sign with RSA and Rabin*. EUROCRYPT (Springer-Verlag, Berlin, Heidelberg, 1996), pp. 399–414
6. P. Britt, Credit Card Skimming: Growing Trend or Media Hype? Transaction World Magazine, September, 2001. Available on-line at: http://www.transactionworld.com/articles/2001/september/riskmanagement1.asp
7. Y. Chen, M.K. Mihcak, D. Kirovski, Certifying authenticity via fiber-infused paper. ACM SIGecom Exch. **5**(3), 29–37 (2005)
8. S. Church, D. Littman, Machine reading of Visual Counterfeit Deterrent Features and Sum-mary of US Research, 1980–1990. Four Nation Group on Advanced Counterfeit Deterrence, Canada, 1991.
9. J. Collins, RFID fibers for secure applications. RFID J. (2004). Available on-line at: http://www.rfidjournal.com/article/articleview/845/1/14.

10. Commission on Engineering and Technical Systems (CETS), *Counterfeit Deterrent Features for the Next-Generation Currency Design*. (The National Academic Press, Washington, DC, 1993)
11. Creo, Inc, Available on-line at: `http://www.creo.com`
12. CrossID, Inc, Firewall Protection for Paper Documents. Available on-line at: `http://www.rfidjournal.com/article/articleview/790/1/44`.
13. G. DeJean, D. Kirovski, in *Radio Frequency Certificates of Authenticity*. IEEE Antenna and Propagation Symposium (IEEE Computer Society Press, Los Alamitos, CA, USA, 2006)
14. G. DeJean, D. Kirovski, in *RF-DNA: Radio-Frequency Certificates of Authenticity*. Cryptographic Hardware and Embedded Systems. Lecture Notes in Computer Science, vol. 4727 (Springer, Berlin, Heidelberg, 2007), pp. 346–363
15. T. Dierks, E. Rescorla, The Transport Layer Security (TLS) Protocol Version 1.2. Internet draft, available on-line at: `http://www.ietf.org/internet-drafts/draft-ietf-tls-rfc4346-bis-09.txt`.
16. P.P. Ewald, Ann. Phys. **49**, 1–56 (1915)
17. B. Gassend, D. Clarke, M. van Dijk, S. Devadas, in *Silicon Physical Random Functions*. ACM Computer and Communication Security Conference (ACM, New York, NY, USA, 2002)
18. J. Guajardo, S. Kumar, G. Schrijen, P. Tuyls, in *FPGA Intrinsic PUFs and Their Use for IP Protection*. Cryptographic Hardware and Embedded Systems (Springer-Verlag, Berlin, Heidelberg, New York, 2007)
19. IEEE 1363-2000: Standard Specifications For Public Key Cryptography, 2000. Available on-line at: `http://grouper.ieee.org/groups/1363`
20. Inkode, Inc, Available on-line at: `http://www.inkode.com`.
21. D. Kirovski, in *Toward an Automated Verification of Certificates of Authenticity*. ACM Electronic Commerce (ACM, New York, NY, USA, 2004) pp. 160–169
22. P. Kocher, J. Jaffe, B. Jun, in *Differential Power Analysis*. CRYPTO, (Springer-Verlag, Berlin, Heidelberg, 1999) pp. 388–397
23. J.-W. Lee, D. Lim, B. Gassend, G.E. Suh, M. van Dijk, S. Devadas, in *A Technique to Build a Secret Key in Integrated Circuits with Identification and Authentication Applications*. IEEE VLSI Circuits Symposium (IEEE Press, 2004)
24. K. Lofstrom, W.R. Daasch, D. Taylor, in *IC Identification Circuit Using Device Mismatch*. IEEE ISSCC, 2000, pp. 372–373
25. Microwave Engineering Europe, CAD benchmark. Oct 2000 – Feb 2001. Available on-line at: `http://i.cmpnet.com/edtn/europe/mwee/pdf/CAD.pdf`
26. C.W. Oseen, Uber die Wechrelwirkung zwischen zwei elektrischen Dipolen und uber die Drehung der Polarisationsebene in Kristallen und Flussigkeiten. Ann. Phys. **48**, 1–56 (1915)
27. R. Pappu, *Physical One-Way Functions*. Ph.D. thesis, MIT, 2001
28. R. Pappu et al., Physical one-way functions. Science **297**(5589), 2026–2030 (2002)
29. R.L. Rivest, A. Shamir, L. Adleman, A method for obtaining digital signatures and public-key cryptosystems. Commun. ACM **21**(2), 120–126 (1978)
30. RF SAW, Inc, Available on-line at: `http://www.rfsaw.com/tech.html`
31. G.J. Simmons, in *Identification of Data, Devices, Documents and Individuals*. IEEE International Carnahan Conference on Security Technology (IEEE Press, 1991) pp. 197–218
32. B. Škorić. On the entropy of keys derived from laser speckle, In Journal of optics A: Pure and Applied Optics Create an alert, vol. 10, no. 5 (IOP Publishing, UK, Mar 2008)
33. A.N. Tikhonov, V.A. Arsenin, *Solution of Ill-posed Problems*. (Winston & Sons, Washington, 1977)
34. P. Tuyls, B. Škorić, Strong authentication with physical unclonable functions. in *Security, Privacy and Trust in Modern Data Management*, ed. by M. Petković, W. Jonker. Data-Centric Systems and Applications (Springer, 2007)
35. E. Wolf, A generalized extinction theorem and its role in scattering theory. in *Coherence and Quantum Optics*, ed. by L. Mandel, E. Wolf. (Plenum, New York, NY, 1973)

Part V
Hardware Security in Contactless Tokens

Anti-counterfeiting, Untraceability and Other Security Challenges for RFID Systems: Public-Key-Based Protocols and Hardware

Yong Ki Lee, Lejla Batina, Dave Singelee, Bart Preneel, and Ingrid Verbauwhede

1 Introduction

Recently, the use of RFID (radio frequency identification) technology has expanded enormously. It was developed in the middle of the twentieth century and is today being applied in many areas: supply chains, access control, electronic passports, health care, road pricing, etc. The advantage of RFID over bar-code technology is that it does not require direct line-of-sight reading and that tags can be interrogated at greater distances. The technology also enables the automation of some control processes, which results in a significant gain in terms of time and cost.

However, the widespread use of RFID tags has raised privacy and security concerns. Fortunately, these problems can be solved by using cryptographic techniques. The tags have to implement cryptographic protocols to provide security services such as authentication, confidentiality, data integrity, non-repudiation, anti-counterfeiting, and anti-cloning. This is however not sufficient. RFID applications also require the enforcement of strong privacy policies to protect the customer. The privacy of RFID tags has been a hot issue recently [40, 41]. Unfortunately, privacy features have initially been close to non-existent in the design of conventional authentication systems and therefore there is still a lot of work to be done in the context of RFID applications.

In this chapter, we summarize security and privacy issues in RFID systems and state-of-the-art solutions in the literature, and we also introduce our novel solutions, which are presented in [30, 31]. Our solutions rely on asymmetric (i.e., public-key) cryptography. In particular, we use ECC (elliptic curve cryptography). The advantage of ECC lies in a drastic reduction in bit-lengths of the certificates (compared to conventional public-key cryptographic techniques e.g., RSA). Also it has been

Y.K. Lee (✉)
Department of Electrical Engineering, University of California, Los Angeles, CA 90095-1594, USA
e-mail: jfirst@ee.ucla.edu

A.-R. Sadeghi, D. Naccache (eds.), *Towards Hardware-Intrinsic Security,* Information Security and Cryptography, DOI 10.1007/978-3-642-14452-3_11,
© Springer-Verlag Berlin Heidelberg 2010

shown that this could be the option for even extremely low-cost platforms, such as RFID tags and sensor nodes [21, 33]. The novelty of our work is in using ECC primitives exclusively. Namely, in authentication protocols for RFID, one often uses some other symmetric-key primitives, such as hash functions and/or message authentication codes (MACS) to defend against security and privacy threats. We solve the issue by using only certain mathematical properties of ECC. We also introduce the search protocol, a novel scheme which allows an RFID reader (or a server) to efficiently query for a specific tag, without compromising the latter's privacy. The authentication protocols and the search protocol are all made of the same building blocks, but meet different security requirements. This feature allows for a simplified realization of the protocols on a real tag. We demonstrate this by presenting an RFID hardware architecture of an elliptic curve processor over $GF(2^{163})$ that realizes the presented protocols. The results show the plausibility of meeting the security, privacy, and efficiency requirements even in a passive RFID tag.

The remainder of this chapter is organized as follows. Section 2 addresses the security and privacy requirements for RFID systems, including scalability, anti-cloning, and protection against tracking and impersonation attacks. In Sect. 3, relevant previous works are discussed and divided into three main categories of cryptographic techniques for RFID. We present our authentication protocols and a search protocol in Sects. 4, 5 and 6. We show the implementation results of these protocols for a particularly designed hardware architecture in Sect. 7. We conclude this chapter in Sect. 8.

2 Security and Privacy Requirements

When designing cryptographic authentication protocols for RFID, both security and privacy have to be addressed. These two design objectives should clearly be distinguished from each other: While *security* addresses the soundness of a protocol, *privacy* addresses the resistance against unauthorized identification, tracking, or linking tags.

2.1 Security Objectives

The main security requirement of an RFID authentication protocol is *tag authentication*. By carrying out the protocol, the reader (or the server) is assured of the identity of the RFID tag. In some scenarios, *mutual authentication* is required, in which the reader has to authenticate itself to the tag. Without loss of generality, we only consider tag authentication in the remainder of this chapter. Tag authentication makes it possible to verify the authenticity of a product (i.e., ensure anti-counterfeiting) or a user (carrying a token that contains the tag). When authentication claims explicitly use the location of a tag, one needs distance-bounding protocols [6, 20, 43], a special class of RFID authentication protocols. In order to be used for authentication purposes, the protocol has to be resistant to an *impersonation attack*, where the attacker maliciously pretends to be the tag. In general, one should consider several types of adversarial attacks (which could result in the impersonation of a tag when conducted successfully), including the following ones:

Eavesdropping: By passively eavesdropping the communication between a tag and a reader, the adversary can try to obtain secret information (such as secret keys), which can then be used to identify itself as a legitimate tag.

Replay attacks: An active adversary can try to use a tag's response to a reader's challenge to impersonate the tag.

Man-in-the-middle attacks: This is an active attack where the adversary is interposed between the reader and the tag and can intercept all messages going between the two parties. He can modify or delete them, or insert new messages.

Cloning: An active adversary can try to crack a particular tag and extract the secret key. He can then try to use this revealed secret to impersonate other tags. To prevent this *cloning attack*, a secret key should be pertinent only to a single tag so that a revealed secret key cannot be used for any other tag. Sometimes one also requires that an adversary cannot extract a secret key from a tag (to make identical copies of that particular tag). We will denote this strong requirement as *full anti-cloning resistance*.

Side-channel attacks: These attacks are based on the information gained from the physical implementation of the cryptographic protocol and do not exploit weaknesses in the theoretical design of the protocol. Examples are timing attacks [28], power analysis [29], fault injection attacks [24], etc. Since this chapter focusses on the design of authentication schemes, these attacks are out of scope.

Another security requirement is *availability*. An adversary can try to tamper with the tag such that it does no longer function. Other types of *denial-of-service attacks* focus on the network itself, and block the communication between the tag and the reader (e.g., by jamming the network). Techniques to prevent this type of attacks are not discussed in this chapter.

2.2 Privacy Objectives

The prevalence of RFID technology introduces several privacy risks. The threat to privacy especially grows when a tag's identity is combined with personal information [25]. There are two important notions of privacy: anonymity and untraceability. An authentication protocol offers *anonymity* if the identity (e.g., the unique EPC code [15]) remains unknown for any unauthorized third party. The protocol offers *untraceability* if it is not possible for any unauthorized third party to determine the (in)equality of two tags. Untraceability is a stronger privacy requirement than anonymity. In the rest of this chapter, we will mainly focus on untraceability. One often uses two other privacy-related concepts: backward and forward privacy (these notions originate from the definition of backward and forward secrecy [13]). A protocol is *forward private* when an adversary that compromises a tag, i.e., she learns its current state and secret keys, is unable to identify the previous outputs of this tag. A protocol is *backward private* when an adversary that compromises a tag, i.e., she learns its current state and secret keys, is unable to identify the future outputs of this tag.

To track an RFID tag, an attacker can carry out the same passive and active attacks as described in Sect. 2.1. One should note that security does not automatically imply privacy. This can be illustrated by the following example. The Schnorr protocol [42] is a well-known cryptographic authentication protocol whose security properties can be formally proven [4]. However, it is not private since a tag (the prover in the Schnorr protocol) can be traced by an eavesdropper as shown in [31].

Several theoretical frameworks to address the privacy of RFID systems have been proposed in the literature [2, 27, 36, 46]. Each of these frameworks describes a formal adversarial model, which defines the means of the adversary and his goals. To analyze the privacy properties of the authentication protocols discussed in this chapter, we import two adversarial characteristics from the theoretical framework of Vaudenay [46]: wide (or narrow) attackers and strong (or weak) attackers. If an adversary has access to the result of the verification (accept or reject) in a server, she is a *wide* adversary. Otherwise she is a *narrow* adversary. If an adversary is able to extract a tag's secret and reuse it in other instances of the protocol, she is a *strong* adversary. Otherwise she is a *weak* adversary. A *wide-strong* adversary is hence the most powerful. If a protocol is untraceable against a *narrow-strong* adversary, we say the protocol is *narrow-strong private*.

2.3 General Objectives

When designing a cryptographic authentication protocol for RFID, one should always take into account the specific characteristics and constraints of this technology. We will briefly focus on two general requirements that have an impact on the security design. First of all, an authentication protocol for RFID should be *cheap to implement*, *lightweight*, and *efficient*. Passive RFID tags, which require an external power source to provoke signal transmission, only have a very low gate complexity. This limits the choice of cryptographic building blocks that can be used. For example, authentication protocols which make use of hash functions (such as [3, 14, 22, 34, 35, 38, 44, 47] are beyond the capability of such low-cost devices. Both the computational and the communication (i.e., the number of rounds) overhead of the protocol should be limited as much as possible. A second important requirement is *scalability*. Millions of RFID tags are deployed, each containing secret information (such as secret keys). Readers have to be able to identify or query for a specific tag, without which results in a significant increase of the complexity of the system.

3 State of the Art

In this section, we survey the state of the art for RFID authentication protocols. In the beginning the main efforts were on designing solutions that rely exclusively on private-key (also called symmetric-key) cryptography. The main reason lies in the common perception of public-key cryptography being too slow, power-hungry, and too complicated for such low-cost environments. However, recent works proved this concept to be wrong, as, for example, the smallest published ECC implementations [21, 33] consume less area than any known secure cryptographic hash function (e.g., the candidate algorithms proposed in the SHA-3 competition [37]). One alternative is, therefore, to pursue protocols that use only public-key cryptography. In addition, we also mention RFID authentication protocols that are based on physical unclonable functions (PUFs) [45]. It was shown that those solutions can also prevent counterfeiting in on-line and off-line scenarios (even providing full anti-cloning resistance) and are feasible for active tags. However, they were not designed to offer any privacy protection, and hence have to be combined with other privacy-preserving RFID authentication protocols.

Note that we only consider RFID authentication protocols on the logical level. Danev et al. [10] have shown that one can also identify RFID tags with a high accuracy from a small distance (e.g., less than 1 m), based on their physical-layer fingerprints. This technique automatically prevents cloning attack. However, the downside of this solution is the requirement that the distance between RFID tag and reader should be small, in order to have a high accuracy. On the other hand, allowing a large distance between reader and tag, as is the case for RFID authentication protocols on the logical level, gives more freedom to the attacker and hence makes him more powerful (e.g., she can carry out man-in-the-middle attacks).

Next we give a (non-exhaustive) overview of protocols that are proposed for RFID authentication, spanning a large set of properties fulfilled and diverse feasibility issues. We divide the protocols into several groups on the basis of main cryptographic blocks that were used in the constructions.

3.1 Authentication Protocols Based on Private-Key Cryptography

There have been many attempts to design authentication protocols for RFID tags by means of symmetric-key primitives. One of the first was the work of Feldhofer [16] that proposed a challenge-response protocol based on the AES block-cipher. The implementation consumes a chip area of 3,595 gates and has a current consumption of 8.15 μA at a frequency of 100 kHz. Of other notable solutions for authentication protocols we mention here the HB^+ protocol [26] that was presented as an extremely cheap solution but still secure against active adversaries. It meets even the cost requirements for the tags of 5–10 c range. The authors have built upon the basic "human authentication" protocol, due to Hopper and Blum (HB) [23].

Other variants of HB followed, as a result of attacks that appeared, such as the work of Gilbert et al. [18]. This attack is a man-in-the-middle attack against HB^+

but it requires many failed authentications to extract HB^+ keys. As a fix a new protocol called HB^{++} from Bringer et al. [8] was proposed. HB^{++} is claimed to be secure against man-in-the-middle attacks (as in [18]) but it requires additional secret key material and universal hash functions to detect the attacks. In the follow-up work Bringer and Chabanne [7] proposed a new HB^+ variant (so-called Trusted-*HB*) that builds upon Krawczyk's hash-based authentication schemes using special LFSR constructions (via Toplitz matrix). The new protocol is also resistant to man-in-the-middle attacks. Many more attacks followed, of which the most recent one is the work of Frumkin and Shamir [17]. In their paper, Frumkin and Shamir discuss several weaknesses of Trusted-*HB*.

A novel authentication and forward private RFID protocol is proposed by Berbain et al. [5]. The protocol is using pseudo-random number generators and universal hash functions as basic building blocks, which makes it suitable for low-footprint solutions. The security of their scheme is proven in the standard model but it remains unclear whether it can withstand physical attacks (i.e., tampering with the tag, such that the tag can be cloned).

Many RFID authentication protocols based on private-key cryptography suffer from scalability issues. To authenticate the tag, the reader (or the server) often has to perform an exhaustive search through all the shared symmetric keys and compute for each of these keys the corresponding response, in order to find a match. The complexity of this task grows linearly with the number of RFID tags being deployed. Using symmetric group keys (i.e., sharing symmetric keys among several RFID tags and the reader) solves the scalability issue, but causes cloning attacks, where an attacker uses a revealed secret (obtained from cracking a tag) to impersonate other RFID tags.

3.2 Authentication Protocols Based on PUFs

Here we introduce some solutions that require a physical unclonable function. As other chapters in this book deal with the issue extensively, we only briefly mention some important works specifically claimed for RFID tags or similar low-power devices.

Tuyls and Batina were the first to propose PUFs for the purpose of counterfeiting, e.g., RFIDs [45]. However, the constructions discussed were too complex and required a hash function as well as ECC. Hammouri and Sunar [19] proposed a protocol that combines the advantages of PUFs and the HB protocol. Their solution is secure against an active adversary and offers resistance against active tampering due to the use of a PUF.

3.3 Authentication Protocols Based on Public-Key Cryptography

Recently, the research community also focussed on RFID authentication protocols based on public-key cryptography. This approach solves the scalability issues in general, that burden symmetric-key solutions and also prevents cloning attacks. However, several protocols suffer from privacy vulnerabilities. In [31], it is shown that some conventional public-key-based authentication protocols, such as the Schnorr protocol [42] and the Okamoto protocol [39], do not resist tracking attacks. Accordingly, the EC-RAC (elliptic curve based randomized access control) protocol has been proposed in the same paper to address the established privacy threat. However, in [9, 11], it is shown that EC-RAC is also vulnerable to tracking attacks and replay attacks, and in addition [9], the randomized Schnorr protocol has been proposed as an alternative for EC-RAC. EC-RAC has been gradually revised in [30, 32] in order to countermeasure the known attacks as in [9, 12]. The remainder of this chapter is focussed on authentication protocols based on public-key cryptography, more specifically on ECC.

4 Untraceable Authentication Protocols Based on ECC

We introduce a few interesting untraceable authentication protocols. All protocols presented in this section are narrow-strong private, i.e., the protocols are untraceable even if secret keys of a tag are known to an attacker. However, protocols must be designed more elaborately in order to achieve wide privacy, i.e., the protocols are untraceable even if the decision (accept/reject) of the server is known to an attacker. However, the design of wide privacy-preserving RFID authentication protocols remains an open research problem.

4.1 Notation

Let us first introduce some notations. We denote P as the base point, y and $Y(= yP)$ are the server's private-key and public-key pair, and x_1 and $X_1(= x_1 P)$ are a tag's private-key and public-key pair. A tag's public-key is also called a *verifier*. One should note, although the name suggests that it can be publicly known, that the public-key of the tag should be kept secret in the server. Revealing this key causes tracking attacks.

4.2 EC-RAC II

In the revised EC-RAC protocols [32], also denoted as EC-RAC II and which solve the weaknesses of the original EC-RAC protocol proposed in [31], the tag authentication is enabled by conducting the ID-transfer scheme and the password-transfer scheme. The two schemes are based on the same design concept, and therefore, we introduce the ID-transfer scheme only.

The ID-transfer scheme of EC-RAC II is shown in Fig. 1. In this scheme, a tag generates a random number r_{t1} and a point T_1, and transfers T_1 to the server. Then, the server responds with a random challenge r_{s1}, and a tag produces and

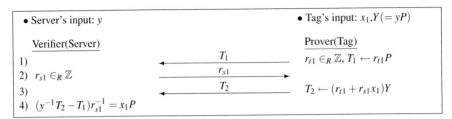

Fig. 1 ID-transfer scheme of EC-RAC II [32]

transfers another point T_2 to the server. After receiving T_2, the server calculates a tag's ID-verifier $x_1 P (= X_1)$, which is used to check whether the corresponding tag is registered in the server.

4.3 Randomized Schnorr Protocol

Another solution suggested to prevent tracking attacks is the randomized Schnorr protocol [9], which is shown in Fig. 2.

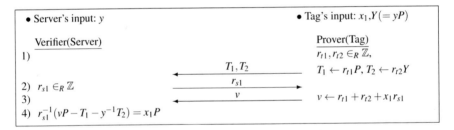

Fig. 2 Randomized Schnorr protocol [9]

In this protocol, a tag generates two random numbers r_{t1} and r_{t2}, and computes and transmits the two corresponding messages T_1 and T_2. After receiving a challenge r_{s1} from the server, a tag computes and transmits an authentication code T_2 to the server. Then, the server derives the tag's ID-verifier $(x_1 P)$ and checks if it is registered in the server.

4.4 Man-in-the-Middle Attacks

In [32], the security proof of the ID-transfer scheme is done by reducing it to well-known hard cryptographic problems. In order to show the security against replay attacks, the ID-transfer scheme is reduced to the Schnorr protocol, and in order to show the resistance against tracking attacks, it is reduced to the decisional Diffie–Hellman (DDH) problem. However, in the attacker's model, an attacker's ability is limited to observing the exchanged messages and forging messages to impersonate

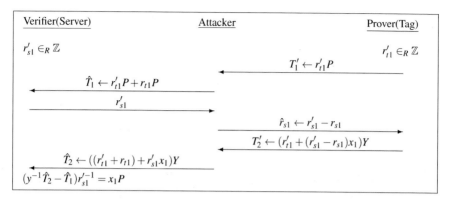

Fig. 3 Man-in-the-middle attack on EC-RAC II

a server or a tag. In other words, an attacker does not know when the server accepts or rejects (forced) messages, i.e., the attacker is assumed to be narrow.

In [12], Deursen and Radomirović demonstrate that man-in-the-middle attacks can be carried out on the revised EC-RAC protocols when an attacker is wide, as shown in Fig. 3. An attacker utilizes messages exchanged in a previous session of the protocol $(T_1(= r_{t1}P), r_{s1}, T_2(= (r_{t1} + r_{s1}x_1)Y)$ and the messages received in the current session $(T_1'(= r_{t1}'P)), r_{s1}', T_2'(= (r_{t1}' + (r_{s1}' - r_{s1})x_1)Y))$ in order to generate $\hat{T}_1(= T_1 + T_1'), \hat{r}_{s1}(= r_{s1}' - r_{s1})$, and $\hat{T}_2(= T_2 + T_2')$. By checking whether the server accepts the forged messages, an attacker can know if the currently communicating tag is the tag that generated the previous messages. As a result, a tag can be traced by a wide attacker.

The randomized Schnorr protocol has a similar problem as the ID-transfer scheme, as noted in [12]. A man-in-the-middle attack that allows to track the tag is shown in Fig. 4.

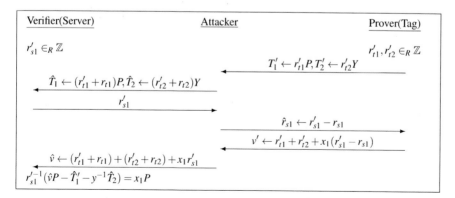

Fig. 4 Man-in-the-middle attack on the randomized Schnorr protocol

One can note that the ID-transfer scheme is more efficient than the randomized Schnorr protocol while they have similar security properties. The ID-transfer scheme requires two EC point multiplications on both the server and a tag, while the randomized Schnorr protocol requires three and two EC point multiplications, respectively.

5 EC-RAC IV

The man-in-the-middle attack may not be useful for most of the RFID applications. Some of the applications may not show the authentication results, which make this attack useless. Moreover, an attacker should be able to block and retransmit the messages properly in the communication between a reader and a tag, which will be difficult due to a short communication distance in most of the RFID applications. In addition, if a reader is placed far away from a tag, the issue for an attacker is on the accessibility to an authentic reader while keeping the communication with a tag. This can cause a large delay resulting in communication loss between a reader and a tag. If a reader uses a threshold for the response delay of a tag or employs distance-bounding protocols, man-in-the-middle attacks can be prevented.

The man-in-the-middle attacks described above can also be prevented by introducing more computation in a server and a tag. The attack shown in Fig. 3 exploits the linear property of the EC group operation. Therefore, if we can break linear relations among the exposed messages, this attack can be prevented.

The solution is shown in Fig. 5 where $x(r_{s1}P)$ is the x-coordinate of $r_{s1}P$. This introduces a non-linear operation (note that the resulting point is used as a scalar value), and the EC point multiplication is a one-way function. Therefore, it can prevent an attacker from forging a server's challenges (r_{s1}). For this, a server and a reader need to perform an extra EC point multiplication. The resulting protocol, denoted as EC-RAC IV in the literature, solves the man-in-the-middle attack which could be applied on EC-RAC II and the randomized Schnorr protocol. More details can be found in [30]. As in EC-RAC II, this protocol relies exclusively on ECC primitives to benefit from the advantages such as short certificates and low-cost implementations [21, 33].

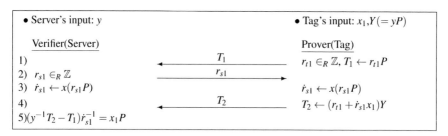

Fig. 5 EC-RAC IV [30]

6 Search Protocol

The search protocol for an RFID system aims to find a specific tag in a pool of many tags. If one of the secure authentication protocols presented in this chapter is used to search for a specific tag, the server must authenticate each tag one by one in a random order. In this case, the computation complexity will increase linearly with the number of the tags. A large library can be a good example for this use case. Suppose each book is equipped with a tag. A book can be easily misplaced, e.g., because of a visitor's negligence or a librarian's mistake. If we just use a randomized authentication protocol to find a specific book, the server should authenticate half of the books in the library on average before finding the required one. Therefore, designing an efficient, secure search protocol is an essential requirement in an RFID system.

In an efficient search protocol, the server would expect the response only from a designated tag. Otherwise, the server should handle responses from multiple tags. On the other hand, a tag should not respond before properly authenticating the server since a query may not be from an authentic server but from an attacker who wants to track a tag. Therefore, the protocol should be a one-round protocol, and a tag should authenticate the server without giving any challenge. Also note that we should consider the possibility of replay attacks because an attacker that can reuse past messages to hear responses from a specific tag can perform tracking attacks. Moreover, the messages from the server should be understandable or verified only by a designated tag to again prevent tracking attacks.

Note that the search protocol cannot achieve narrow-strong privacy, since an attacker that knows the secret keys of the tag can interpret the server's messages as much as the tag itself. As a result, this type of search protocol can be only wide-weak private.

The specific requirements for the search protocol can be summarized as follows:

1. One-round authentication: The protocol should be done in one round. Therefore, the server should generate messages without receiving a challenge from a tag.
2. Dedicated authentication: Only the designated tag should be able to verify that the messages are originally generated by the server.
3. Secure and private authentication (see requirements of the RFID authentication protocols discussed in this chapter): This implies protection against replay attacks.
4. Wide-weak privacy.

6.1 Protocol Description

In this protocol, we suppose that a tag has two pairs of private-keys and public-keys. Two private-keys of a tag are x_1 and x_2, and their public-keys are $X_1 (= x_1 P)$ and $X_2 (= x_2 P)$, respectively. A tag stores x_1 and x_2, and the server stores x_1 and X_2.

In order to prevent replay attacks, the server should somehow utilize a challenge from a tag, which requires at least two rounds. So, we first design a two-round protocol and reduce it to a one-round protocol. A two-round protocol can be considered as a function $f(c)$ in the server, which outputs authentication messages with an input of a challenge c from a tag, as follows:

$$f(c) = \{rP, r(x_1 + c)x_2P\},\tag{1}$$

where r is a random number.

In order to reduce to a one-round protocol, we change the protocol such that the server generates a challenge instead of receiving from a tag. Therefore, the server will generate and transmit the following three messages:

$$\{c, rP, r(x_1 + c)x_2P\},\tag{2}$$

In this case, we need to make sure that c cannot be used twice to prevent replay attacks. A tag can keep a counter and update with the received one only when the messages are valid and the received counter is bigger than the stored one. Another way is to use a time stamp. Since the time is incremental like a counter, a tag can use the same mechanism as the counter. Note that even if the time stamp covers up to 1,000 years and down to nanoseconds, the resolution can be covered with 65 bits which is much less than a full word size, e.g., 163 bits for a reasonable security level of ECC. The final search protocol is given in Fig. 6.

After verifying the messages from the server, a tag can respond to the server. In order to make sure that a proper tag is responding to the server, tag authentication can be realized by carrying out one of the protocols discussed in Sect. 4. For the search protocol, the server and a tag need to perform two EC point multiplications each.

- **Server Input:** $x_1, X_2(= x_2P), c_s$ (server counter).
- **Tag Input:** x_1, x_2, c_t (tag counter).

A. Server $\rightarrow$ Tag (Message Generation): $c_s, rP, r(x_1 + c_s)X_2$

1. Increase the counter c_s.
2. Generate a random number r, $1 \leq r \leq n - 1$.
3. Calculate $S_1 \leftarrow rP$ and $S_2 \leftarrow r \cdot (x_1 + c_s) \cdot X_2$.
4. Broadcast three messages: c_s, S_1, S_2.

B. Tag (Message Verification)

1. If $c_t \geq c_s$, then a tag ignores the messages and halts.
2. Verify whether S_1 has the desired prime order n.
   If $S_1 = O$ or $n \cdot S_1 \neq O$, then halts.
3. Otherwise, verify whether $S_2 = (x_1 + c_s) \cdot x_2 \cdot S_1$.
4. If S_2 is valid, then updates the counter as $c_t \leftarrow c_s$, and responds to the server.

Fig. 6 Search protocol

6.2 Search Protocol Analysis

We show that the proposed protocol satisfies all the four conditions for the search protocol. The first two conditions can be easily shown. The search protocol is definitely a one-round authentication protocol. Moreover, only the valid server can generate the messages since it requires x_1 and X_2, and only a specific tag can verify them since it requires x_1 and x_2.

6.2.1 Security Against Replay Attacks

For this attack, an attacker should be able to generate $r(x_1 + c)x_2 P = rx_1 x_2 P + rcx_2 P$ for a new value of c using some of the previously exchanged protocol messages. Since x_1 and x_2 are fixed, independent random values, $x_2 P$ and $x_1 x_2 P$, can be considered as two independent public-keys of a tag. Therefore, by the transmission of rP, the server and a tag can share two independent shared-secrets of $rx_2 P$ and $rx_1 x_2 P$, which are indistinguishable from a random point, assuming the hardness of the decisional Diffie–Hellman problem. Therefore, $r(x_1 + c)x_2 P$ can be considered for an attacker as follows:

$$rx_1 x_2 P + rcx_2 P = R_1 + cR_2,$$

where R_1 and R_2 are random points.

Note that R_1 and R_2 are unknown to an attacker and are independently generated each time of the protocol. This can be reduced to ECDSA (Fig. 7 [1]) where the signature is done on c as shown in Theorem 1. Therefore, as long as ECDSA is a secure signature algorithm, an attacker should not be able to generate another valid message for a different c.

Theorem 1 *The search protocol can be cryptographically reduced to the ECDSA assuming the hardness of the decisional Diffie–Hellman problem.*

Proof In ECDSA the transmitted messages, i.e., the signature on m, are as follows:

$$\{r, s\} = \left\{ x_1 \ (\text{mod } n), \ k^{-1}e + k^{-1}rd \ (\text{mod } n) \right\}, \tag{3}$$

where $(x_1, y_1) = kP$, k is a random number, $e = SHA1(m)$, and d is the private-key of the signature generator.

We can consider a new signature algorithm which is stronger than ECDSA as follows:

$$\{r, s\} = \{r_1, \ r_2 e + r_3\}, \tag{4}$$

where r_1, r_2, and r_3 are independent random numbers.

In order to verify the signature in Eq. (4), a verifier should receive r_1, r_2, and r_3 securely, which are not solved, and therefore, the algorithm may not properly

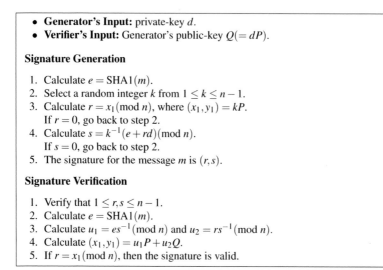

- **Generator's Input:** private-key d.
- **Verifier's Input:** Generator's public-key $Q(=dP)$.

Signature Generation

1. Calculate $e = \text{SHA1}(m)$.
2. Select a random integer k from $1 \leq k \leq n-1$.
3. Calculate $r = x_1 (\text{mod } n)$, where $(x_1, y_1) = kP$.
   If $r = 0$, go back to step 2.
4. Calculate $s = k^{-1}(e + rd)(\text{mod } n)$.
   If $s = 0$, go back to step 2.
5. The signature for the message m is (r, s).

Signature Verification

1. Verify that $1 \leq r, s \leq n-1$.
2. Calculate $e = \text{SHA1}(m)$.
3. Calculate $u_1 = es^{-1}(\text{mod } n)$ and $u_2 = rs^{-1}(\text{mod } n)$.
4. Calculate $(x_1, y_1) = u_1 P + u_2 Q$.
5. If $r = x_1 (\text{mod } n)$, then the signature is valid.

Fig. 7 EC digital signature algorithm

work. At this moment, we just care whether an attacker can get any advantage in the exchanged message. Nevertheless, it is evident that the signature in Eq. (4) is at least as secure as the signature in Eq. (3).

Now let us have a closer look to the search protocol:

$$\{c, rP, r(x_1 + c)x_2 P\} = \{c, rP, rx_1 x_2 P + crx_2 P\}.$$

Here, c is comparable with the message e being signed, and rP, $rx_1 x_2 P$, and $rx_2 P$ are comparable with three random values r_1, r_2, and r_3. rP is an actual random point, and $rx_1 x_2 P$ and $rx_2 P$ are undistinguishable from real random points for an attacker as long as the decisional Diffie–Hellman problem is hard. Therefore, the search protocol can be reduced to ECDSA with an assumption of the hardness of the decisional Diffie–Hellman problem.

6.2.2 Wide-Weak Privacy

There are three exchanged messages in the protocol: c, rP, and $r(x_1 + c)x_2 P$. Among these messages, rP is a random point and $r(x_1 + c)x_2 P$ is indistinguishable from a random point as long as the decisional Diffie–Hellman problem is hard. Therefore, an attacker has no benefit from these two messages to track a tag. Therefore, the protocol is at least narrow-weak private. A wide attacker knows whether a set of messages are accepted or not. In order to utilize this decision, an attacker should be able to forge a set of messages related with a valid message set and check if it is accepted by a tag, similar to the attacks shown in Figs. 3 and 4. However, a success of this attack means that an attacker can generate a valid digital signature, noting that the search protocol can be reduced to the EC digital signature algorithm. Therefore, the proposed search protocol is wide-weak private.

Note that c does not involve any secret information of a tag. However, if c is a counter being queried for a specific tag, it could cause some leakage. This can be solved by increasing c in a different way. The server may keep only one counter and use it for all the tags. Since a tag will accept the counter as long as it is larger than the saved value, the protocol will work and the revealed counter will not indicate how many times a certain tag has been queried. An alternative solution is the use of a time stamp. Since the time is incremental like a counter, it prevents reusing the value c. Note that the tag does not need to generate the value c (it only has to check that the value is larger than the saved value), and hence does not need to have a timer. By using a time stamp, c will no longer be a counted number of queries. Even if the time stamp covers up to 1,000 years with a precision down to nanoseconds, this resolution can be covered with 65 bits, which is much less than a full word size (e.g., 163 bits) for a reasonable security level.

6.3 Combining Authentication Protocols

After the search protocol, the authentication of a tag should follow, e.g., the ID-transfer scheme, in order to make sure that a proper tag is responding to the server. When two protocols are combined, we need to check security and privacy again if there is vulnerability due to the combination. The exchanged messages in the ID-transfer scheme are as follows:

$$r_{t1}P, \ r_{s1}, \ (r_{t1} + \dot{r}_{s1}x_1)Y. \tag{5}$$

The exchanged messages in the search protocol are as follows:

$$c, \ rP, \ (rx_1 + rc)x_2P. \tag{6}$$

$(r_{t1} + \dot{r}_{s1}x_1)Y$ is the only message using the base point Y, and therefore, it will be independent of the messages in the search protocol. Moreover, the other messages in the ID-transfer scheme, $r_{t1}P$ and r_{s1}, are random values that are not used in the search protocol. Therefore, they are independent, and the combination will inherit the weaker security and privacy properties between the two protocols.

7 Implementation

In order to show the feasibility of the proposed protocols for RFID tags, we analyze a hardware implementation of our solutions. The EC processor that we designed has an architecture similar to the one presented by Lee et al. [33]. However, further optimizations are performed in our work.

7.1 Overall Architecture

The overall architecture is shown in Fig. 8. The processor is composed of a micro-controller, a bus manager, and an EC processor (ECP). It is connected with a front-end module, a random number generater (RNG), ROM, and RAM as shown in the overall architecture of Fig. 8. The solid arrows are for data exchange, the dash arrows are for addressing, and control signals are omitted in this figure. The ROM stores program codes and data. The program is executed by the micro controller and the data may include a tag's private key, the server's public key, and system parameters. The program is basically an authentication protocol. The microcontroller is able to perform general modular arithmetic operations (additions and multiplications) in a byte-serial fashion. It also gives commands for the execution of the ECP via the bus manager. The ECP loads a key (k) and an EC point (P) from ROM or RAM and executes the EC scalar multiplication (kP). After finishing the scalar multiplication, it stores the results in RAM.

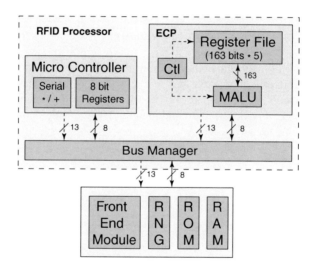

Fig. 8 RFID processor architecture

7.2 New MALU Design

The main differences when compared with [33] are in the register file and the MALU (Modular ALU). The original EC processor uses a MALU, which performs modular additions and multiplications, and it reuses the logic of the modular multiplication for modular square operations. On the other hand, the new MALU we designed includes a specialized squarer logic. Since the modular squaring can be completed in

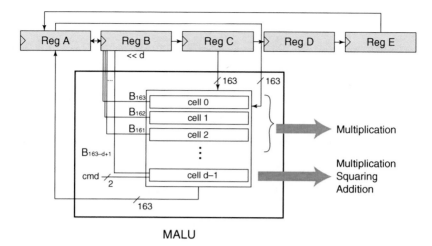

MALU

Fig. 9 MALU architecture with the register file

one cycle on a dedicated squarer while the modular multiplication is performed in a digit-serial fashion, the performance can be substantially increased with an overhead of the square logic. Moreover, the size of register file is reduced to 5×163 bits from 6×163 bits. This reduction is possible since the specialized squarer requires only one operand while the reuse of a multiplier for squaring requires two operands of the same value. As a result, the overall circuit area can be reduced further even after including the squarer in the MALU while achieving a much better performance.

The MALU operations can be described with Eq. (7), where $A(x)$, $B(x)$, and $C(x)$ represent the values of RegA, RegB, and RegC in Fig. 9:

$$
\begin{aligned}
A(x) &= A^2(x) \bmod P(x) && \text{if cmd} = 2, \\
A(x) &= B(x) \cdot C(x) \bmod P(x) && \text{if cmd} = 1, && (7) \\
A(x) &= A(x) + C(x) \bmod P(x) && \text{if cmd} = 0,
\end{aligned}
$$

where $A(x) = \Sigma a_i x^i$, $B(x) = \Sigma b_i x^i$, $C(x) = \Sigma c_i x^i$, and $P(x) = x^{163} + x^7 + x^6 + x^3 + 1$.

Its architecture is shown in Fig. 9, where the squaring and the addition use only the last cell (cell $d - 1$) while the multiplication uses all the cells. At the completion of each operation, only register RegA is updated while registers RegB and RegC hold the same data as at the beginning of the operation. (We make the shift of d-bits of RegB a circular shift so that the value goes back to the original after finishing a multiplication). Therefore, RegB and RegC can be used to store not only field operands but also some intermediate values.

7.3 Performance Evaluation

The performance comparison is summarized in Table 1 where both architectures have the digit size of 4 in the MALU. This work achieves about 24% better performance with a smaller circuit area, and the energy consumption is much smaller. Moreover, this work includes the coordinate conversion to affine coordinates from Z-coordinates while the work of [33] gives output in Z-coordinates.

The performance results of the protocols are summarized in Table 2 where a $0.13\,\mu$m CMOS technology is used, and the gate area does not include RNG, ROM, and RAM which are required to store or run programmed protocols. The area specifies a complete EC processor with required registers.

According to [16], the current consumption for a security part should not exceed $15\,\mu$A, which corresponds to $22.5\,\mu$W for 1.5 V in our CMOS library. Therefore, the power consumption of $13.8\,\mu$W in our design will be low enough even if we count extra power consumption in the required memories for the design.

Table 1 Performance comparison

Criteria	[33]	This work
Circuit area (Gage equivalent)	15,356	14,566
Cycles for EC scalar multiplication	78,544	59,790
Frequency (kHz)	323	700
Power (μW)	12.1	13.8
Energy for EC scalar multiplication (μJ)	2.94	1.18

Table 2 Performance results of protocols

Protocols	Cycles	Time (ms)	ROM for program	ROM for data	RAM	Non-volatile RAM
EC-RAC IV (Fig. 5)	328,074	469	80	126	107	–
Search protocol (Fig. 6)	120,505	172	61	105	128	21

- Gate equivalent area: 14,566 GE; frequency: 700 kHz; power: 13.8 μW; tech.: 0.13 μm.
- All the required memories are in bytes.
- Search protocol needs extra 21 bytes of non-volatile RAM for a counter.
- Gate area and power do not include memories.
- Data transmission in protocols is counted 1 cycle per byte.

8 Conclusions

In this chapter, we addressed the security and privacy requirements for RFID systems. We presented various RFID authentication protocols which are all made of the same building blocks, but meet different security and privacy requirements. Further on, the search protocol is presented as a novel scheme where a server (or a reader) can efficiently query for a specific tag, without compromising the tag's privacy.

In addition, we presented a hardware architecture that can realize the proposed protocols. The performance results show the feasibility of the protocols even for a passive tag and outperform other secure and private protocols proposed in the literature.

Acknowledgments This work was partially supported by the US National Science Foundation CCF-0541472, by the IAP Programme P6/26 BCRYPT of the Belgian State (Belgian Science Policy), by K.U. Leuven-BOF (OT/06/40), by the FWO project G.0300.07, and by the Flemish IBBT projects.

References

1. ANSI. X9.62 The Elliptic Curve Digital Signature Algorithm (ECDSA). http://www.ansi.org
2. G. Avoine, Adversarial Model for Radio Frequency Identification. Cryptology ePrint Archive, Report 2005/049, 2005. http://eprint.iacr.org/
3. G. Avoine, P. Oechslin, in *A Scalable and Provably Secure Hash-Based RFID Protocol*. Proceedings of the 3rd IEEE International Conference on Pervasive Computing and Communications Workshops (PERCOMW '05) (IEEE Computer Society, Washington, DC, 2005)
4. M. Bellare, A. Palacio, GQ and Schnorr identification schemes: Proofs of security against impersonation under active and concurrent attacks. in *Advances in Cryptology - CRYPTO'02*. Lecture Notes in Computer Science, vol. 2442 (Springer, Berlin, 2002), pp. 162–177
5. C. Berbain, O. Billet, J. Etrog, H. Gilbert, in *An Efficient Forward Private RFID Protocol*. CCS '09: Proceedings of the 16th ACM conference on Computer and Communications Security (ACM, New York, NY, 2009), pp. 43–53
6. S. Brands, D. Chaum, Distance-bounding protocols. in *Advances in Cryptology - EURO-CRYPT '93*. Lecture Notes in Computer Science, vol. 765 (Springer, Berlin, Heidelberg, 1994), pp. 344–359
7. J. Bringer, H. Chabanne, Trusted-HB: A low-cost version of HB^+ secure against man-in-the-middle attacks. IEEE Trans. Inf. Theory **54**(9), 4339–4342 (2008)
8. J. Bringer, H. Chabanne, E. Dottax, in HB^{++}: *A Lightweight Authentication Protocol Secure Against Some Attacks*. Security, Privacy and Trust in Pervasive and Ubiquitous Computing - SecPerU (IEEE Computer Society, Washington, DC, 2006)
9. J. Bringer, H. Chabanne, T. Icart, in *Cryptanalysis of EC-RAC, a RFID Identification Protocol*. International Conference on Cryptology and Network Security - CANS'08. Lecture Notes in Computer Science (Springer, Heidelberg, 2008)
10. B. Danev, T.S. Heydt-Benjamin, S. Čapkun, in *Physical-Layer Identification of RFID Devices*. Proceedings of the 18th USENIX Security Symposium (USENIX Security '09) (USENIX, Montreal, 2009), pp. 125–136
11. T. Deursen, S. Radomirović, Attacks on RFID Protocols, in Cryptology ePrint Archive: listing for 2008 (2008/310), 2008
12. T. Deursen, S. Radomirović, Untraceable RFID protocols are not trivially composable: Attacks on the revision of EC-RAC, in Cryptology ePrint Archive: Report 2009/332, 2009
13. W. Diffie, P.C. Van-Oorschot, M.J. Weiner, Authentication and authenticated key exchanges. Designs, Codes Cryptogr. **2**(2), 107–125 (1992)
14. T. Dimitriou, in *A Secure and Efficient RFID Protocol that Could Make Big Brother (Partially) Obsolete*. Proceedings of the 4th Annual IEEE International Conference on Pervasive Computing and Communications (PERCOM '06) (IEEE Computer Society, Washington, DC, 2006), pp. 269–275

15. EPC global. Class 1 Generation 2 UHF Air Interface Protocol Standard version 1.2.0. http://www.epcglobalinc.org/home, 2008. 108 pages

16. M. Feldhofer, S. Dominikus, J. Wolkerstorfer, in *Strong Authentication for RFID Systems using the AES Algorithm*. ed. by M. Joye, J.J. Quisquater. Cryptographic Hardware and Embedded Systems - CHES'04. Lecture Notes in Computer Science, vol. 3156 (Springer, Heidelberg, 2004), pp. 357–370

17. D. Frumkin, A. Shamir, in *Un-Trusted-HB: Security Vulnerabilities of Trusted-HB*. Proceedings of RFIDSec09, Leuven, Belgium, 2009

18. H. Gilbert, M. Robshaw, H. Sibert, An active attack against HB^+ - a provably secure lightweight authentication protocol. IEE Process. Lett. **41**(21), 1169–1170 (2005)

19. G. Hammouri, B. Sunar, in *PUF-HB: A Tamper-Resilient HB Based Authentication Protocol*. ed. by S. Bellovin, R. Gennaro. Applied Cryptography and Network Security: 6th International Conference, ACNS 2008. Lecture Notes in Computer Science, vol. 5037 (Springer, Heidelberg, 2008)

20. G.P. Hancke, M.G. Kuhn, in *An RFID Distance Bounding Protocol*. Proceedings of the 1st International Conference on Security and Privacy for Emerging Areas in Communications Networks (SECURECOMM '05) (IEEE Computer Society, Washington, DC, 2005), pp. 67–73.

21. D. Hein, J. Wolkerstorfer, N. Felber, in *ECC Is Ready for RFID - A Proof in Silicon*. Selected Areas in Cryptography. Lecture Notes in Computer Science, vol. 5381 (Springer, Heidelberg, 2009), pp. 401–413.

22. D. Henrici, P. Müller, in *Hash-Based Enhancement of Location Privacy for Radio-Frequency Identification Devices Using Varying Identifiers*. Proceedings of the Second IEEE Annual Conference on Pervasive Computing and Communications Workshops (PERCOMW '04) (IEEE Computer Society, Washington, DC, 2004), pp. 149–153

23. N.J. Hopper, M. Blum, in *Secure Human Identification Protocols*. ASIACRYPT '01: Proceedings of the 7th International Conference on the Theory and Application of Cryptology and Information Security (Springer-Verlag, Berlin, Heidelberg, New York, NY, 2001), pp. 52–66

24. M. Hutter, T. Plos, J.-M. Schmidt, in *Contact-Based Fault Injections and Power Analysis on RFID Tags*. Proceedings of the 19th IEEE European Conference on Circuit Theory and Design (ECCTD '09) (IEEE Computer Society, 2009), pp. 409–412

25. A. Juels, RFID security and privacy: A research survey. IEEE J. Sel. Areas Commun. **24**(2), 381–394 (2006)

26. A. Juels, S.A. Weis, in *Authenticating Pervasive Devices with Human Protocols*. Proceedings of CRYPTO'05. Lecture Notes in Computer Science, vol. 3126 (IACR, Springer-Verlag, Berlin, Heidelberg, New York, NY, 2005), pp. 293–308

27. A. Juels, S.A. Weis, *Defining Strong Privacy for RFID*. Cryptology ePrint Archive, Report 2006/137, 2006. http://eprint.iacr.org/

28. P. Kocher, in *Timing Attacks on Implementations of Diffie-Hellman, RSA, DSS and Other systems*. In N. Koblitz, editor, *Advances in Cryptology: Proceedings of CRYPTO'96*, number 1109 in Lecture Notes in Computer Science, pages 104–113. Springer-Verlag, Berlin, Heidelberg, New York, NY, 1996.

29. P. Kocher, J. Jaffe, B. Jun, in *Differential Power Analysis*. ed. by M. Wiener. Advances in Cryptology: Proceedings of CRYPTO'99. Lecture Notes in Computer Science, vol. 1666, (Springer-Verlag, Berlin, Heidelberg, New York, NY, 1999), pp. 388–397

30. Y.K. Lee, L. Batina, D. Singelee, I. Verbauwhede, in *Low-Cost Untraceable Authentication Protocols for RFID*. ACM Conference on Wireless Network Security - WiSec '10 (ACM, New York, NY, USA, 2010), pp. 55–64

31. Y.K. Lee, L. Batina, I. Verbauwhede, in *EC-RAC (ECDLP Based Randomized Access Control): Provably Secure RFID Authentication Protocol*. IEEE International Conference on RFID (IEEE, 2008), pp. 97–104

32. Y.K. Lee, L. Batina, I. Verbauwhede, in *Untraceable RFID Authentication Protocols: Revision of EC-RAC*. IEEE International Conference on RFID (IEEE, 2009), pp. 178–185

33. Y.K. Lee, K. Sakiyama, L. Batina, I. Verbauwhede, Elliptic curve based security processor for RFID. IEEE Trans. Comput. **57**(11), 1514–1527 (Nov 2008)

34. J. Lim, H. Oh, S. Kim, in *A New Hash-Based RFID Mutual Authentication Protocol Providing Enhanced User Privacy Protection*. Proceedings of the 4th International Conference on Information Security Practice and Experience (ISPEC '08). Lecture Notes in Computer Science, vol. 4991 (Springer-Verlag, Berlin, Heidelberg, New York, NY, 2008), pp. 278–289

35. D. Molnar, A. Soppera, D. Wagner, in *A Scalable, Delegatable Pseudonym Protocol Enabling Ownership Transfer of RFID Tags*. Proceedings of the 12th Annual International Workshop of Selected Areas in Cryptography (SAC '05). Lecture Notes in Computer Science, vol. 3897 (Springer-Verlag, Berlin, Heidelberg, New York, NY, 2005), pp. 276–290.

36. C.Y. Ng, W. Susilo, Y. Mu, R. Safavi-Naini, in *RFID Privacy Models Revisited*. European Symposium on Research in Computer Security (ESORICS'08). Lecture Notes in Computer Science, vol. 5283 (Springer-Verlag, Berlin, Heidelberg, New York, NY, 2008), pp. 251–266

37. NIST National Institute of Standards and Technology. Cryptographic Hash Algorithm Competition. http://csrc.nist.gov/groups/ST/hash/sha-3/index.html.

38. M. Ohkubo, K. Suzuki, S. Kinoshita, RFID privacy issues and technical challenges. Commun. ACM **48**(9), 66–71 (2005)

39. T. Okamoto, Provably secure and practical identification schemes and corresponding signature schemes. *Advances in Cryptology - CRYPTO'92*, ed. by E.F. Brickell. Lecture Notes in Computer Science, vol. 740 (Springer-Verlag, Berlin, Heidelberg, New York, NY, 1992), pp. 31–53

40. T. Phillips, T. Karygiannis, R. Kuhn, Security standards for the RFID market. Secur. Priv. **3**(6), 85–89 (2005)

41. A. Razaq, W. Luk, K. Shum, L. Cheng, K. Yung, Second–generation RFID. Secur. Priv. **6**(4), 21–27 (2008)

42. C.-P. Schnorr, Efficient identification and signatures for smart cards. *Advances in Cryptology - CRYPTO'89*, ed. by G. Brassard. Lecture Notes in Computer Science, vol. 435 (Springer-Verlag, Berlin, Heidelberg, New York, NY, 1989), pp. 239–252

43. D. Singelée, B. Preneel, *Distance Bounding in Noisy Environments*. Proceedings of the 4th European Workshop on Security and Privacy in Ad Hoc and Sensor Networks (ESAS '07). Lecture Notes in Computer Science, vol. 4572 (Springer-Verlag, Berlin, Heidelberg, New York, NY, 2007), pp. 101–115

44. B. Song, C.J. Mitchell, in *RFID Authentication Protocol for Low-Cost Tags*. Proceedings of the First ACM Conference on Wireless Network Security (WISEC '08) (ACM, New York, NY, USA, 2008), pp. 140–147

45. P. Tuyls, L. Batina, *RFID-Tags for Anti-Counterfeiting*. ed. by D. Pointcheval. Topics in Cryptology - CT-RSA - The Cryptographers' Track at the RSA Conference, San Jose, CA, USA. Lecture Notes in Computer Science, vol. 3860 (Springer, Heidelberg, Feb 13–17 2006), pp. 115–131

46. S. Vaudenay, in *On Privacy Models for RFID*. Advances in Cryptology (ASIACRYPT'07). Lecture Notes in Computer Science, vol. 4833 (Springer-Verlag, Berlin, Heidelberg, New York, NY, 2007), pp. 68–87

47. S.A. Weis, S. Sarma, R. Rivest, D. Engels, in Security and Privacy Aspects of Low-Cost Radio Frequency Identification Systems. Proceedings of the 1st International Conference on Security in Pervasive Computing (SPC '03). Lecture Notes in Computer Science, vol. 2802 (Springer-Verlag, Berlin, Heidelberg, New York, NY, 2003), pp. 454–469

Contactless Security Token Enhanced Security by Using New Hardware Features in Cryptographic-Based Security Mechanisms

Markus Ullmann and Matthias Vögeler

1 Introduction

1.1 Benefits of Contactless Smart Cards

Contact-based smart cards are widely accepted. What are then the reasons for focussing on contactless cards in recent times? First, because of the abrasion of the physical contacts, contact-based smart cards have a shorter lifetime compared to contactless cards. Second, contactless interfaces do not need to comply with mechanical form factors. Third, the usage of contactless cards is more comfortable compared to contact-based cards. And furthermore, with the upcoming dispersion of mobiles with near-field communication interfaces (NFC) the communication with PICCs is possible with mobiles, too.

1.2 Security Limitation of Supposed Security Mechanisms for an Authenticated Connection Establishment Between Terminals and Contactless Cards

1.2.1 Security Attacks

From a security perspective, one benefit of contact-based cards is that the operation of the cards is only possible if they are mechanically connected to a smart card terminal. Since this is not required for contactless smart cards, one significant security thread is the communication between an attacker's terminal and the card without

M. Ullmann (✉)
Bonn-Rhine-Sieg University of Applied Sciences, Sankt Augustin, Germany; Bundesamt für Sicherheit in der Informationstechnik, Bonn, Germany
e-mail: Markus.Ullmann@bsi.bund.de

A.-R. Sadeghi, D. Naccache (eds.), *Towards Hardware-Intrinsic Security,* Information Security and Cryptography, DOI 10.1007/978-3-642-14452-3_12,
© Springer-Verlag Berlin Heidelberg 2010

the awareness of the cardholder, e.g., while he carries the contactless card in his pocket. This attack is possible, even with passive contactless smart cards within the activation distance of the contactless card. In this context "passive" means that the smart card has no own electrical power supply (e.g., battery). According to ISO/IEC 14443 [11], the real activation distance of contactless smart cards (PICC) and terminals (PCD) depends on technical quantities such as terminal power, terminal antenna coil, and antenna diameters of terminal and card [17]. Measurement results are published in [13].

Besides that, an adversary might just be interested in eavesdropping of an existing radio frequency data transmission between a terminal and a contactless card. Again, in the case of contactless smart cards and terminals with an ISO/IEC 14443 interface the real range for eavesdropping a communication depends on technical quantities such as magnetic field strength, signal to noise ratio, noise class, see [17]. Considering noise class business according to [9] the maximum range for eavesdropping the communication of a contactless smart card (PICC) is theoretically below 4.5 m for a forced bit error rate of 0.1%, see [17].

1.2.2 Security Requirements

To counter the mentioned security threats, specific security mechanisms are needed. Therefore, contactless security tokens have to fulfill the following security requirements:

- Authentication of terminals.
- Strong session key agreement between authenticated terminal and contactless card (for the establishment of a secure channel).
- Forward secrecy of session keys.

In order to address these requirements, new password-based cryptographic protocols are discussed for an authenticated connection establishment between terminals and contactless cards, see the password authenticated connection establishment (PACE) [7] or alternatively terminal card AMP (TC-AMP) [21]. TC-AMP is a simplified version of TP-AMP, [14, 15], adapted to contactless smart cards.

Today for technical reasons, only static passwords can be used in these protocols. Typically, the static password of a contactless card is printed on the front side or back side of the card. If an attacker has once seen a contactless card he knows the according static password. As a consequence, the attacker can successfully initialize protocol runs with this card. Expecting a very powerful attacker, the possibility of tracking contactless security tokens based on their static password cannot be excluded.

Here, we suppose the enhancement of current contactless cards with a flexible display component, which enables system architects to apply dynamic passwords in password-based protocols instead of static ones (see Fig. 1).

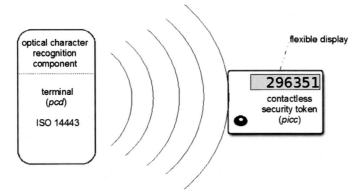

Fig. 1 Passive contactless security token

1.3 Security Limitation of Device Authentication Protocols Based on Irrevocable Authentication Certificates

1.3.1 Authentication Protocol Based on Irrevocable Authentication Certificates

Typically, certificates with a long validity period are generated and used for entity authentication. Lost or broken keys shall be revoked. The common mechanism to invalidate certificates and keys are certificate revocation lists. Usually, objectives like key pair generation and certificate generation as well as the publishing of certificate revocation lists are activities of a certification authority.

If entities of a public key infrastructure are always online, revocation lists (CRLs) are a very useful concept, because entities are able to check the validity of keys and certificates online right before their usage.

But in general, this mechanism fails in case of ubiquitous devices like contactless smart cards or RFIDs. The reason is that ubiquitous devices would require a secure online connection to get fresh CRLs to check the validity of keys and certificates.

If CRLs cannot be established, one can alternatively issue certificates with a short validity period. One consequence is a frequent generation and distribution of new certificates for all entities. In the following, we are considering validity periods of certificates of hours or days instead of years. Then the issue of a certificate revocation list can be omitted, because invalid keys and certificates can only be used during a very short time frame. Moreover, entities which become untrusted just do not get any new authentication certificate.

First, authentication certificates with a short validity period were used in the extended access control protocol (EAC), see [7]. They are supposed to grant access to an authorized terminal to sensitive biometric data of electronic passports. EAC consists of two cryptographic protocols: the chip authentication protocol (CA) and the terminal authentication protocol (TA).

Moreover, CA and TA are in general usable for a client $\Leftrightarrow$ server authentication (contactless smart card $\Leftrightarrow$ remote terminal/Internet server) based on server authentication certificates with a short validity period shown later in Sect. 5.1.

1.3.2 Motivation for Secure Time Synchronization Protocols

A contactless smart card (PICC) accepts an authentication certificate only if it holds the following properties:

1. the electronic signature of an authentication certificate is valid and
2. the current local time of the smart card t_{PICC} is lower or equal as the expiration time t_{CertEx} of an authentication certificate ($t_{PICC} \leq t_{CertEx}$)

Therefore, a contactless card requires a notion of time to check the validity of an authentication certificate. Initially, the local time of a smart card t_{PICC} is set to the global time t_{glob} during the personalization process.

Today, PICCs do not have a hardware clock. This means the local time t_{PICC} is a static date. But it is obviously necessary to update t_{PICC} at least right before the verification of an authentication certificate. Therefore, the local time can be autonomically updated by a PICC using the most recent certificate effective time t_{CertGe} contained in a valid authentication certificate presented during an authentication operation of a remote terminal. If a PICC is frequently used its local time t_{PICC} does not deviate too much from the real time.

But if a PICC is not used for a long time period, the local time t_{PICC} is not updated. Consequently, a PICC accepts authentication certificates over longer time periods as originally intended. Moreover, stolen keys and certificates have a bigger security impact. So, overaged time values t_{PICC} cause a specific security vulnerability. This downside shall be overcome in the future.

One solution, to remedy this security vulnerability, we want to propose here is to equip contactless smart cards with a real-time clock (RTC). One can think of RTCs, which are based on quartz technology or based on the progress of a physical process, e.g., chemical reaction or radioactive decay. While the first solution needs to be powered by a battery, the second one does not need a power supply.

Unfortunately, hardware clocks have a certain clock drift which depends on environmental conditions like temperature, pressure. Beyond that, batteries or timer can fail. So in any case, a time synchronization procedure is needed to update the value of the local time t_{PICC}.

The application of authentication certificates with a short validity period for ubiquitous devices allow the synchronization of the local time t_{PICC} become a security relevant process. That is the reason why we propose a secure time synchronization procedure in chapter "Strong PUFs: Models, Constructions, and Security Proof".

Abbreviation	Semantics
$<G>$	cyclic group
$\mathbb{Z}_n^*$	multiplicative group of n
G, G'	elliptic curve base point
$\Gamma, A, B, K, M, P, Q, T, X_1, X_2, Y_1, Y_2$	elliptic curve point
$A_x, B_x, M_x, Q_x, X_{1,x}, K_x$	x-coordinate of the curve point
π	shared short secret (password) with limited entropy
$x, x_1, x_2, y, y_1, y_2, r_{PICC}$	random value $\in \mathbb{Z}_n^*$
s	random secret
k_i, sk_C, sk_S, μ	symmetric key
k_{Enc}	symmetric key for encryption
$h(), h_i()$	strong hash function
ENC()	symmetric encryption algorithm
DEC()	symmetric decryption algorithm
MAC()	Message Authentication Code calculation
k_{MAC}	symmetric key for MAC calculation
$SK_{TS}|PK_{TS}$	key pair (secret key—public key) of a TS
PICC	contactless security token
PCD	local terminal
TS	time server
CA	chip authentication protocol
TA	terminal authentication protocol
t_{glob}	current global time
t_{PICC}	current local time of a contactless smart card
t_{CertGe}	authentication certificate generation time
t_{CertEx}	authentication certificate expiration time
CV	card verifiable authentication certificate
CV_{Chain}	chain of CV authentication certificates

Fig. 2 Used abbreviations

2 Contactless Security Token

Here we suppose an extension of common passive contactless smart cards with further technical hardware components to form a platform for a contactless security token. Contactless security token consists of

- a common chip card controller with cryptographic coprocessor, random number generator, and ISO/IEC 14443 contactless interface (e.g., NXP SmartMX) and
- a flexible display component (first suggested in [20]), a real-time clock, a battery and buttons as shown in Fig. 3. The internal battery is only needed to power the real-time clock.

2.1 Flexible Display Technology

Several flexible display technologies are ready for the integration in contactless smart cards. Here we give only a brief overview:

Fig. 3 Components of
contactless security token

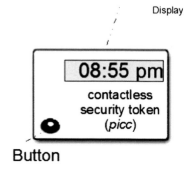

Display

Button

- Flexible liquid crystal display (LCD)
- Organic light emitting diode display (OLED)
- Electrophoretic displays, see [5]

All display types have different properties. Electrophoretic display technology, for example, combines high reflexivity with excellent readability in direct sunlight and a very low energy consumption. There is no need of a backlight, which is the main energy consumer in most displays. Energy is only needed during alteration of the displayed information.

Electrophoretic display technology is also known as e-Ink technology. Today, only segment displays (e.g., 11-segment display) are available for the integration in PICCs. In future we will see matrix displays in OLED technology.

2.2 Real-Time Clock

A RTC usually consists of an oscillator and a counter register, which is incremented by hardware after a certain number of oscillator pulses. Oscillators have a slight random deviation from their normal frequency, called drift. This can be due to impure crystals, for example. In addition, the clock drift depends on several environmental conditions, like pressure, temperature. The clock drift is expressed in parts per million (ppm). A typical clock drift for an existing PICC RTC is about 100 ppm. This sums up to a clock inaccuracy of about 1 h over a time period of 1 year.

Today, first prototypes of timer and micro-batteries are ready for integration into contactless smart card systems. For example, the RTC IC (NXP PCF8564) has a 32 kHz micro-crystal and needs 250 nA for operation. Unfortunately, the lifetime of the batteries is only about 3 years. This is the main reason why RTCs are unsuitable for contactless smart cards with a long validity period of about 10 years, today.

2.3 Buttons

Buttons open the possibility for user interaction. This is very useful for the confirmation/interruption of transaction performed on a contactless security token.

Further technical issues of contactless security tokens are not discussed in this chapter.

3 Authenticated Connection Establishment

3.1 Password-Based Cryptographic Protocols

The basic idea of password-based cryptographic protocols is to combine a strong session key agreement with an implicit entity authentication based on a shared secret with limited entropy, called a password, in one cryptographic protocol. The initial idea goes back to Bellovin and Merret and their publication of the encrypted key exchange protocol (EKE) [3]. Beyond the security requirements mentioned in Sect. 1.2.2, password-based protocols have to fulfill specific security properties.

First, if passwords with low entropy (e.g., passwords with six numeric characters) are considered, one thread is the possibility that an adversary can search through all secret passwords in a reasonable time. So, off-line and online dictionary attacks should not be feasible.

1. Security against off-line dictionary attacks: A passive eavesdropper Eve who records protocol runs shall not be able to get any information concerning passwords used for these protocol runs; in particular, Eve cannot calculate the password based on the protocol transcripts. But also, if Eve guessed an used password, she shall not be able to calculate the used session key.
2. Security against online dictionary attacks: There are two types of online dictionary attacks:

   - Type (1): An active adversary cannot abuse the protocol, so as to eliminate a significant number of possible passwords.
   - Type (2): An active adversary can only test at most one password per protocol run by attempting to masquerade using this password.

For more information, see [21]. A good compendium on published password-based protocols is to be found in [12].

3.2 Password Authenticated Connection Establishment (PACE)

The PACE protocol, see [7], is adaptable for prime fields and elliptic curves. Here, in order to increase the performance we have chosen to use the elliptic curves variant.

terminal (PCD)	security token (PICC)	Step
	choose $0 \le s < 2^m$ randomly	(a)
$\mu = h(\pi\|1) \bmod n$	$\mu = h(\pi\|1) \bmod n$	(b)
	$z = \mathrm{ENC}(\mu, s)$	(c)
	$\xleftarrow{\quad z \quad}$	
$s = \mathrm{DEC}(\mu, z)$		(d)
choose $x_1 \in \mathbb{Z}_n^*$ randomly		(e)
$X_1 = x_1 * G$		(f)
	$\xrightarrow{\quad X_1 \quad}$	
	abort if $X_1 \notin < G >^*$	
	choose $y_1 \in \mathbb{Z}_n^*$ randomly	(g)
	$Y_1 = y_1 * G$	(h)
	$\xleftarrow{\quad Y_1 \quad}$	
abort if $Y_1 \notin < G >^*$		
$P = x_1 * Y_1$	$P = y_1 * X_1$	(i)
$G' = s * G + P$	$G' = s * G + P$	(j)
choose $x_2 \in \mathbb{Z}_n^*$ randomly		(k)
$X_2 = x_2 * G'$		(l)
	$\xrightarrow{\quad X_2 \quad}$	
	abort if $X_2 \notin < G >^*$	
	choose $y_2 \in \mathbb{Z}_n^*$ randomly	(m)
	$Y_2 = y_2 * G'$	(n)
	$\xleftarrow{\quad Y_2 \quad}$	
abort if $Y_2 \notin < G >^*$		
$K = x_2 * Y_2$	$K = y_2 * X_2$	(o)
$k_{MAC} = h(K_x\|2)$	$k_{MAC} = h(K_x\|2)$	(p)
$t_{PCD} = \mathrm{MAC}(k_{MAC}, Y_{2,x})$		(q)
	$\xrightarrow{\quad t_{PCD} \quad}$	
	$t_{PICC} = \mathrm{MAC}(k_{MAC}, X_{2,x})$	(r)
	$\xleftarrow{\quad t_{PICC} \quad}$	
	$t'_{PCD} = \mathrm{MAC}(k_{MAC}, Y_{2,x})$	(s)
	abort if $t'_{PCD} \neq t_{PCD}$	(t)
$t'_{PICC} = \mathrm{MAC}(k_{MAC}, X_{2,x})$		(u)
abort if $t'_{PICC} \neq t_{PICC}$		(v)

Fig. 4 Password authenticated connection establishment

The operations are then performed in the cyclic group $< G >:= \{t * G | t \in \mathbb{N}\}$, $n := | < G > |$. In the following, $< G >^*$ denotes the cyclic group $< G >$ without the point at infinity. (Please refer to Fig. 2 for a complete overview of the notation used throughout this chapter.) For security reasons and simplicity, we recommend the usage of published secure domain parameter of a trusted authority, see [6].

The protocol consists of the following steps, the complete implementation is shown in Fig. 4. Before the protocol starts, the communication partners (terminal and smart card) of course have to agree on an elliptic curve E and a base point G.

1. The protocol starts with the selection of a random number s, $0 \le s < 2^m$, by the smart card in step (a). m is defined as the block size of the blockcipher used for the encryption of s. Both the smart card and the terminal derive a key μ using

a key derivation function, here $h(\pi|1)$ is used. In the next step s is encrypted using a blockcipher with key μ, $z = \mathrm{ENC}(\mu, s)$, and z is then transmitted to the terminal, which decrypts z.

2. The terminal and the card map the nonce s to a new base point G'. Here the following randomized mapping is used:

   a. An anonymous Diffie–Hellman key agreement based on G is used to calculate a random point $P \in< G >^*$ (steps (e)–(i)).
   b. Thereupon, P and s are exclusively used to calculate a new base point $G' = s * G + P$ in step (j) for the subsequent Diffie–Hellman key agreement.

3. An anonymous Diffie–Hellman key agreement based on the new base point G' is performed to calculate a common secret curve point K (steps (k)–(o)).

4. Then, two different keys $k_{\mathrm{ENC}} = h(K_x|1)$ for encryption and $k_{\mathrm{MAC}} = h(K_x|2)$ for calculation of message authentication codes (MAC) are derived from K. First, k_{MAC} is used for a MAC calculation in step (p) and (q) performed as mutual authentication of terminal and card in steps ((r)–(u)).

After a successful PACE protocol run, Secure Messaging is started using the derived keys k_{ENC} and k_{MAC}.

3.3 Security Token Operation

Every time a PACE protocol run is initiated, the PICC generates a fresh password π and displays it. Only this PCD which is able to optically read this fresh displayed π can successfully perform the PACE protocol and shares a session key K with the contactless security token, afterward.

3.4 Security Analysis of PACE Using Fresh Passwords

There is an exhaustive security analysis of PACE available. A cryptographic proof can be found in [4]. The idea of the proof is derived from Michel Abdalla, Pierre-Alain Fouque, and David Pointcheval [1] which is a refinement of the proof idea of Mihir Bellare, David Pointcheval, and Phillip Rogaway, see [2]. Beyond that, a formal proof of the mentioned security properties is performed and will be published soon. Here, the proof obligations are proven using induction based on an algebraic approach performed with an interactive proof system. This approach is more or less derived from the theoretical foundations of the AVISPA system [23].

It is important to note that in our setting π is a dynamic secret with low entropy. Dynamic means that the PICC generates a fresh password and displays it at any time a PACE protocol run is initialized.

Although technically challenging, the tracking of PICCs is in principle possible, if static passwords are used (see Sect. 1.2.1). The introduction of dynamic passwords for every run of the PACE protocol completely prevents this attack.

Obviously an adversary Eve can always try to guess a password π for a PACE protocol run, even when a cardholder carries his contactless security token in his pocket. Now the probability for Eve guessing the used password is in our setting with dynamic passwords:

$$s_n = a + aq + aq^2 + \ldots + aq^n = a(1 + q + q^2 + \ldots + q^n) = \frac{a(1 - q^{(n+1)})}{1 - q}. \quad (1)$$

With the precondition that π is a six-figure sum:

$$a = \frac{1}{1,000,000}, q = \frac{999,999}{1,000,000}. \quad (2)$$

This means, Eve has to guess $n = 693, 146$ passwords to find the right one with a probability of $s_n = 0.5$ (50%). As shown in [21] one unsuccessful PACE run last nearly 1 s on the SmartMX. So, Eve needs 19,254 h time for this attack to guess a password with a probability of $s_n = 0, 5$ (50%).

But now, Eve has not any advantage knowing a former used password, because the contactless security token uses a fresh one during any new PACE protocol run.

3.5 Brute-Force Online-Attacks on Passwords

Due to the low entropy of the used passwords, online dictionary attacks (brute-force attacks) cannot be ignored, especially in the contactless setting, where the PICC can be activated without notice of the owner. In order to prevent such attacks a smart card operating system has the following options:

1. It can implement an error counter. If the error counter then reaches a certain value, the smart card operating system permanently disables the smart card. For a contactless operating smart card the protocol might be executed by an attacker, if he is able to place at close range a smart card terminal. The disadvantage of this simple countermeasure is that it could be exploited for denial of service (DoS) attacks if the attacker might just have the goal to disable the smart card.
2. It can implement a time delay between a session where the protocol failed and a new session so that a guessing attack would consume too much time. If the time delay on the other hand is too big a DoS attack would be possible again. A standard technique for realizing a time delay is to use an EEPROM memory cell to realize a delay counter. Of course on the one hand that would stress the lifetime of a dedicated EEPROM cell a lot and on the other hand even more important, an adversary might be able to detect the EEPROM-writing routine and disable the power supply for the smart card before the EEPROM writing has been finished. The SmartMX of NXP offers an interesting hardware solution that does not involve the usage of EEPROM, here, the smart card operating system is able to set a dedicated bit, called *Delay Latch*, which will be automatically erased again after a couple of minutes independently of the power supply of the smart card.

4 Secure Time Synchronization

4.1 Time Values

If time is used in a distributed system a joint reference time is needed. Here, this reference time is labeled as global time (t_{glob}) and is represented by the coordinated universal time (UTC). UTC is a high-precision atomic time standard with uniform seconds defined by the international atomic time [18]. This time is the reference time for all necessary timed background processes, like the setting of t_{PICC} during the personalization of a PICC or the setting of t_{CertGe} and t_{CertEx} as part of a certificate generation. As already mentioned, t_{PICC} should not deviate much from t_{glob}. In this chapter we assume, that time values are technically implemented using a 32 bit integer in accordance to UNIX time, see [24]. Unix time is a system for describing points in time. It is the number of seconds elapsed since midnight of January 1, 1970, UTC. Unix time is widely used in technical systems.

Figure 5 shows the considered system structure for time synchronization. In this scenario a PICC is connected to a remote terminal (time server (TS)) with the help of a local terminal (PCD) that is connected to a internet PC. In this setting these

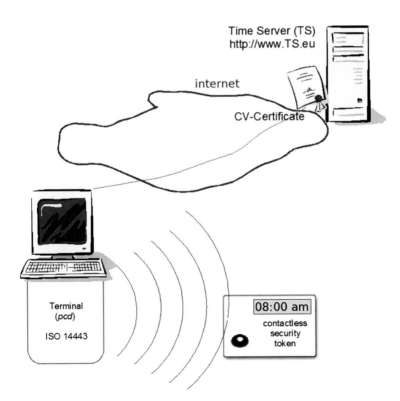

Fig. 5 Time server-ba'sed time synchronization

components play no relevant role during a secure time synchronization process. The local PCD and the PC only act as a relay station for the PICC.

4.2 Time Server-Based Synchronization Protocols

A timer can easily be synchronized by an up-to-date time stamp. Therefore, a PICC and the time server (TS) have to establish a communication channel, in order to exchange time stamps, see Fig. 5. In this case the PICC generates a time request and sends it to the time server (TS) through the communication channel. Next, the time server responses with a time stamp expressed in UTC. Obviously, there is a propagation delay between the sending and the receiving of the time stamp, because the time stamp needs a specific transmission time to cross the communication channel between the time server and the PICC. Unfortunately, neither the time server nor the PICC can determine the accurate propagation delay. This propagation delay briefly results in a synchronization failure of t_{PICC}.

To overcome such synchronization failures specific mechanisms have been developed. A well-known protocol, which especially addresses this kind of synchronization failures is the network time protocol (NTP), see [10]. The principle behavior of NTP is demonstrated in Fig. 6. In this protocol, the propagation delay is measurable by the PICC, because it assumes that the propagation delay of the request and the response (time stamp) is equal. If this assumption holds, t_{PICC} can be computed by the following formula:

$$t_{\mathrm{PICC}} = t_{\mathrm{TS}} + \Delta T/2. \tag{3}$$

This equation idealizes real technical behavior, beside the assumption of a symmetric propagation delay. Further delays going back to the involved entities are not considered. For example, a time server needs some time to pass the generated time stamp from the processing unit to the transmission unit. As well the PICC needs

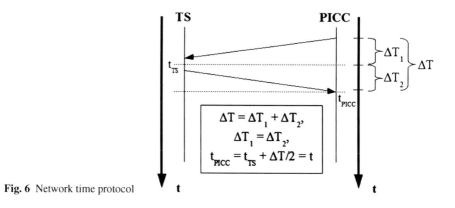

Fig. 6 Network time protocol

some time to pass the received time stamp from its receiver unit to its processing unit, and so on.

Technically, the equality of the propagation delays of the request and the response transmission cannot be guaranteed, if the Internet is used. One reason is unequal communication routes of the request and the response transmission. Although the TCP protocol formally offers an option to force the same communication route for both directions, Internet service provider do not support this TCP feature, because it can be abused for denial of service attacks.

Furthermore we showed in [22], that NTP is very vulnerable to delay attacks. If an attacker is able to independently delay both communication directions he can decide whether the clock of the PICC is ahead or behind the correct time. And in addition, the extent of the failure is not limited. It is important to emphasize that no protection is possible to counter this kind of attack if only one single time server is used.

Therefore, we address this problem by using only single time stamps, without any correction by an approximated transmission delay. We accept, that this time synchronization approach leads to a specific time failure, caused by the propagation delay of the time stamp. On the other hand, we gain a deterministic behavior, that is we know for sure, that the PICC's time t_{PICC} is always behind the global time t_{TS}.

In order to limit the extent of the failure, the PICC rejects all time stamps if for the communication delay $\Delta T > \Delta T_{max}$ holds, where ΔT_{max} is a predefined constant. Then, since $\Delta T_1, \Delta T_2 > 0$ the absolute extent of the failure is by guarantee lower than ΔT_{max}.

4.3 Security Requirements for Time Synchronization

Besides the discussed attack on NTP, common time synchronization protocols like the network time protocol have further weaknesses. A brief overview can be found in [19].

At first, if time stamps are generated without authentication information a PICC cannot be sure that a received time stamp is generated from the correct time server (TS). Therefore, classical time stamps must be enhanced with authentication information like a digital signature.

But if an attacker is able to access the communication channel, for example, if he records signed time stamps, he can replay old time stamp messages. A signature cannot prevent this attack. In this case, we have an authentic time stamp without any possibility for the PICC to detect this kind of attack. In order to avoid such replay attacks the PICC has to challenge the time server (TS) with a nonce.

Summing up, a secure time synchronization protocol has to fulfill the following security requirements:

1. the time synchronization has to be performed on the basis of authentic time server time values,
2. the time synchronization has to resist replay attacks,

3. the time synchronization has to analyze and consider runtime delay attacks and
4. the time synchronization has to safeguard a deterministic time behavior.

The last requirement means that it must be clear whether t_{PICC} is a very short time ahead or behind t_{TS} after a successful time synchronization process. We assume that the time server as such is a trustworthy entity, which generates genuine and correct time stamps.

4.4 Secure Time Synchronization Protocols

Our ambition is not to specify totally new protocols. Rather, we would like to adapt existing and already used cryptographic protocols for contactless smart cards, see [7], to fulfill the mentioned security requirements in Sect. 4.3. The considered protocols are the chip- and terminal authentication protocol which are originally used for ID cards and e-passports. One advantage using these protocols is that a security analysis of these protocols already exists.

If contactless smart cards are broadly equipped with a hardware clock in the future, time synchronization processes will be performed very often. Then calculation efforts of the time server become an important issue in respect to possible time server denial of service attacks. So in addition, we are looking for security mechanisms which need only low-calculation efforts on the time server. From this point of view, calculation efforts of the PICCs are less critical.

It is important to emphasize that the suggested time synchronization protocols work only with a single time stamp to fulfill the fourth security requirement (deterministic time behavior).

The chip- and terminal authentication protocols, see Sect. 5.1 or [7], are adaptable for prime fields and elliptic curves. Here, in order to increase the performance on the timer server side, we have also chosen to use the elliptic curves variant.

To apply the adaptations of the chip and terminal authentication protocols the time server TS has to be equipped with a static key pair $(SK_{TS}|PK_{TS})$ and a specific card verifiable certificate, according to [7].

Besides the supposed protocols, password-based protocols [12] have been analyzed concerning an adaption for a secure time synchronization as well, see [21]. But, no secure and efficient adaption has been found.

4.4.1 Message Authentication Code-Based Time Synchronization (MACT)

The adaption of the chip authentication protocol [7] results in an ephemeral static Diffie–Hellman key agreement protocol with a subsequent implicit unilateral authentication of TS based on the agreed common secret curve point K. Before the protocol starts the communication partners TS and PICC have to agree on an elliptic curve E and a base point G. The protocol consists of the following steps.

1. The protocol starts with a choice of a random value x_1 in step (a) and the generation of a random curve point X_1 in step (b).

Time Server (TS)	Security Token (PICC)	Step

$$\xrightarrow{CV_{Chain}}$$

	choose $x_1 \in \mathbb{Z}_n^*$ randomly	(a)
	$X_1 = x_1 * G, K = x_1 * PK_{TS}$	(b)
	$t = 0$	(c)

$$\xleftarrow{X_1}$$

abort if $X_1 \notin <G>^*$		(d)
$K = SK_{TS} * X_1$		(e)
$k_{MAC} = h(K_x|2)$	$k_{MAC} = h(K_x|2)$	(f)
$m_{TS} = MAC(k_{MAC}, X_{1,x}||t_{TS})$		(g)

$$\xrightarrow{m_{TS},t_{TS}}$$

| | $m'_{TS} = MAC(k_{MAC}, X_{1,x}||t_{TS})$ | (h) |
| | abort, if $m'_{TS} \neq m_{TS} \vee t \geq \Delta T_{max}$ | (i) |

Fig. 7 Message authentication code-based time synchronization (MACT)

2. The PICC measures the runtime of the protocol starting with protocol step (c).
3. The protocol proceeds with the transmission of X_1 to the time server.
4. Next, the time server calculates the ephemeral static Diffie–Hellman key K using his secret key (SK_{TS}) in step (e).
5. Then, a key $k_{MAC} = h(K_x|2)$ for the calculation of message authentication codes (MAC) is derived from K. k_{MAC} is used for a MAC calculation in step (g) and the according MAC verification of the PICC in step (h). A successful MAC verification in step (h) authenticates TS and in a consequence the time stamp t_{TS}.
6. The PICC aborts the protocol in step (i) if the runtime t of the protocol is $\geq \Delta T_{max}$.

Figure 7 illustrates the adaption of the chip authentication protocol.

We consider t_{TS} as a non confidential value. On the other hand, if it is desired to keep t_{TS} undisclosed, the communication partners should additionally derive an encryption/decryption key $k_{ENC} = h(K_x|1)$ for the encryption of t_{TS} before transmission.

4.4.2 Signature-Based Time Synchronization (SigT)

The adaptation of the terminal authentication protocol is a two-move challenge–response protocol that provides explicit unilateral authentication of a TS using a digital signature.

Before the protocol starts the communication partners TS and PICC have to agree on an elliptic curve E and a base point G.

Figure 8 presents the adaption of the terminal authentication protocol. The protocol consists of the following steps:

1. The protocol starts with the selection of a random number r_{PICC} by the smart card in step (a).
2. The PICC measures the runtime of the protocol starting with protocol step (b).

Time Server (TS)	**Security Token (PICC)**	Step

$$\xrightarrow{\quad CV_{Chain} \quad}$$

	choose $r_{PICC} \in \mathbb{Z}_n^*$ randomly	(a)
	$t = 0$	(b)

$$\xleftarrow{\quad r_{PICC} \quad}$$

$Sig = Sign(SK_{TS}, h(r_{PICC}\|t_{TS}))$ | | (c) |

$$\xrightarrow{\quad Sig, t_{TS} \quad}$$

	$Verify(PK_{TS}, Sig, h(r_{PICC}\|t_{TS}))$	(d)
	abort, if verify fails $\vee\, t \geq \Delta T_{max}$	(e)

Fig. 8 Signature-based time synchronization (SigT)

3. In the next step r_{PICC} as well as the hash value of the current time stamp t_{TS} are signed using ECDSA, see [8] with the key SK_{TS}. Then the signature and the time stamp t_{TS} are transmitted to the smart card.
4. If the verification of the signature succeeds in step (d), the time server TS and the time stamp t_{TS} are authenticated.
5. The PICC aborts the protocol in step (e) if the runtime t of the protocol is $\geq \Delta T_{max}$.

4.5 Security and Performance Analysis

4.5.1 Security Analysis of the Proposed Time Synchronization Protocols

In this section we evaluate the protocols against our security requirements from Sect. 4.3. The first requirement of authentic time stamps from the time server is addressed, because the time server provides a certificate chain. Since the PICC knows the public key of the root instance (CVCA), it is able to verify the certificate chain. The certificate chain also provides the public key of the time server. In order to proof that the time server knows the according private key SK_{TS}, the PICC has to challenge the time server. In protocol MACT this is realized by performing an ephemeral static Diffie–Hellman key exchange, which only succeeds if the time server knows the private key SK_{TS}. The resulting common secret is used to derive a MAC key, which is used by the time server to append a MAC to the time stamp, which can be verified by the PICC. The verification of the MAC authenticates the time stamp with high probability, if and only if the time server knows the private key SK_{TS}. We know that the security of MACT strongly depends on the computational Diffie–Hellman assumption and the nondisclosure of the secret key SK_{TS} of the time server.

Protocol SigT traces a different approach: It uses the public key of the time server to verify a signature of a random challenge that has been sent to the time server. Since the signature generation also involves the time stamp, the PICC is able to verify the authenticity of the time stamp with high probability. The security of SigT relies on the intractability assumption of the discrete logarithmic problem, the nondisclosure of the secret key SK_{TS} of the time server and a secure hash function.

In order to fulfill the second security requirement, we have to show, that the PICC withstands replay attacks. Both protocols cryptographically link a random value with high entropy to a time stamp. Thus, the PICC is able to conclude that the received time stamp belongs to the current session.

The third security requirement is respected, because the PICC measures the time between the time request and the concerning time response (time stamp) in both protocols. If t exceeds a certain value ΔT_{max} the time stamp is not accepted (MACT: step (i), SigT: step (e)). This procedure guarantees that the time stamp is not older than ΔT_{max}, see also the analysis about induced transmission delays in Sect. 4.2.

Finally, we demand a deterministic time behavior, which means that an attacker is not able to manipulate the protocol such, that according to his attack, the clock of the PICC is ahead or behind the real time. Since the PICC just adopts the time stamp t_{TS} this requirement is easily fulfilled. After synchronization, the time of the PICC is always behind the real clock, at most by ΔT_{max}.

4.5.2 Performance and Runtime Analysis of the Proposed Time Synchronization Protocols

There are two major factors that determine the running time of the protocols. The first one is the number of transmissions and the amount of data to be transmitted, the second one is the computation time that is needed to perform the protocols.

The two protocols perform three transmissions such as:

1. The certificate chain CV_{Chain} is transmitted from the time server to the smart card.
2. A challenge is sent from the smart card to the time server. While a random elliptic curve point is sent in the one protocol, the other one sends a random number.
3. A verifiable time information is transmitted from the time server to the smart card.

In addition the amount of data is of the same magnitude. Thus, the communication effort of the protocols is of the same grade.

For protocol MACT the server performs a scalar elliptic point multiplication (step (e)) followed by a key derivation (step (f), one way hash function) and a MAC calculation (step (g)). The effort for the scalar elliptic point multiplication is many times over the remaining effort. Protocol SigT requires to perform an elliptic curve signature generation consisting of a scalar elliptic point multiplication, a few more modulo arithmetic and of hashing the signature text. Again the scalar elliptic point multiplication is dominating. The reason why we do not rate the effort for the two protocols for the time server as equivalent is that a scalar elliptic point multiplication with a constant elliptic point can speed up easily (by doing some pre-computations) by about a factor of 4. Thus, protocol SigT has essential performance advantages.

For the smart card, as a first step, a verification of the certificate chain CV_{Chain} has to be performed for both protocols. Then protocol MACT needs to compute two scalar elliptic point multiplications (step (b)), followed by a key derivation (hash function, step (f)) and a MAC calculation (step (h)). The calculation time is determined by the two scalar elliptic point multiplications. Protocol SigT requires to perform an elliptic curve signature verification, which consists of two dominating scalar point multiplication. In conclusion, the two protocols for the smart card require a similar computational effort.

We provide a rough landmark for performance relations on a normal PC. The time consumption of one 256-bit scalar elliptic point multiplication with a random point is about the same as 700 SHA-256 bit hashes of 256 bytes or about the same as 1400 AES-128-CBC computations of 256 bytes. We have achieved this landmark by performing the speed command of the OpenSSL command-line tool.

5 Applications

Contactless security tokens can be used for different (security) purposes, as authentication token, secure signature creation device, prepaid card with the ability to display the holding/credit balance, etc. A further application is proposed by Nithyanand et al. for checking of terminal revocation, see [16].

On the other hand common client systems are vulnerable to phishing and Trojan horses. Phishing attacks are performed by misrouting users to Web pages under the attackers control to achieve valid login data and transaction identifiers. Trojan horses are used by attackers to spy out passwords or to directly manipulate user transactions, like bank transfers.

We depict, how a contactless security token can help to discover and avoid successful attacks.

5.1 Authentication of Internet Services

Figure 9 exhibits an authentication of an Internet service, here www.bank.eu. Password-based protocols were originally supposed for a client server authentication. As a consequence, PACE could also be used for a mutual authentication of an Internet service (e.g., bank server) and a PICC based on a common shared static password. Obviously, this approach does not assure a strong server authentication, because the shared password could be disclosed by an attacker, anyway. As a consequence, a server authentication procedure based on asymmetric keys is needed. Therefore, an Internet service (e.g., bank server) needs an asymmetric key pair and an according certificate. Here, for technical reasons, CV certificates according to [7] are used. If the URL is an ingredient of the CV certificate of the Internet server the PICC can display the authentic URL of the certificate after a successful authentication of the bank server. Then the user can be sure that he is connected to the intended bank server. If a phishing attack is performed during a server authentication procedure, the authentication fails. As consequence, no URL is displayed and the user becomes aware of this situation.

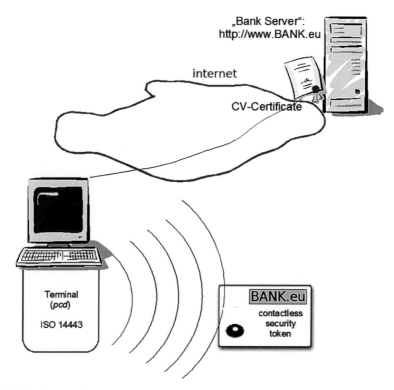

Fig. 9 Mutual authentication bank-server security token

After a successful terminal and chip authentication protocol both components can establish a common secure channel with the agreed session key K. Now, the local PCD and the PC act only as transceivers and cannot influence transactions between Internet service and PICC, except denial of service attacks. This includes the capabilities of Trojan horses, too.

At that point the secure channel with the contactless security token can be used to perform secure user interactions by using the display and buttons of the PICC. For example, the bank server can use the PICC's display to repeat the credit transfer data and ask for confirmation via the PICC's button, then a transaction number (TAN) can be submitted and displayed, which is explicitly linked to this credit transfer data and cannot be misused by PC malware for another transaction. PC malware (like Trojan horses) cannot influence the operation of a security token.

Figure 10 describes the terminal- and chip authentication protocols. Thereby the steps (a)–(f) represent the terminal authentication protocol. In the present context, this protocol is used for authentication of Internet services.

Steps (g)–(m) describe the chip authentication protocol to authenticate the PICC and to agree in a fresh session key K with derived keys for encryption K_{ENC} and MAC generation K_{MAC}.

'"Bank Server"'		**Security Token (PICC)**	Step
	$\xrightarrow{CV_{chain}}$		
choose $x_1 \in \mathbb{Z}_n^*$ random			(a)
$X_1 = x_1 * G, h = h(X_{1,x})$			(b)
	$\xrightarrow{h}$		
		choose $r_{PICC} \in \mathbb{Z}_n^*$ random	(c)
	$\xleftarrow{r_{PICC}}$		
$Sig = Sign(SK_{PCD}, r_{PICC}|h(X_{1,x}))$			(d)
	$\xrightarrow{Sig}$		
		$Verify(PK_{PCD}, Sig, r_{PICC}|h)$	(e)
		abort, if $Verify$ fails	(f)
	$\xrightarrow{X_1}$		
		abort, if $h(X_{1,x}) \neq h$	(g)
$K = x_1 * PK_{PICC}$		$K = SK_{PICC} * X_1$	(h)
$k_{ENC} = h(K_x|1)$		$k_{ENC} = h(K_x|1)$	(i)
$k_{MAC} = h(K_x|2)$		$k_{MAC} = h(K_x|2)$	(j)
		$m_{PICC} = MAC(k_{MAC}, X_{1,x})$	(k)
	$\xleftarrow{m_{PICC}}$		
$m'_{PICC} = MAC(k_{MAC}, X_{1,x})$			(l)
abort, if $m'_{PICC} \neq m_{PICC}$			(m)

Fig. 10 Internet server authentication

6 Conclusion

This chapter supposes the integration of new hardware components into contactless smart cards like a flexible display, a real-time clock (RTC), and a battery to power the RTC to form a novel security token (PICC). We have depicted how the new features of this components can be used to significantly enhance the security. First, the security benefit for password-based cryptographic protocols for an authenticated connection establishment between terminals (PCD) and contactless cards has been shown. The presented password-based protocol PACE has only been chosen as an example to realize the secure protocols, other password-based protocols for PICCs, like TC-AMP, can be employed as well. Second, the uses for a secure authentication of Internet services and secure transactions performed on the contactless security token are shown.

If security mechanisms rely on a valid time, secure time synchronization procedures are necessary. So, cryptographic protocols for synchronizing the RTC of the contactless security token using Internet time server are presented and analyzed, too.

References

1. M. Abdalla, P.A. Fouque, D. Pointcheval, in *Password-Based Authenticated Key Exchange in the Three-Party Setting*. Proceedings Public Key Cryptography (PKC 2005). Lecture Notes in Computer Science, vol. 3386 (Springer, Heidelberg, 2005), pp. 65–84
2. M. Bellare, D. Pointcheval, P. Rogaway, in *Authenticated Key Exchange Secure Against Dictionary Attacks*. Proceedings Eurocrypt 2000. Lecture Notes in Computer Science, vol. 1807 (Springer, Heidelberg, 2000), pp. 139–155

3. S.M. Bellovin, M. Merritt, in *Augmented Encrypted Key Exchange: Password-Based Protocol Secure Against Dictionary Attacks*. Proceedings of the Symposium on Research in Security and Privacy (IEEE Computer Society Press, Oakland, CA, 1992)
4. J. Bender, M. Fischlin, D. Kügler, in *Security Analysis of the PACE Key-Agreement Protocol*. Proceedings Information Security Conference 2009. Lectures Notes in Computer Science, vol. 5735 (Springer, Berlin, Heidelberg, 2009), pp. 33–48
5. T. Bert, H.D. Smert, F. Beunis, K. Neyts, Complete electrical and optical simulation of electronic paper. Displays J. **27**(2), 50–55 (2006)
6. Brainpool: ECC Brainpool Standard Curves and Curve Generation, Version 1.0, available online at http://www.ecc-brainpool.org/ecc-standard.htm, 2005
7. BSI: Technical Guideline TR-03110: Advanced Security Mechanisms for Machine Readable Travel Documents Extended Access Control (EAC) and Password Authenticated Connection Establishment (PACE). Version 2.0, February 2008
8. BSI: Technical Guideline TR-03111, Elliptic Curve Cryptography, Version 1.10, available online at: www.bsi.bund.de, 2008
9. European Radiocommunication Committee (ERC) within the European Conference of Postal, Telecommunications Administrations: Propagation Model and Interference Range Calculation for Inductive Systems 10 KHz–30 MHz. Marbella, Feb 1999
10. IETF: Network Time Protocol (Version 3), Mills, D., March 1992
11. ISO/IEC: ISO/IEC 144443 contactless Integrated Circuits Cards, Part 1–4: Physical Characteristics (1), Radio Frequency Power and Signal Interface (2), Initialization and Anticollision (3) and, Transmission Protocol (4), 2000
12. D. Jablon, List of Research Paper on Password-Based Cryptography, available online at www.jablon.org/passwordlinks.html, 2008
13. Z. Kfir, A. Wool, in *Picking Virtual Pockets Using Relay Attacks on Contact-less Smartcard Systems*. Proceedings of the 1st International Conference on Security and Privacy for Emerging Areas in Communication Networks (IEEE Computer Society Press, Silver Spring, MD, 2005)
14. T. Kwon, Practical Authenticated Key Agreement Using Passwords, long paper available online: http://dasan.sejong.ac.kr/tkwon/research/tpampfull.pdf, 2003
15. T. Kwon, in *Practical Authenticated Key Agreement using Passwords*. Information Security. Lecture Notes in Computer Science, vol. 3225 (Springer, Berlin, Heidelberg, Sept 2004), pp. 1–12
16. R. Nithyanand, G. Tsudik, E. Uzun, Readers Behaving Badly: Reader Revocation in PKI-Based RFID Systems. Cryptology ePrint Archive, Report 2009/465, 2009. http://eprint.iacr.org
17. NXP: ISO/IEC 14443 Eavesdropping and Activation Distance, 13,56 MHz Proximity Smart Cards, Application Note. Rev. 1.01, Jan 2008
18. Physikalisch-Technische Bundesanstalt: Coordinated Universal Time, available online at http://www.ptb.de
19. J. Tsang, K. Beznosov, in *A Security Analysis of the Precise Time Protocol (Short Paper)*. Proceedings of the 8th International Conference on Information and Communication Security (ICICS 2006), Raleigh, NC, USA, Nov 2006, pp. 50–59
20. M. Ullmann, in *Flexible Visual Display Unit as Security Enforcing Component for Contactless Smart Card Systems*. 1st International EURASIP Workshop on RFID Technology (RFID 2007), Vienna, Austria, 2007, pp. 87–90
21. M. Ullmann, D. Kügler, H. Neumann, S. Stappert, M. Vögeler, in *Password Authenticated Key Agreement for Contactless Smart Cards*. 4th Workshop on RFID Security (RFIDSec 2008), Budapest, Hungary, 2008, pp. 140–161
22. M. Ullmann, M. Vögeler, in *Delay Attacks - Implication on NTP and PTP Time Synchronization*. Proceedings of the International IEEE Symposium on Precision Clock Synchronization for Measurement, Control and Communication, ISPCS 2009, Brescia, Italy, 2009, pp. 97–102
23. L. Viganò, Automated security protocol analysis with the AVISPA tool. Electr. Notes Theor. Comput. Sci. **155**, 61–86 (2006)
24. Wikipedia: Unix time, available online at http://en.wikipedia.org/wiki/unix_time

Enhancing RFID Security and Privacy by Physically Unclonable Functions

Ahmad-Reza Sadeghi, Ivan Visconti, and Christian Wachsmann

1 Introduction

Radio frequency identification (RFID) is a technology that enables RFID *readers* to perform fully automatic wireless identification of objects that are labeled with RFID *tags*. Initially, this technology was mainly used for electronic labeling of pallets, cartons, and products to enable seamless supervision of supply chains. Today, RFID technology is widely deployed to many other applications as well, including animal and product identification [2, 42], access control [2, 47], electronic tickets [47] and passports [27], and even human implantation [30].

As pointed out in previous publications (see, e.g., [30, 68]), this prevalence of RFID technology introduces various risks, in particular concerning the privacy of its users and holders. The most deterrent privacy risk concerns the tracking of users, which allows the creation and misuse of detailed user profiles. Thus, an RFID system should provide *anonymity* (confidentiality of the tag identity) as well as *untraceability* (unlinkability of the communication of a tag) even in case the state (e.g., the secret) of a tag has been disclosed. Despite these privacy risks, classical threats to authentication and identification systems must be considered as well. Indeed, potential threats to RFID systems are attacks, where the adversary tries to impersonate or copy a legitimate tag. By legitimate we mean a tag created by an accredited tag issuer. Thus, appropriate countermeasures must be provided (*authentication* and *unclonability*). However, there are some other risks such as denial-of-service attacks, where an adversary unnoticeably interacts with tags and exploits deficiencies of the underlying protocols to permanently disable legitimate tags remotely [7], which must also be prevented (*availability*). In addition to the privacy and security requirements discussed above, RFID systems in practice must achieve various functional goals, including fast verification of cost-efficient tags (*efficiency*) and support of a huge number of tags (*scalability*). However, depending on the underlying application scenario and the given technological constraints, practical realizations

C. Wachsmann (✉)
Horst Görtz Institute for IT-Security (HGI), Ruhr-University Bochum, Bochum, Germany
e-mail: christian.wachsmann@trust.rub.de

A.-R. Sadeghi, D. Naccache (eds.), *Towards Hardware-Intrinsic Security,* Information
Security and Cryptography, DOI 10.1007/978-3-642-14452-3_13,
© Springer-Verlag Berlin Heidelberg 2010

may not be able to fulfill all of these requirements. In particular, the security and functional requirements often contradict the privacy requirements.

Most currently used RFID systems do not offer privacy at all (see, e.g., [47, 48, 62, 63]). This is mainly because current cost-efficient tags do not provide the necessary computational resources to run privacy-preserving protocols [2, 47], which heavily rely on public-key cryptography. Moreover, as pointed out in Sect. 3.1, privacy-preserving solutions without public-key cryptography do not fulfill important security or functional requirements and thus are inapplicable to real-world applications.

The design of a secure privacy-preserving RFID scheme requires a careful analysis in an appropriate formal security and privacy model. Existing security and privacy models for RFID (see, e.g., [3, 7, 8, 33]) often do not consider important aspects like adversaries with access to auxiliary information (on whether the identification of a tag was successful or not) or the privacy of corrupted tags (whose secrets have been disclosed). Recently, a comprehensive security and privacy model that generalizes and improves many previous works in a single concise framework has been proposed in [67] and refined in [45, 53]. In the following, we refer to the privacy model of [67] as the *V-Model* (Vaudenay Model). The V-Model [67] introduces eight privacy notions, which correspond to adversaries of different strength. The strongest *achievable* privacy notion in this model (*narrow-strong privacy*) allows the adversary to arbitrarily corrupt tags but does not capture the availability of auxiliary information. If auxiliary information is of concern, the weaker notions of *destructive* and *forward privacy* must be considered while *weak privacy* does not adequately model the capabilities of real-world adversaries since weak privacy does not allow tag corruption. However, [67] showed that narrow-strong privacy requires the use of public-key cryptography [67], which in general clearly exceeds the capabilities of current cost-efficient RFIDs [2, 47]. Moreover, it has been shown that forward privacy can be achieved but at the cost of using public-key cryptography while the feasibility of the stronger notion of destructive privacy currently is an open question [67].

1.1 Contribution

In this chapter, we propose a new privacy-preserving tag authentication protocol for RFID that can be proven to be destructive private in the V-Model [67]. This means that our protocol provides untraceability of tags against adversaries that permanently destroy a tag by physically attacking (i.e., corrupting) it. Our protocol is based on the weak private protocol proposed in [67] and uses physically unclonable functions (PUFs) as tamper-evident key storage in a similar way as described in [65]. This means that the tag authentication key is not stored on the tag but reconstructed from the physical characteristics of the RFID chip each time it is needed. The properties of the PUF ensure that any attempt to physically tamper with the PUF to obtain the authentication secret of the tag result in destruction of the PUF and the tag secret, which corresponds to the definition of a destructive adversary in the V-Model [67].

2 High-Level RFID System and Requirement Analysis

We first informally analyze the general scenario of radio frequency identification (RFID) on a very high level.

2.1 System Model

An RFID system consists of at least an operator $\mathcal{I}$, a reader $\mathcal{R}$ and a tag $\mathcal{T}$ [17]. The operator $\mathcal{I}$ is the entity that enrolls and maintains the RFID system. Hence, $\mathcal{I}$ initializes each tag $\mathcal{T}$ and reader $\mathcal{R}$ before it is deployed in the system. A tag $\mathcal{T}$ or reader $\mathcal{R}$ that has been initialized by the operator $\mathcal{I}$ is called *legitimate*. A tag $\mathcal{T}$ is a hardware token with constrained computing and memory capabilities that is equipped with a radio interface [2, 17, 47]. All information (e.g., secrets and data) that is stored on a tag $\mathcal{T}$ is denoted as the *state* of $\mathcal{T}$. Usually, tags are attached to objects or carried by the users of the RFID system [16, 46]. A reader $\mathcal{R}$ is a stationary or mobile computing device that interacts with all tags within its reading range to authenticate them. Depending on the specific use case (e.g., electronic passports [27]), the reader $\mathcal{R}$ may obtain additional information like the tag identity or some data stored on the tag $\mathcal{T}$. Readers can have a sporadic or permanent online connection to some backend system $\mathcal{D}$, which typically is a database maintaining detailed information on all tags in the system [15]. The backend is initialized and maintained by the operator $\mathcal{I}$ and can be read and updated by the readers $\mathcal{R}$.

2.2 Trust and Adversary Model

The operator $\mathcal{I}$ is the entity that maintains the RFID system and thus can be considered to be honest. However, $\mathcal{I}$ may be curious since he may collect user information (see, e.g., [29, 68]) while in general at the same time nobody can blame him for cheating.

Since RFID tags and readers communicate over a radio link, every entity can eavesdrop or manipulate this communication, even from outside the nominal reading range [37]. Thus, the adversary can be every (potentially unknown) entity that needs not to be a member of the RFID system. Besides the communication between a tag $\mathcal{T}$ and a reader $\mathcal{R}$, an adversary can also obtain useful auxiliary information (e.g., by visual observation) on whether the reader $\mathcal{R}$ accepted the tag $\mathcal{T}$ [33, 67]. Most commercial RFID tags are cost-efficient devices without (expensive) protection mechanisms against physical tampering [2, 47]. Hence, an adversary in practice can physically attack (*corrupt*) a tag to access its state (e.g., its secrets) [26, 38, 39, 41].

RFID readers are embedded devices that can be integrated into mobile devices (e.g., mobile phones or PDAs) or laptops and personal computers. The resulting complexity exposes readers to sophisticated hard- and software attacks (e.g., viruses or Trojans). Hence, an adversary in practice can get full control of (*corrupt*) an RFID reader [5]. This problem aggravates for mobile readers that can easily be lost or stolen.

2.3 Security and Privacy Threats

The most deterrent privacy risk concerns the *tracking* of users, which allows the creation and misuse of detailed profiles of a user of the RFID system [30]. For instance, tracking or identification of a tag enables the creation of detailed movement profiles, which can leak sensitive information on the personal habits and interests of the tag user.

A major security risk concerns adversaries who trick an honest reader to accept illegitimate tags. The main threats are to create faked (illegitimate) tags that are accepted by legitimate readers (*forgery*) and to simulate (*impersonate*) or to copy (*clone*) legitimate tags. Another threat concerns attacks that permanently prevent users from using the RFID system (*denial-of-service*) [7].

2.4 Security and Privacy Objectives

Based on the discussion in the previous paragraphs, we consider RFID systems that provide *anonymity* as well as *untraceability* even when the state of (i.e., the data stored on) a tag has been disclosed. Anonymity means the confidentiality of the tag identity whereas untraceability refers to the unlinkability of the tag communication. To distinguish tracing in past or future protocol runs, the notions of *forward untraceability* and *backward untraceability* are defined in [40]. In use cases like electronic passports, where tags store privacy-sensitive data, *reader authentication* is an additional goal to prevent disclosure of this data to illegitimate readers.

The major security objective of an RFID system is to ensure that only legitimate tags are accepted by legitimate readers (*tag authentication*). Hence, the reader must be able to distinguish between legitimate and illegitimate tags. Most use cases additionally require the reader to be capable of determining the (authentic) tag identity (*tag identification*).

3 Related Work

3.1 Privacy-Preserving RFID Protocols

A general problem with privacy-preserving authentication of low-cost tags that are incapable of public-key operations is how to inform the reader which key should be used for the authentication. Indeed, a tag cannot disclose its identity before the reader has been authenticated since this would violate untraceability. However, a reader cannot authenticate a tag unless it knows the identity (i.e., the key) of that tag. Essentially there are three approaches that address this problem.

The first approach is that the reader performs an exhaustive search for the secret key that is used by the authenticating tag [68]. Solutions to optimize this approach (see, e.g., [42, 64]) suffer from inefficiency since tag verification depends on the total number of tags in the system. Clearly, this violates the efficiency and scalability requirements of most practical RFID systems. A prominent family of lightweight authentication protocols in this context is the HB protocols (see, e.g., [32, 34–36]). These protocols are subject to man-in-the-middle attacks [18, 20, 21, 52], require the reader to perform an exhaustive search for the authentication secret of the authenticating tag and usually require many rounds of interaction [55]. Moreover, tag corruption is usually not considered in the security evaluation of the HB protocols.

In the second approach, a tag updates its identity after each interaction such that its new identity is unlinkable and only known to the tag and the authorized readers, which allows readers to identify tags in constant time (see, e.g., [12, 24, 40, 51, 61]). However, this approach requires each tag to be always synchronized with all readers in the system. In general, it is easy to mount denial-of-service attacks that desynchronize the tag and the readers (see, e.g., [12, 24]).

Another approach to enhance the privacy of RFID systems without lifting the computational requirements on tags are anonymizer-enabled protocols, where external devices (*anonymizers*) are in charge of providing anonymity of tags (see, e.g., [1, 22, 31, 56, 57, 59]). The main concept of anonymizer-enabled protocols is that each tag stores a ciphertext that encrypts the information carried by the tag (e.g., the tag identifier) under the public key of the reader. This ciphertext is transmitted to the reader in the tag authentication protocol. Since this ciphertext is static data that can be used to track and to identify the tag, it must be frequently changed to provide anonymity and untraceability. However, current RFIDs [2, 47] are not capable of updating this public-key encrypted ciphertext on their own and thus, privacy in these protocols relies on third parties, called anonymizers, that frequently refresh the ciphertexts stored on the tags. Most anonymizer-enabled RFID systems are subject to impersonation attacks since tag authentication is only based on the ciphertext that the tag sends to the reader. Moreover, existing security models do not capture RFID systems that use anonymizers. The authors of [58] address these issues and propose an anonymizer-enabled RFID system that provides untraceability, tag authentication, and basic availability along with a general security and privacy framework for anonymizer-enabled RFID systems that are based on the security and privacy odel of [67].

For a broad overview about privacy issues in RFID systems, see also [58].

3.2 RFID Protocols Based on Physically Unclonable Functions

To prevent cloning of a tag it must be infeasible to determine its authentication secret by both attacking the corresponding authentication protocols as well as by physically attacking the tag. One solution to counterfeit cloning attacks is to employ physical protection mechanisms that aggravate reading out the memory of a tag [43, 60]. However, this would dramatically increase the price of tags and render them inappropriate for most commercial applications. A more economic solution to prevent cloning can be implemented by using physically unclonable functions (PUFs) [19, 65].

A PUF consists of an inherently unclonable noisy function P that is embedded into a physical object [66]. The unclonability of a PUF comes from randomness generated during its manufacturing processes. A PUF maps challenges to responses. A *challenge* c is a stimulus signal input to the PUF that makes the PUF to return a *response* $r' = P(c)$ that is specific for that PUF with respect to the stimulus c. This response r' relies on the physical properties of the corresponding physical object, which, however, is subject to environmental noise (e.g., temperature or supply voltage variations). Thus, the PUF will always return slightly different responses r' to the same stimulus c. These slight deviations can be removed by a small circuit, called *fuzzy extractor* that (up to a certain threshold) maps different responses r' to a unique value r for each specific challenge c [14]. The fuzzy extractor needs some additional input w (called *helper data*) to remove the effects of noise on the PUF. Moreover, two different PUFs that are challenged with the same stimulus will return seemingly independent responses with overwhelming probability.

A PUF can be embedded into a microchip, e.g., by exploiting statistical variations of delays of gates and wires within the chip [19]. These deviations are unique for every sample from a set of chips (even from the same lot or wafer) that implement the same circuit.

Physically unclonable functions are a very interesting and promising approach to increase the security of existing RFID systems. Moreover, they open new directions toward cost-efficient privacy-preserving protocols based on physical assumptions. They provide cost-effective and practical tamper-evident storage for cryptographic secrets that even cannot be learned or reproduced by the manufacturer of the corresponding PUF (as long as the manufacturer produced the PUF following the prescribed procedure).

One of the first proposals of using PUFs in RFID systems is introduced by [54]. It proposes the manufacturer of a tag to store a set of challenge–response pairs in a database, which can later be used by RFID readers that are connected to this database to identify a tag. The idea is that the reader chooses a challenge from the database, queries the tag and checks whether the database contains a tuple that matches the response received from the tag. One problem of this approach is that challenge–response pairs cannot be reused since this would enable replay attacks and allow tracing of tags. Hence, the number of tag authentications is limited by the database and the time required to measure the reference responses for the database. This scheme has been implemented by [11] who provide a realization of PUF-enabled RFID tags and analyze their security and usability. The authors of [25] propose a similar approach based on the physical characteristics of SRAM cells. The advantage of this approach is that SRAM-PUFs can be implemented using the existing SRAM memory cells of the RFID chip without the need for additional hardware.

In [65], the authors propose to use a PUF as secure key storage for the secret authentication key of the RFID tag. This means that instead of storing the key in some protected memory, a PUF is used to reconstruct the key whenever it is needed. Since the key is inherently hidden within the physical structure of the PUF, obtaining this secret by hardware-related attacks is supposed to be intractable for real-world adversaries [19]. According to [65], a PUF-based key storage can be implemented with less than 1,000 gates. However, their authentication scheme relies on public-key cryptography, which is still much too expensive for current low-cost RFID tags.

The authors of [6] follow the approach of frequently updating the identity of tags to provide privacy (see Sect. 3.1) and suggest to use PUFs instead of pseudorandom functions. They propose to equip each tag with a PUF P that is used to derive new tag identifiers. Since readers cannot recompute these identifiers, the authors propose the readers to access a database that stores a tuple $(T_0, T_1, \ldots, T_m)$ for each legitimate tag T where T_0 is a random tag identifier and $T_{i+1} = P(T_i)$ for $i \in \{0, \ldots, m-1\}$. To authenticate to a reader, a tag first sends its current identifier T_i and then updates its identity to $T_{i+1} \leftarrow P(T_i)$. The reader then checks whether there is a tuple that contains a value T_i in the database. In case the reader finds T_i, it accepts the tag and invalidates all previous database entries T_j where $j \leq i$ to prevent replay attacks. A major drawback of this scheme is that a tag can only be authenticated m times without being re-initialized, which, as the authors mention, allows an adversary to perform denial-of-service attacks.

3.3 Privacy Models for RFID

One of the first privacy models for RFID [50] defines anonymity and backward untraceability based on a security game where an adversary must distinguish a random value from the output of a tag. However, it does not consider forward untraceability. A privacy model specific for RFIDs that cannot perform any cryptographic operations [29] is based on assumptions on the number of queries an adversary can make to a tag but does not capture adversaries who can corrupt tags. Thus, it does not cover backward and forward untraceability, which is required to realistically model adversaries against cost-efficient tags in practice. Another privacy model [3, 4] provides various flexible definitions for different levels of privacy based on a security experiment where an adversary must distinguish two known tags. This model is extended in [33] by the notion of auxiliary information. In [8], a *completeness* and *soundness* requirement is added to the definition of [33], which means that a reader must accept *all* but *only* valid tags. The definition of [33] has been further improved in [23] to cover backward untraceability. Another privacy model [7] is based on the universal composability (UC) framework and claims to be the first model that considers availability. However, it does not allow the adversary to corrupt tags and does not capture backward untraceability. Recently, [67] presented a privacy model that generalizes and classifies previous RFID privacy models by defining eight levels of privacy that correspond to real-world adversaries of different strength. The strongest privacy notion of [67] captures anonymity, backward and forward untraceability and adversaries with access to auxiliary information. Moreover, it provides a security definition equivalent to [8] that covers tag authentication. The model of [67] has been extended in [53] to consider reader authentication whereas [45] aims at reducing the mentioned eight privacy classes to three privacy classes. Recently in [9, 10] other privacy notions have been considered along with denial of service attacks. The authors of [44] use the framework of [53, 67] to classify and to examine the privacy properties of various existing symmetric key-based authentication protocols for RFID and show several impossibility results for this class of protocols.

4 RFID Security and Privacy Model of Vaudenay [67]

In this section, we review the RFID security and privacy model proposed by Vaudenay (V-Model) [67], which is one of the most comprehensive RFID privacy and security models up to date. We start by setting the notation that will be used later and then give a fairly detailed and at the same time more formal specification of the V-Model [67].

4.1 General Notation

For a finite set S, $|S|$ denotes the size of set S whereas for an integer (or a bitstring) n the term $|n|$ means the bit-length of n. The term $s \in_R S$ means the assignment of a uniformly chosen element of S to variable s. With $\emptyset$ we denote both the empty set and the empty string. Let A be a probabilistic algorithm. Then $y \leftarrow A(x)$ means that on input x, algorithm A assigns its output to variable y. The term $[A(x)]$ denotes the set of all possible outputs of A on input x. $A_K(x)$ means that the output of A depends on x and some additional parameter K (e.g., a secret key). The term $\mathsf{Prot}[\mathsf{A} : x_\mathsf{A}; \mathsf{B} : x_\mathsf{B}; * : x_{pub}] \to [\mathsf{A} : y_\mathsf{A}; \mathsf{B} : y_\mathsf{B}]$ denotes an interactive protocol Prot between two probabilistic algorithms A and B. Hereby, A (resp. B) gets a private input x_A (resp. x_B) and a public input x_{pub}. While A (resp. B) is operating, it can interact with B (resp. A). After the protocol terminates, A (resp. B) returns y_A (resp. y_B).

Let E be some event (e.g., the result of a security experiment), then $\Pr[E]$ denotes the probability that E occurs. Probability $\varepsilon(l)$ is called *negligible* if for all polynomials $f(\cdot)$ it holds that $\varepsilon(l) \leq 1/f(l)$ for all sufficiently large l. Probability $1 - \varepsilon(l)$ is called *overwhelming* if $\varepsilon(l)$ is negligible.

4.2 Pseudorandom Function (PRF)

Let $l \in \mathbb{N}$ be a security parameter, $\kappa, \alpha, \beta \in \mathbb{N}$ be polynomially bounded in l and $F : \{0, 1\}^{\kappa+\alpha} \longrightarrow \{0, 1\}^\beta$ be a family of functions. Consider the following security experiment $\mathsf{Exp}^{\mathsf{prf}\text{-}b}_{\mathcal{A}_{\mathsf{prf}}}$, where an adversary $\mathcal{A}_{\mathsf{prf}}$ interacts with a PRF-challenger $\mathcal{C}_{\mathsf{prf}}$:

When initialized with l, κ, α, β and $b \in_R \{0, 1\}$, $\mathcal{C}_{prf}$ chooses $K \in_R \{0, 1\}^\kappa$ and initializes an oracle $\mathcal{O}^{F_K}$ that on input $x \in \{0, 1\}^\alpha$ returns $y \leftarrow F_K(x)$ if $b = 1$ and $y \in_R \{0, 1\}^\beta$ otherwise. After a polynomial number of queries to oracle $\mathcal{O}^{F_K}$, $\mathcal{A}_{prf}$ then must return a bit b'. $\mathcal{A}_{prf}$ wins the security experiment if $b = b'$.

Definition 1 (Pseudorandom function [49]) A pseudo random function (PRF) is a family of functions F with the following properties:

1. Each function $F_K \in F$ can be identified by a unique index $K \in \{0, 1\}^\kappa$.
2. There is a polynomial time algorithm that gives an index $K \in \{0, 1\}^\kappa$ and input $x \in \{0, 1\}^\alpha$ computes $F_K(x)$.
3. Each probabilistic polynomial time adversary $\mathcal{A}_{prf}$ has at most negligible advantage: $\mathrm{Adv}^{prf}_{\mathcal{A}_{prf}} = \left| \Pr\left[\mathrm{Exp}^{prf-1}_{\mathcal{A}_{prf}} = 1\right] - \Pr\left[\mathrm{Exp}^{prf-0}_{\mathcal{A}_{prf}} = 1\right] \right|$.

4.3 Physically Unclonable Function (PUF)

A physically unclonable function is an inherently unclonable noisy function that is embedded into a physical object (e.g., an integrated circuit) [66]. When challenged with a stimulus (*challenge*), a PUF generates an output (*response*) that depends on both the challenge and the physical properties of the object containing the PUF. However, the physical object is subject to noise (e.g., temperature and/or supply voltage variations) and hence, when queried with the same challenge multiple times, the PUF will always return slightly different responses. To eliminate these output variations *Fuzzy Extractors* [13, 14] can be used.

To the best of our knowledge, currently there is no widely accepted security model for PUFs. Moreover, setting up a model that realistically reflects the properties of real PUFs requires precise physical evaluation results to determine the capabilities of an adversary against PUFs in practice. However, industry considers this data as trade secret while academia usually is restricted to prototype implementations of PUFs (e.g., on FPGAs) that do not reflect the properties of real product-quality PUF implementations (e.g., on ASICs). Hence, in this chapter, we fall back to an idealized model of PUFs that does *not* reflect *real* PUF implementations but captures the *desired* properties of an *ideal* PUF component.

Let $l \in \mathbb{N}$ be a security parameter, $\gamma, \kappa \in \mathbb{N}$ be polynomially bounded in l and $\mathsf{P} : \{0, 1\}^\gamma \longrightarrow \{0, 1\}^\kappa$ be a function. Consider the following security experiment $\mathrm{Exp}^{puf-b}_{\mathcal{A}_{puf}}$ that is similar to $\mathrm{Exp}^{prf-b}_{\mathcal{A}_{prf}}$ described above. The difference is that, when initialized with l, γ, κ and $b \in_R \{0, 1\}$, the PUF-challenger $\mathcal{C}_{puf}$ initializes an oracle $\mathcal{O}^\mathsf{P}$ that on input $x \in \{0, 1\}^\gamma$ returns $y \leftarrow \mathsf{P}(x)$ if $b = 1$ and $y \in_R \{0, 1\}^\kappa$ otherwise. After a polynomial number of queries to $\mathcal{O}^\mathsf{P}$, $\mathcal{A}_{puf}$ must return a bit b'. $\mathcal{A}_{puf}$ wins the security experiment if $b = b'$.

Definition 2 (Ideal PUF) An ideal PUF is a function P with the following properties:

1. For all $c \in \{0, 1\}^\gamma$ and all tuples $(r_i, r_j) \in \left[\mathsf{P}(c)\right]^2$, probability $\Pr[r_i = r_j] = 1$.

2. In the above experiment, any probabilistic polynomial time adversary $\mathcal{A}_{\text{puf}}$ has at most negligible advantage: $\text{Adv}^{\text{puf}}_{\mathcal{A}_{\text{puf}}} = \left| \Pr\left[\text{Exp}^{\text{puf-1}}_{\mathcal{A}_{\text{puf}}} = 1 \right] - \Pr\left[\text{Exp}^{\text{puf-0}}_{\mathcal{A}_{\text{puf}}} = 1 \right] \right|$.

3. Any attempt to physically tamper with the object implementing P results in destruction of P, i.e., P cannot be evaluated any more.

Note that the second property of Definition 2 is similar to the pseudo-randomness property of a PRF (see Definition 1). Hence, the output of an ideal PUF is pseudo-random, which can be achieved in practice by using Fuzzy Extractors [13, 14]. In addition, the second property of Definition 2 implies that the adversary cannot compute the output of the PUF for an adaptively chosen challenge even after adaptively querying the PUF for a polynomial number of times. In return, this means that the adversary cannot emulate (i.e., impersonate or clone) the PUF based on its input/output behavior. According to the third property of Definition 2, the adversary cannot obtain any information about the PUF by physical means, which entirely prevents cloning of the PUF. Moreover, the capabilities of the adversary are not limited concerning the creation and querying of other PUFs, which means that different ideal PUFs are independent pseudorandom functions.

4.4 System Model

As most RFID privacy models, the V-Model [67] considers RFID systems that consist of one single operator $\mathcal{I}$, one single reader $\mathcal{R}$, and a polynomial number of tags $\mathcal{T}$. The reader $\mathcal{R}$ is assumed to be capable of performing public-key cryptography and of handling multiple instances of the tag identification protocol with different tags in parallel. Tags are passive devices, i.e., they do not have own power supply but are powered by the electromagnetic field of the reader $\mathcal{R}$. Hence, tags cannot initiate communication and have a narrow communication range (of a few centimeters to meters). Tags are assumed to be capable of performing basic cryptographic operations like hashing, random number generation, and symmetric key encryption.

4.4.1 Trust and Adversary Model

In the V-Model [67], the issuer $\mathcal{I}$, the backend $\mathcal{D}$, and the readers $\mathcal{R}$ are assumed to be trusted. Therefore, these entities will behave as intended. All the readers $\mathcal{R}$ and the backend $\mathcal{D}$ are subsumed to *one single* reader entity $\mathcal{R}$. This implies that all readers $\mathcal{R}$ are assumed to be tamper-resistant devices that have a permanent secure online connection to a database $\mathcal{D}$. Tags are considered to be untrusted, which means that the adversary can obtain their state (i.e., all the data stored on them).

The adversary can eavesdrop and manipulate the communication channel between a tag $\mathcal{T}$ and the reader $\mathcal{R}$. The V-Model [67] defines eight adversary classes that differ in their ability to corrupt tags and the availability of auxiliary information (see Sect. 4.5). Hence, depending on the adversary class, the adversary is subject to different restrictions concerning tag corruption. At this point we would like to stress that the V-Model [67] does not pose any limitation regarding corruption of a tag $\mathcal{T}$ while the tag $\mathcal{T}_{ID}$ is involved in the authentication protocol with the reader $\mathcal{R}$. However, the adversary is not allowed to corrupt the reader $\mathcal{R}$.

4.4.2 Security and Privacy Objectives

The main security goal of the V-Model [67] is tag authentication. More precisely, a legitimate reader $\mathcal{R}$ should only accept legitimate tags and must be able to identity them. Reader authentication, availability, and protection against cloning are not captured by the V-Model [67]. The privacy objectives are anonymity and unlinkability.

4.4.3 Protocol Definitions

The operator $\mathcal{I}$ sets up the reader $\mathcal{R}$ and all tags $\mathcal{T}$. Hence, there are two setup protocols where $\mathcal{R}$ and the tags $\mathcal{T}$ are initialized and their system parameters (e.g., keys) are generated and defined. A third protocol between a tag $\mathcal{T}$ and $\mathcal{R}$ covers tag authentication. More formally, the RFID system model of [53] is defined as follows:

Definition 3 (RFID system [53]) An RFID system RFID is a tuple of probabilistic polynomial time algorithms $(\mathcal{R}, \mathcal{T}, \mathsf{SetupReader}, \mathsf{SetupTag}, \mathsf{Ident})$ that are specified as follows:

SetupReader$(1^l) \rightarrow (sk_{\mathcal{R}}, pk_{\mathcal{R}}, \mathrm{DB})$ On input of a security parameter $l \in \mathbb{N}$, this algorithm initializes the reader algorithm $\mathcal{R}$ by creating some public parameters $pk_{\mathcal{R}}$ that are known to all entities and some secret parameters $sk_{\mathcal{R}}$ that are only known to $\mathcal{R}$. This algorithm also initializes a credentials database DB that can only be accessed by $\mathcal{R}$ and that stores the identities and the authentication secrets of all legitimate tags.

SetupTag$_{pk_{\mathcal{R}}}(\mathrm{ID}) \rightarrow (K, S)$ Creates a tag $\mathcal{T}_{ID}$, which is an instance of the tag algorithm $\mathcal{T}$. Hereby, the public key $pk_{\mathcal{R}}$ of $\mathcal{R}$ is used to generate a secret K and an initial tag state S. $\mathcal{T}_{ID}$ is initialized with S and (ID, K) is stored in the credentials database DB of $\mathcal{R}$.

Ident$[\mathcal{T}_{ID}: S;\ \mathcal{R}: sk_{\mathcal{R}}, \mathrm{DB};\ *: pk_{\mathcal{R}} \rightarrow \mathcal{T}_{ID}: -;\ \mathcal{R}: out_{\mathcal{R}}]$ This is an interactive protocol between a tag $\mathcal{T}_{ID}$ and the reader $\mathcal{R}$. $\mathcal{T}_{ID}$ takes as input its current state S while $\mathcal{R}$ has as input its secret key $sk_{\mathcal{R}}$ and the credentials database DB. The common input to all parties is the public key $pk_{\mathcal{R}}$ of $\mathcal{R}$. After the protocol terminates, $\mathcal{R}$ returns either the identity ID of $\mathcal{T}_{ID}$ or $\perp$ to indicate that $\mathcal{T}_{ID}$ is not a legitimate tag.

4.5 Adversary Model

In the V-Model [67], the privacy and security objectives are defined as security experiments, where a polynomially bounded adversary can interact with a set of oracles that model the capabilities of the adversary. These oracles are

CreateTag^b(ID) This oracle allows the adversary to set up a tag $\mathcal{T}_{\text{ID}}$ with identifier ID by internally calling SetupTag$_{pk_\mathcal{R}}$(ID) to create (K, S) for $\mathcal{T}_{\text{ID}}$. If input $b = 1$, the adversary chooses $\mathcal{T}_{\text{ID}}$ to be legitimate, which means that (ID, K) is added to the credentials database DB of $\mathcal{R}$. For input $b = 0$, the adversary chooses $\mathcal{T}_{\text{ID}}$ to be illegitimate and (ID, K) is *not* added to DB.[1] This models the fact that an adversary can obtain (e.g., buy) legitimate tags and create forgeries.

Draw(δ) $\rightarrow$ ($vtag_1, b_1, \ldots, vtag_n, b_n$) Initially, the adversary cannot interact with any tag but must query the Draw oracle to get access to a set of tags that has been chosen according to a given tag distribution δ. This models the fact that the adversary can only interact with the tags within his reading range. The adversary usually only knows the tags it can interact with by some temporary tag identifiers $vtag_1, \ldots, vtag_n$. The Draw oracle manages a secret look-up table Γ that keeps track of the real-tag identifier ID_i that is associated with each temporary tag identifier $vtag_i$ (i.e., $\Gamma[vtag_i] = \text{ID}_i$). Moreover, the Draw oracle also provides to the adversary the information on whether the tags are legitimate ($b_i = 1$) or not ($b_i = 0$).

Free($vtag$) Contrary to Draw, the Free oracle makes a tag $vtag$ inaccessible to the adversary. This means that the adversary cannot interact with the tag $vtag$ any longer until it is made accessible again (under a new temporary identifier $vtag'$) by another Draw query. This models the fact that a tag can get out of the reading range of the adversary.

Launch( ) $\rightarrow$ π Makes the reader $\mathcal{R}$ to start a new instance π of the Ident protocol, which allows the adversary to start different concurrent Ident protocol instances with the reader $\mathcal{R}$.

SendReader(m, π) $\rightarrow$ m' Sends a message m to the instance π of the Ident protocol that is running on the reader $\mathcal{R}$. The reader $\mathcal{R}$ interprets m as a protocol message of instance π of the Ident protocol and responds with a message m'. This allows the adversary to perform active attacks on the Ident protocol.

SendTag($m, vtag$) $\rightarrow$ m' Sends a message m to the tag $\mathcal{T}_{\text{ID}}$ that is known as $vtag$ to the adversary. $\mathcal{T}_{\text{ID}}$ interprets m as a protocol message of the Ident protocol and responds with a message m'. This allows the adversary to perform active attacks on the Ident protocol.

Result(π) Returns 1 if the instance π of the Ident protocol has been completed and the tag $\mathcal{T}_{\text{ID}}$ that participated in this instance π has been accepted by the reader $\mathcal{R}$. In case $\mathcal{R}$ identified an illegitimate tag, Result returns 0. This allows the adversary to obtain auxiliary information on whether the authentication of $\mathcal{T}_{\text{ID}}$ was successful or not.

---

[1] Note that illegitimate tags created by the CreateTag oracle are initialized in the same way as legitimate tags with the only difference that their identifier ID and secret K is not added to the credentials database DB of $\mathcal{R}$. As shown in [67], an adversary can use such tags to violate the privacy objectives.

Corrupt(*vtag*) → *S* Returns the current state *S* of the tag $\mathcal{T}_{\text{ID}}$ that is known as *vtag* to the adversary. This model (physical) attacks on $\mathcal{T}_{\text{ID}}$ that discloses the current tag state *S* (i.e., all information stored on or used by $\mathcal{T}_{\text{ID}}$ at the time of corruption) to the adversary.

The V-Model [67] distinguishes the following adversary classes, which differ in their ability to corrupt tags and the availability of auxiliary information (i.e., the ability to call the Corrupt and the Result oracle):

- *Weak adversaries* cannot corrupt tags and are limited to eavesdropping and manipulating the communication between the tags and the reader.
- *Forward adversaries* can obtain the state of the tags only as the last interaction with the oracles defined above. This means that after having corrupted a tag for the first time, a forward adversary can no longer observe any protocol execution or interact with any tag or reader. However, he can still corrupt all remaining non-corrupted tags.
- *Destructive adversaries* cannot reuse a tag after corrupting it. This means that a destructive adversary cannot observe or interact with a corrupted tag nor can he impersonate the corrupted tag to the reader. However, he can still observe and interact with any non-corrupted tag.[2]
- *Strong adversaries* are not restricted in their ability to corrupt tags.

Moreover, the V-Model [67] defines *narrow* variants of the four adversary classes described above (i.e., narrow-weak, narrow-forward, narrow-destructive, and narrow-strong). In addition to the restrictions concerning tag corruption of the corresponding adversary class, a narrow adversary cannot obtain auxiliary information from the communication between the tags and the reader.

Definition 4 (Adversary classes [67]) An adversary is a probabilistic polynomial time algorithm that has arbitrary access to all of the oracles described in Sect. 4.5. Weak adversaries cannot access the Corrupt oracle. Forward adversaries can no longer query any other oracle than Corrupt after they made the first query to the Corrupt oracle. Destructive adversaries cannot query any oracle for *vtag* again after they made a Corrupt(*vtag*) query. Strong adversaries have no restrictions on the use of the Corrupt oracle. Narrow adversaries cannot access the Result oracle.

According to the above notation and definitions, we now recall the definitions of correctness, security, and privacy of the V-Model [67].

---

[2] Note that, in case of PUF-enabled RFID tags, a destructive adversary can corrupt the tag and read out its memory whereas the properties of the PUF ensure that the PUF is destroyed and the adversary does not obtain any information on the PUF.

4.6 Definition of Correctness, Security, and Privacy

4.6.1 Correctness

Correctness describes the honest behavior of legitimate tags $\mathcal{T}$ and the reader $\mathcal{R}$. With overwhelming probability, the reader $\mathcal{R}$ returns $out_{\mathcal{R}} = \text{ID}$ when interacting with a legitimate tag $\mathcal{T}_{\text{ID}}$ and $out_{\mathcal{R}} = \bot$ otherwise. More formally

Definition 5 (Correctness [53]) An RFID system RFID as defined in Definition 3 is correct if for every $l \in \mathbb{N}$, every $(sk_{\mathcal{R}}, pk_{\mathcal{R}}, \text{DB}) \in [\text{SetupReader}(1^l)]$ and every $(K, S) \in [\text{SetupTag}_{pk_{\mathcal{R}}}(\text{ID})]$ it holds with overwhelming probability that

$$\text{Ident}[\mathcal{T}_{\text{ID}}:S; \ \mathcal{R}:sk_{\mathcal{R}}, \text{DB}; \ *:pk_{\mathcal{R}}] \to [\mathcal{T}_{\text{ID}}: -; \ \mathcal{R}:\text{ID}].$$

4.6.2 Security Definition

The security definition given by the V-Model [67] focuses on attacks where the adversary aims to impersonate or forge a legitimate tag. It does *not* capture security against cloning and availability.

The definition of tag authentication is based on a security experiment $\text{Exp}_{\mathcal{A}_{\text{sec}}}^{\mathcal{T}\text{-auth}}$ where a strong adversary $\mathcal{A}_{\text{sec}}$ must make the reader $\mathcal{R}$ to identify some tag $\mathcal{T}_{\text{ID}}$ in some instance π of the Ident protocol. To exclude trivial attacks (e.g., relay attacks), $\mathcal{A}_{\text{sec}}$ is not allowed to simply forward all the messages from $\mathcal{T}_{\text{ID}}$ to $\mathcal{R}$ in instance π nor to corrupt $\mathcal{T}_{\text{ID}}$. This means that at least some of the protocol messages that made $\mathcal{R}$ to return ID must have been partly computed by $\mathcal{A}_{\text{sec}}$ without knowing the secrets of $\mathcal{T}_{\text{ID}}$. With $\text{Exp}_{\mathcal{A}_{\text{sec}}}^{\mathcal{T}\text{-auth}} = 1$ we denote the case where $\mathcal{A}_{\text{sec}}$ wins the security experiment.

Definition 6 (Tag authentication [53]) An RFID system (Definition 3) achieves tag authentication if for every strong adversary $\mathcal{A}_{\text{sec}}$ $\Pr[\text{Exp}_{\mathcal{A}_{\text{sec}}}^{\mathcal{T}\text{-auth}} = 1]$ is negligible.

Note that tag authentication is a critical property and hence must be preserved even against strong adversaries.

4.6.3 Privacy Definition

The privacy definition of the V-Model [67] is very flexible and, dependent on the adversary class considered (see Definition 4), it covers different notions of privacy. It captures anonymity and unlinkability and focuses on the privacy leakage of the communication of tags with the reader $\mathcal{R}$. It is based on the existence of a simulator $\mathcal{B}$, called *blinder*, that can simulate the tags and the reader $\mathcal{R}$ without knowing any of their secrets such that an adversary $\mathcal{A}_{\text{prv}}$ cannot distinguish whether it is interacting with the real or the simulated RFID system. The rationale behind this simulation-based definition is that the communication of the tags with the reader $\mathcal{R}$ does not leak any information about the tags. Hence, everything the adversary $\mathcal{A}_{\text{prv}}$ observes from the interaction with the tags and the reader $\mathcal{R}$ appears to be

Strong	$\Rightarrow$	Destructive	$\Rightarrow$	Forward	$\Rightarrow$	Weak
$\Downarrow$		$\Downarrow$		$\Downarrow$		$\Downarrow$
Narrow-Strong	$\Rightarrow$	Narrow-Destructive	$\Rightarrow$	Narrow-Forward	$\Rightarrow$	Narrow-Weak

Fig. 1 Privacy notions defined in the PV-Model [53] and their relations

independent of the tags and consequently, $\mathcal{A}_{prv}$ cannot distinguish different tags based on their communication, which corresponds to unlinkability.

In the following, we express this privacy definition in a more formal way by a privacy experiment $\text{Exp}_{\mathcal{A}_{prv}}^{prv\text{-}b}$. Let $\mathcal{A}_{prv}$ be an adversary according to Definition 4, $l \in \mathbb{N}$ be a given security parameter and $b \in_R \{0, 1\}$. In the first phase of the experiment, the reader $\mathcal{R}$ is initialized with $(sk_{\mathcal{R}}, pk_{\mathcal{R}}, \text{DB}) \leftarrow \text{SetupReader}(1^l)$. The public key $pk_{\mathcal{R}}$ is given to $\mathcal{A}_{prv}$ and to the simulator $\mathcal{B}$. Now, $\mathcal{A}_{prv}$ is allowed to arbitrarily interact with all oracles defined in Sect. 4.5. Hereby, $\mathcal{A}_{prv}$ is subject to the restrictions of its corresponding adversary class (see Definition 4). If $b = 1$, all queries to the **Launch, SendReader, SendTag**, and **Result** oracles are redirected to and answered by the simulator $\mathcal{B}$. Hereby, $\mathcal{B}$ can observe all queries $\mathcal{A}_{prv}$ makes to all the other oracles that are not simulated by $\mathcal{B}$ and the corresponding responses ("$\mathcal{B}$ sees what $\mathcal{A}_{prv}$ sees"). After a polynomial number of oracle queries, the second phase of the experiment starts. In this second stage, $\mathcal{A}_{prv}$ can no longer interact with the oracles but is given the hidden table Γ of the **Draw** oracle. Finally, $\mathcal{A}_{prv}$ returns a bit b', which we denote with $\text{Exp}_{\mathcal{A}_{prv}}^{prv\text{-}b} = b'$.

Definition 7 (Privacy [67]) Let C be one of the adversary classes according to Definition 4. An RFID system (Definition 3) is C-private if for every adversary $\mathcal{A}_{prv}$ of C there is a probabilistic polynomial time algorithm $\mathcal{B}$ (blinder) such that

$$\text{Adv}_{\mathcal{A}_{prv}}^{prv} = \left| \Pr\left[\text{Exp}_{\mathcal{A}_{prv}}^{prv\text{-}0} = 1\right] - \Pr\left[\text{Exp}_{\mathcal{A}_{prv}}^{prv\text{-}1} = 1\right] \right|$$

is negligible. $\mathcal{B}$ simulates the **Launch, SendReader, SendTag**, and **Result** oracles to $\mathcal{A}_{prv}$ without having access to $sk_{\mathcal{R}}$ and DB. Hereby, all oracle queries $\mathcal{A}_{prv}$ makes and their corresponding responses are also sent to $\mathcal{B}$.

All privacy notions defined in the PV-Model [53] are summarized in Fig. 1, which also shows the relations among them. It has been shown that strong privacy is impossible [67] while the technical feasibility of destructive privacy has been an open problem.

5 A PUF-Based Destructive-Private RFID Protocol

In this section, we address an open problem of [67] by presenting the first destructive-private RFID protocol. Our protocol is based on the weak-private protocol of [67], which is a simple challenge–response protocol. To achieve destruc-

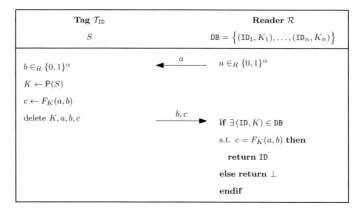

Fig. 2 Destructive-private PUF-based RFID protocol

tive privacy, in our protocol, the tag $\mathcal{T}$ does not directly use its state S as authentication key K. Instead, K is derived by evaluating a physically unclonable function P on input S each time K is needed. Hence, the properties of the PUF ensure that the adversary cannot access the tag secret K but destroys the tag $\mathcal{T}$ by any attempt to corrupt it.

Let $l \in \mathbb{N}$ be a given security parameter, $\alpha, \beta, \gamma, \kappa \in \mathbb{N}$ be polynomial in l and $F : \{0, 1\}^{\kappa} \times \{0, 1\}^{2\alpha} \longrightarrow \{0, 1\}^{\beta}$ be a family of pseudorandom functions. Each tag $\mathcal{T}$ is equipped with a PUF $P : \{0, 1\}^{\gamma} \longrightarrow \{0, 1\}^{\kappa}$ and is initialized by a random state $S \in_R \{0, 1\}^{\gamma}$. The credentials database DB of the reader $\mathcal{R}$ contains a tuple (ID, K) for each legitimate tag $\mathcal{T}_{\text{ID}}$ where $K \leftarrow P(S)$.

Our destructive-private tag authentication protocol is illustrated in Fig. 2. The reader $\mathcal{R}$ starts by sending a random challenge a to the tag $\mathcal{T}_{\text{ID}}$, which first chooses a random value b and then queries its PUF with its state S to reconstruct its tag authentication secret K. Next, the tag $\mathcal{T}_{\text{ID}}$ evaluates $F_K(a, b)$, sends the result c and b to the reader $\mathcal{R}$, and immediately erases K, a, b and c from its temporary memory. On receipt of c the reader $\mathcal{R}$ recomputes $F_K(a, b)$ for each tuple (ID, K) in its credential database DB until it finds a match. If the reader $\mathcal{R}$ finds a matching (ID, K), it accepts the tag $\mathcal{T}_{\text{ID}}$ by returning ID. Otherwise, the reader $\mathcal{R}$ rejects the tag $\mathcal{T}_{\text{ID}}$ and returns $\bot$.

5.1 Correctness

Clearly, if both tag $\mathcal{T}_{\text{ID}}$ and reader $\mathcal{R}$ are legitimate, then the correctness of the Ident protocol shown in Fig. 2 follows directly from the properties of the PRF F (see Definition 1) and the correctness of the PUF P (see Definition 2).

6 Security Analysis

6.1 Tag Authentication

Theorem 1 *The RFID protocol illustrated in Fig. 2 achieves tag authentication (Definition 6).*

Proof Assume by contradiction that the protocol shown in Fig. 2 does not achieve tag authentication. This means that there is an adversary $\mathcal{A}_{sec}$ who can generate, with non-negligible probability p, a protocol message $(\tilde{b}, \tilde{c})$ for a given $\tilde{a}$ such that $\tilde{c} = F_{\tilde{K}}(\tilde{a}, \tilde{b})$ where $(\tilde{\text{ID}}, \tilde{K}) \in$ DB without having made a **Corrupt** or **SendTag**$(\tilde{a}, \cdot)$ query to the tag $\mathcal{T}_{\tilde{\text{ID}}}$. In the following, we show that $\mathcal{A}_{sec}$ can be transformed into a probabilistic polynomial time algorithm $\mathcal{A}_{prf}$ that contradicts the security property of the underlying PRF F (Definition 1). Hence, the pseudorandomness of F ensures that there is no such adversary $\mathcal{A}_{sec}$.

The construction of $\mathcal{A}_{prf}$ is as follows: Given the security parameters l, κ, α, β and a description of the PRF F from the PRF-challenger $\mathcal{C}_{prf}$, $\mathcal{A}_{prf}$ initializes the RFID system by first choosing γ polynomial in l and then setting $sk_{\mathcal{R}} \leftarrow \emptyset$, $pk_{\mathcal{R}} \leftarrow (l, \gamma, \kappa, \alpha, \beta, F)$ and DB $\leftarrow \emptyset$. Then $\mathcal{A}_{prf}$ guesses the identifier $\tilde{\text{ID}}$ of the tag $\mathcal{T}_{\tilde{\text{ID}}}$ that will be impersonated by $\mathcal{A}_{sec}$. Note that the probability of correctly guessing $\tilde{\text{ID}}$ is polynomial since $\mathcal{A}_{sec}$ can create at most a polynomial number of tags. Next, $\mathcal{A}_{prf}$ initializes $\mathcal{A}_{sec}$ with $(l, \gamma, \kappa, \alpha, \beta, F)$ and simulates all the oracles defined in Sect. 4.5 to $\mathcal{A}_{sec}$:

- **CreateTag**(ID) : If there already is a tuple (ID, $\cdot$, $\cdot$) $\in$ DB or if ID $= \tilde{\text{ID}}$, then $\mathcal{A}_{prf}$ aborts. Otherwise, $\mathcal{A}_{prf}$ chooses $S \in_R \{0, 1\}^\gamma$ and $K \in_R \{0, 1\}^\kappa$ and updates DB $\leftarrow$ DB $\cup \{(\text{ID}, K, S)\}$.
- **Draw, Free, Launch**: The simulation of the **Draw**, **Free**, and **Launch** oracle is straightforward. Note that $\mathcal{A}_{prf}$ knows the secret look-up table Γ of the **Draw** oracle.
- **SendTag**($a, vtag$): If $\Gamma[vtag] = \tilde{\text{ID}}$, then $\mathcal{A}_{prf}$ responds with $b \in_R \{0, 1\}^\alpha$ and $c \leftarrow \mathcal{O}^{F_{\tilde{K}}}(a, b)$. Else, $\mathcal{A}_{prf}$ gets $(\Gamma[vtag], K, S)$ from DB and responds with $b \in_R \{0, 1\}^\alpha$ and $c \leftarrow F_K(a, b)$.
- **SendReader**($\emptyset, \pi$): If π has been previously generated by a **Launch** oracle query and the corresponding protocol transcript is $\text{tr}_\pi = \emptyset$, then $\mathcal{A}_{prf}$ returns $a \in_R \{0, 1\}^\alpha$ and updates $\text{tr}_\pi \leftarrow a$.
- **SendReader** $((b, c), \pi)$: If π has been previously generated by a **Launch** oracle query and the corresponding protocol transcript is $\text{tr}_\pi = a$, then $\mathcal{A}_{prf}$ updates the protocol transcript $\text{tr}_\pi \leftarrow (a, b, c)$ and aborts otherwise.
- **Result**(π): If π has been previously generated by a **Launch** oracle query and the corresponding protocol transcript $\text{tr}_\pi = (a, b, c)$ has been obtained through $a \leftarrow$ **SendReader**($\emptyset, \pi$), then computes $c' \leftarrow F_K(a, b)$ for each tuple (ID, K) in DB. If a $c' = c$ for some (ID, K) then returns 1, otherwise returns 0.

- **Corrupt**($vtag$): If there is a tuple ($\Gamma[vtag]$, K, S) in DB, $\mathcal{A}_{prf}$ returns S. Note that according to Definition 6, $\mathcal{A}_{sec}$ is not allowed to corrupt the tag $T_{\tilde{ID}}$ and hence, $\mathcal{A}_{prf}$ needs not to simulate the **Corrupt** oracle for the tag $T_{\tilde{ID}}$.

With non-negligible probability, after a polynomial number of oracle queries, $\mathcal{A}_{sec}$ returns a protocol message ($\tilde{b}$, $\tilde{c}$) for a given $\tilde{a}$. Next, $\mathcal{A}_{prf}$ sends $x \leftarrow (\tilde{a}, \tilde{b})$ to $\mathcal{C}_{prf}$ who responds with the challenge y, which is either $y \leftarrow F_{\tilde{K}}(x)$ or $y \in_R \{0, 1\}^\beta$. In case $y = \tilde{c}$, $\mathcal{A}_{prf}$ returns 0 and 1 otherwise.

Note that in case $b = 1$, $\mathcal{A}_{prf}$ perfectly simulates all oracles defined in Sect. 4.5 to $\mathcal{A}_{sec}$. Hence, in case $b = 1$, by assumption $\mathcal{A}_{sec}$ generates ($\tilde{b}$, $\tilde{c}$) for any given $\tilde{a}$ such that $\tilde{c} = F_{\tilde{K}}(\tilde{a}, \tilde{b})$ holds with non-negligible probability. In return, this means that $\mathcal{A}_{prf}$ has a non-negligible advantage of distinguishing the output of F and a randomly chosen value. Clearly, this contradicts the pseudorandomness of the PRF F (Definition 1), which proves Theorem 1. $\square$

6.2 Destructive Privacy

Theorem 2 *The RFID protocol illustrated in Fig. 2 achieves destructive privacy (Definition 7).*

Proof According to Definition 7, destructive privacy means that there is a blinder $\mathcal{B}$ that simulates the **Launch, SendTag, SendReader,** and **Result** oracle such that no destructive adversary $\mathcal{A}_{prv}$ (Definition 4) can distinguish between the blinder $\mathcal{B}$ and the real oracles. Hence, to prove Theorem 2, we first give the construction of the blinder $\mathcal{B}$ and then show that every destructive adversary $\mathcal{A}_{prv}$ has at most negligible probability to distinguish the blinder $\mathcal{B}$ from the real oracles.

The blinder $\mathcal{B}$ is initialized with the security parameters l, γ, κ, α, β and the public key $pk_{\mathcal{R}}$ of the reader $\mathcal{R}$ and works as follows:

- **Launch**( ): The simulation of the **Launch** oracle is straightforward.
- **SendTag**(a, $vtag$): Return $b \in_R \{0, 1\}^\alpha$ and $c \in_R \{0, 1\}^\beta$.
- **SendReader** (π): Return $a \in_R \{0, 1\}^\alpha$.
- **SendReader** ((b, c), π): Since oracle queries of this form do not generate any output nor change the state of the tag and the reader, the blinder $\mathcal{B}$ needs not to simulate their responses.
- **Result**(π): If π has been previously generated by a **Launch** oracle query and the corresponding protocol transcript $\text{tr}_\pi = (a, b, c)$ has been generated by $a \leftarrow$ **SendReader** ($\emptyset$, π) and (b, c) $\leftarrow$ **SendTag** (a, $vtag$), return 1 and 0 otherwise.

In the following, we show that if there is a destructive adversary $\mathcal{A}_{prv}$ who can distinguish the blinder $\mathcal{B}$ from the real oracles, then we can use $\mathcal{A}_{prv}$ to construct a polynomial time algorithm that violates the security properties of either the underlying PRF F or the PUF P.

Let game game$^{(0)}$ be the game where the adversary $\mathcal{A}_{prv}$ interacts with the real oracles as defined in Sect. 4.5. Now consider the following hybrid game game$^{(1)}$ that is exactly as game$^{(0)}$ with the only difference that the states S and the authentication secrets K of all tags are simulated by randomly chosen values. In the following, we show that if $\mathcal{A}_{prv}$ can distinguish between game$^{(0)}$ and game$^{(1)}$, then we can use $\mathcal{A}_{prv}$ to construct a polynomial time algorithm $\mathcal{A}_{puf}$ that contradicts the security property of the PUF P (Definition 2).

According to the protocol specification given in Sect. 5, the states and PUFs of different tags are chosen independently. Moreover, $\mathcal{A}_{puf}$ can trivially simulate different tags by following the protocol specifications. Hence, we assume w.l.o.g. that $\mathcal{A}_{prv}$ creates just one single tag $\mathcal{T}_{ID}$ during his attack. To create this tag $\mathcal{T}_{ID}$, $\mathcal{A}_{puf}$ chooses $S \in_R \{0, 1\}^{\gamma}$ and sets $K \leftarrow \mathcal{O}^{P}(S)$. Note that $\mathcal{O}^{P}(S)$ returns either $K \leftarrow P(S)$ as in game$^{(0)}$ or $K \in_R \{0, 1\}^{\kappa}$ as in game$^{(1)}$. Now, $\mathcal{A}_{puf}$ can interact with all the oracles defined in Sect. 4.5 that are simulated by $\mathcal{A}_{puf}$ based on the input of $\mathcal{C}_{puf}$. The simulation of the **Draw**, **Free** and **Launch** oracle is straightforward. Note that the output of the **Result** and **Corrupt** oracle is independent of the PUF of tag $\mathcal{T}_{ID}$ and hence, these oracles can be simulated in a trivial way. Since **SendReader** queries generate no output and do not change the state S of the tag $\mathcal{T}_{ID}$, they need not be simulated by $\mathcal{A}_{puf}$. On a **SendTag**$(a, vtag)$ oracle query, $\mathcal{A}_{puf}$ responds with $b \in_R \{0, 1\}^{\alpha}$ and $c \leftarrow F_K(a, b)$.

Note that $\mathcal{A}_{prv}$ is a destructive adversary and hence, by making a **Corrupt**$(vtag)$ query, $\mathcal{A}_{prv}$ can obtain the state S of the tag $vtag$ but he can no longer send any query that involves the tag $vtag$ afterward. After a polynomial number of oracle queries, $\mathcal{A}_{prv}$ returns a bit b'. In case $b' = 1$ (which indicates that $\mathcal{A}_{prv}$ detected $\mathcal{B}$), with non-negligible probability $\mathcal{O}^{P}$ must have returned a random $K \in_R \{0, 1\}^{\kappa}$. Hence, $\mathcal{A}_{puf}$ can distinguish the between the output of a PUF and a randomly chosen value, which contradicts the security property of the PUF (Definition 2). As a result, it follows that

$$\left| \Pr\left[\text{game}^{(0)} = 1 \right] - \Pr\left[\text{game}^{(1)} = 1 \right] \right| \tag{1}$$

is negligible.

Next, consider the hybrid game game$^{(2)}$ that is exactly as game$^{(1)}$ with the only difference that the **SendTag** oracle is simulated by the blinder $\mathcal{B}$ as described above. In the following, we show that if $\mathcal{A}_{prv}$ can distinguish between game$^{(1)}$ and game$^{(2)}$, then we can use $\mathcal{A}_{prv}$ to construct a polynomial time algorithm $\mathcal{A}_{prf}$ that contradicts the security property of the PRF F (Definition 1).

Let $q \in \mathbb{N}$ be the number of **SendTag** queries made by $\mathcal{A}_{prv}$, which is polynomial in l. Moreover, let $i \in \{0, \dots, q\}$. Now consider the following hybrid game game$_i$ with $\mathcal{A}_{prv}$: The first i **SendTag** queries of $\mathcal{A}_{prv}$ are answered by the blinder $\mathcal{B}$ (as in game$^{(2)}$), while the remaining $q - i$ queries are forwarded and answered by the real **SendTag** oracle (as in game$^{(1)}$). Note that game$_0$ corresponds to game$^{(1)}$ whereas game$_q$ corresponds to game game$^{(2)}$. Hence, and due to the contradicting assumption made at the beginning of the proof, it holds that

$$\text{Adv}_{\mathcal{A}_{prv}}^{prv} = \left| \Pr[\text{game}_0 = 1] - \Pr[\text{game}_q = 1] \right|$$

is non-negligible. In return, this means that there must be some index $i \in \{1, \ldots, q\}$ such that

$$\left| \Pr[\text{game}_{i-1} = 1] - \Pr[\text{game}_i = 1] \right| \tag{2}$$

is non-negligible. Note that Eq. (2) implies that w.l.o.g. $\mathcal{A}_{\text{prv}}$ detects $\mathcal{B}$ in game game_i with non-negligible probability while he has at most negligible probability to detect $\mathcal{B}$ in game game_{i-1}.

We can use $\mathcal{A}_{\text{prv}}$ to construct the following polynomial time algorithm $\mathcal{A}_{\text{prf}}$ that violates the security property of the PRF F (Definition 1). Therefore, $\mathcal{A}_{\text{prf}}$ plays the hybrid game game'_i with $\mathcal{A}_{\text{prv}}$, which is like game_i except that the i-th SendTag$(a, vtag)$ query is answered as follows: $\mathcal{A}_{\text{prf}}$ chooses $b \in_R \{0, 1\}^\alpha$ and sends $x \leftarrow (a, b)$ to the PRF-challenger $\mathcal{C}_{\text{prf}}$, which responds with $y \leftarrow \mathcal{O}^F(x)$ that is either $y = F_K(x)$ or $y \in_R \{0, 1\}^{2\alpha}$ Then, $\mathcal{A}_{\text{prf}}$ sends (b, c) to $\mathcal{A}_{\text{prv}}$. Note that, in case $\mathcal{C}_{\text{prf}}$ sends $y = F_K(x)$ then $\text{game}'_i = \text{game}_{i-1}$ and $\text{game}'_i = \text{game}_i$ otherwise. Hence, if $\mathcal{A}_{\text{prv}}$ returns 1 (which indicates that $\mathcal{A}_{\text{prv}}$ detected $\mathcal{B}$) then $\mathcal{A}_{\text{prf}}$ must have played game_i. Clearly, this allows $\mathcal{A}_{\text{prf}}$ to distinguish the output of the PRF F from a random value, which contradicts the security property of the PRF (Definition 1). Hence, the PRF ensures that Eq. (2) is negligible and, as a consequence, that

$$\left| \Pr\left[\text{game}^{(1)} = 1 \right] - \Pr\left[\text{game}^{(2)} = 1 \right] \right| \tag{3}$$

is negligible.

Next, consider the hybrid game $\text{game}^{(3)}$ that is exactly as $\text{game}^{(2)}$ with the only difference that the Result oracle is simulated by the blinder $\mathcal{B}$ as described above. In the following, we show that if there is an adversary $\mathcal{A}_{\text{prv}}$ who can distinguish between $\text{game}^{(2)}$ and $\text{game}^{(3)}$, then $\mathcal{A}_{\text{prv}}$ can be used to construct a polynomial time algorithm $\mathcal{A}_{\text{sec}}$ that contradicts tag authentication (Definition 6).

In the following, let $p \in \mathbb{N}$ be the number of Result queries made by $\mathcal{A}_{\text{prv}}$, which is polynomial in l. Moreover, let $i \in \{0, \ldots, p\}$. Now consider the following hybrid game game_i^*: The first i Result queries of $\mathcal{A}_{\text{prv}}$ are answered by the blinder $\mathcal{B}$ (as in $\text{game}^{(3)}$), while the remaining $p - i$ queries are forwarded and answered by the real Result oracle (as in $\text{game}^{(2)}$). Note that game_0^* corresponds to $\text{game}^{(2)}$ whereas game_p^* is equivalent to $\text{game}^{(3)}$. Hence, and due to the contradicting assumption made at the beginning of the proof, it holds that

$$\text{Adv}_{\mathcal{A}_{\text{prv}}}^{\text{prv}} = \left| \Pr[\text{game}_0^* = 1] - \Pr[\text{game}_p^* = 1] \right|$$

is non-negligible. In return, this means that there must be some index $i \in \{1, \ldots, p\}$ such that

$$\left| \Pr[\text{game}_{i-1}^* = 1] - \Pr[\text{game}_i^* = 1] \right| \tag{4}$$

is non-negligible. Note that Eq. (4) implies that w.l.o.g. $\mathcal{A}_{prv}$ detects $\mathcal{B}$ in game game$_i^*$ with non-negligible probability while he has at most negligible probability to detect $\mathcal{B}$ in game game$_{i-1}^*$. This means that in game$_i^*$ $\mathcal{A}_{prv}$ runs a protocol instance π where the Result oracle simulated by $\mathcal{B}$ returns a different output then the real Result oracle. According to the description of $\mathcal{B}$ given at the beginning of this proof and the definition of the Result oracle in Sect. 4.5, this can only happen if $\mathcal{A}_{prv}$ generates a protocol transcript $\mathtt{tr}_\pi = (a, b, c)$ such that $c = F_K(a, b)$ where $(\mathtt{ID}, K) \in \mathtt{DB}$ and tag $\mathcal{T}_{\mathtt{ID}}$ have not been corrupted by $\mathcal{A}_{prv}$. However, as shown in the proof of Theorem 1 this can only happen with negligible probability. Hence, tag authentication ensures that Eq. (5) is negligible and thus

$$\left| \Pr\left[\mathrm{game}^{(2)} = 1 \right] - \Pr\left[\mathrm{game}^{(3)} = 1 \right] \right| \tag{5}$$

is negligible as well.

Note that game$^{(3)}$ corresponds to the game where $\mathcal{A}_{prv}$ interacts with a full blinder $\mathcal{B}$. Hence, from Eqs. (1), (3) and (5) it follows that

$$\left| \Pr\left[\mathrm{game}^{(0)} = 1 \right] - \Pr\left[\mathrm{game}^{(3)} = 1 \right] \right|$$

is negligible. This means that $\mathcal{A}_{prv}$ cannot distinguish between the real oracles and the full blinder $\mathcal{B}$, which completes the proof of Theorem 2.   □

7 Conclusion

In this chapter, we have shown that physically unclonable functions are a very interesting and promising approach to improve the security and privacy of existing RFID systems. However, several aspects of PUFs and their deployment to RFID require further research. Since PUFs are bound to the device in which they are embedded, no other entity can verify the output of a PUF to a given challenge without knowing the correct output value in advance. Another problem with PUFs is that their realizations require careful statistical testing before they can be safely deployed to real-security critical products. Moreover, to our knowledge, there is no complete security and adversary model for PUFs yet.

Acknowledgments We wish to thank Frederik Armknecht, Paolo D'Arco, and Alessandra Scafuro for several useful discussions about RFID privacy notions. This work has been supported in part by the European Commission through the FP7 programme under contract 216646 ECRYPT II, 238811 UNIQUE, and 215270 FRONTS, in part by the Ateneo Italo-Tedesco under Program Vigoni and by the MIUR Project PRIN 2008 "PEPPER: Privacy E Protezione di dati PERsonali" (prot. 2008SY2PH4).

References

1. G. Ateniese, J. Camenisch, B. de Medeiros, in *Untraceable RFID Tags via Insubvertible Encryption*. Proceedings of the 12th ACM Conference on Computer and Communications Security, Alexandria, VA, USA, 7–11 Nov 2005 (ACM Press, 2005), pp. 92–101
2. Atmel Corporation. Innovative IDIC solutions. `http://www.atmel.com/dyn/resources/prod_documents/doc4602.pdf`, 2007
3. Gildas Avoine. Adversarial model for radio frequency identification. Cryptology ePrint Archive, Report 2005/049, 2005.
4. G. Avoine, E. Dysli, P. Oechslin, in *Reducing Time Complexity in RFID systems*. 12th International Workshop on Selected Areas in Cryptography (SAC), Kingston, ON, Canada, 11–12 Aug 2005. Lecture Notes in Computer Science, vol. 3897 (Springer, Berlin, 2005), pp. 291–306
5. G. Avoine, C. Lauradoux, T. Martin in *When Compromised Readers Meet RFID*. The 5th Workshop on RFID Security 2009, Leuven, Belgium, 30 June–2 July, 2009
6. L. Bolotnyy, G. Robins, in *Physically Unclonable Function-Based Security and Privacy in RFID systems*. Proceedings of the Fifth IEEE International Conference on Pervasive Computing and Communications, White Plains, NY, USA, 19–23 Mar 2007 (IEEE Computer Society, Washington, DC, 2007)
7. M. Burmester, T. van Le, B. de Medeiros, *Provably Secure Ubiquitous Systems: Universally Composable RFID Authentication Protocols*. Proceedings of Second International Conference on Security and Privacy in Communication Networks (SecureComm), Baltimore, MD, USA, 28 Aug–1 Sept 2006 (IEEE Computer Society, Washington, DC, 2006), pp. 1–9
8. I. Damgård, M. Østergaard, *RFID Security: Tradeoffs Between Security and Efficiency*. Cryptology ePrint Archive, Report 2006/234, 2006
9. P. D'Arco, A. Scafuro, I. Visconti, in *Revisiting DoS Attacks and Privacy in RFID-Enabled Networks*. Proceedings of ALGOSENSORS, Rhodes, Greece, 10–11 July 2009. Lecture Notes in Computer Science (Springer, July 2009)
10. P. D'Arco, A. Scafuro, I. Visconti, in *Semi-Destructive Privacy in DoS-Enabled RFID Systems*. Proceedings of RFIDSec, Leuven, Belgium, 30 June-2 July 2009, July 2009
11. S. Devadas, E. Suh, S. Paral, R. Sowell, T. Ziola, V. Khandelwal, in *Design and Implementation of PUF-Based Unclonable RFID ICs for Anti-counterfeiting and Security Applications*. IEEE International Conference on RFID 2008, Las Vegas, NV, USA, 16–17 April, 2008 (IEEE Computer Society, 2008), pp. 58–64
12. T. Dimitriou, in *A Lightweight RFID Protocol to Protect Against Traceability and Cloning Attacks*. Proceedings of the First International Conference on Security and Privacy for Emerging Areas in Communications Networks (SecureComm) Athens, Greece, 5–9 Sept 2005 (IEEE Computer Society, 2005), pp. 59–66
13. Y. Dodis, L. Reyzin, A. Smith, *Fuzzy Extractors: How to Generate Strong Keys from Biometrics and Other Noisy Data*. International Conference on the Theory and Applications of Cryptographic Techniques, Interlaken, Switzerland, 2–6 May, 2004, Proceedings. Lecture Notes in Computer Science, vol. 3027 (Springer, 2004), pp. 523–540
14. Y. Dodis, L. Reyzin, A. Smith, in Security with Noisy Data, chapter Fuzzy Extractors, (Springer, 2007), pp. 79–99
15. EPCglobal Inc. Object Naming Service (ONS), version 1.0, October 2005
16. EPCglobal Inc. Web site of EPCglobal Inc `http://www.epcglobalinc.org/`, April 2008
17. K. Finkenzeller, *RFID-Handbook* 2nd edn. (Carl Hanser Verlag, Munich, Germany, Apr 2003). Translated from the 3rd German edition by Rachel Waddington, Swadlincote, UK
18. D. Frumkin, A. Shamir, *Un-Trusted-HB: Security Vulnerabilities of Trusted-HB*. Cryptology ePrint Archive, Report 2009/044, 2009
19. B. Gassend, D. Clarke, M. van Dijk, S. Devadas, in *Controlled Physical Random Functions*. Proceedings of the 18th Annual Computer Security Applications Conference, Las Vegas, NV, USA, 9–13 Dec 2002 (IEEE Computer Society, 2002), pp. 149–160

20. H. Gilbert, M. Robshaw, H. Silbert, *An Active Attack Against HB+ — A Provable Secure Leightweight Authentication Protocol*. Cryptology ePrint Archive, Report 2007/237, 2007

21. H. Gilbert, M.J.B. Robshaw, Y. Seurin, in *Good Variants of HB+ Are Hard to Find*. in G. Tsudik. Financial Cryptography and Data Security, 12th International Conference, FC 2008, Cozumel, Mexico, 28–31 Jan 2008, Revised Selected Papers. Lecture Notes in Computer Science, (Springer, 2008), pp. 156–170

22. P. Golle, M. Jakobsson, A. Juels, P. Syverson, in *Universal Re-encryption for Mixnets*. The Cryptographers' Track at the RSA Conference 2004, Proceedings. Lecture Notes in Computer Science, San Francisco, CA, USA, 23–27 Feb 2004 (Springer, 2004), pp. 163–178.

23. J.H. Ha, S.J. Moon, J. Zhou, J.C. Ha, A new formal proof model for RFID location privacy. In Jajodia and Lopez [28], pp. 267–281

24. D. Henrici, P. Müller, in *Hash-Based Enhancement of Location Privacy for Radio-Frequency Identification Devices Using Varying Identifiers*. Proceedings of the Second IEEE Annual Conference on Pervasive Computing and Communications Workshops, Orlando, FL, USA, 14–17 Mar 2004 (IEEE Computer Society, 2004), pp. 149–153

25. D.E. Holcomb, W.P. Burleson, K. Fu, *Initial SRAM State as a Fingerprint and Source of True Random Numbers for RFID Tags*. Conference on RFID Security 2007, Malaga, Spain, 11–13 July 2007

26. M. Hutter, J.-M. Schmidt, T. Plos, *RFID and Its Vulnerability to Faults*. 10th International Workshop on Cryptographic Hardware and Embedded Systems (CHES) 2008, Washington, DC, USA, 10–13 Aug 2008, Proceedings. Lecture Notes in Computer Science, vol. 5154 (Springer, 2008), pp. 363–379

27. I.C.A. Organization. Machine Readable Travel Documents, Doc 9303, Part 1 Machine Readable Passports, 5th edn., 2003

28. S. Jajodia, J. Lopez (eds.), *Computer Security — ESORICS 2008*. Lecture Notes in Computer Science, Malaga, Spain, 6–8 Oct 2008, vol. 5283 (Springer, 2008)

29. A. Juels, in *Minimalist Cryptography for Low-Cost RFID Tags (Extended Abstract)*. 4th International Conference on Security in Communication Networks (SCN) 2004, Revised Selected Papers. Lecture Notes in Computer Science, Amalfi, Italy, 8–10 Sep 2004, vol. 3352 (Springer, 2004), pp. 149–164

30. A. Juels, in *RFID Security and Privacy: A Research Survey*. J. Select. Areas Commun. **24**(2), 381–395 (Feb 2006)

31. A. Juels, R. Pappu, in *Squealing Euros: Privacy Protection in RFID-Enabled Banknotes*. 7th International Conference on Financial Cryptography (FC) 2003, Revised Papers. Lecture Notes in Computer Science, Gosier, Guadeloupe, FWI, 27–30 Jan 2003, vol. 2742 (Springer, 2003), pp. 103–121

32. A. Juels, S.A. Weis, Authenticating pervasive devices with human protocols. in *Advances in Cryptology — CRYPTO 2005*, ed. by V. Shoup. 25th Annual International Cryptology Conference, Santa Barbara, CA, USA, 14–18 Aug 2005, Proceedings. Lecture Notes in Computer Science, vol. 3621 (Springer, 2005), pp. 293–308

33. A. Juels, S.A. Weis, *Defining Strong Privacy for RFID*. Cryptology ePrint Archive, Report 2006/137, 2006

34. J. Katz, in *Efficient Cryptographic Protocols Based on the Hardness of Learning Parity with Noise*. in S.D. Galbraith. Cryptography and Coding, 11th IMA International Conference, Cirencester, UK, 18–20 Dec 2007, Proceedings. Lecture Notes in Computer Science, vol. 4887 (Springer, 2007), pp. 1–15

35. J. Katz, J.S. Shin, Parallel and concurrent security of the HB and HB+ protocols. in *Advances in Cryptology — EUROCRYPT 2006*, ed. by S. Vaudenay. 24th Annual International Conference on the Theory and Applications of Cryptographic Techniques, St. Petersburg, Russia, 28 May –1 June 2006, Proceedings. Lecture Notes in Computer Science, vol. 4004 (Springer, 2006), pp. 73–87

36. J. Katz, A, Smith, *Analyzing the HB and HB+ Protocols in the "Large Error" Case*. Cryptology ePrint Archive, Report 2006/326, 2006

37. I. Kirschenbaum, A. Wool, *How to Build a Low-Cost, Extended-Range RFID Skimmer*. Cryptology ePrint Archive, Report 2006/054, 2006
38. O. Kömmerling, M.G. Kuhn, in *Design Principles for Tamper-Resistant Smartcard Processors*. Proceedings of the USENIX Workshop on Smartcard Technology on USENIX Workshop on Smartcard Technology, Chicago, IL, 10–11 May 1999
39. P.C. Kocher, in *Timing Attacks on Implementations of Diffie-Hellman, RSA, DSS, and Other Systems*. 16th Annual International Cryptology Conference, Santa Barbara, CA, USA, Proceedings, 18–22 Aug 1996. Lecture Notes in Computer Science, vol. 1109 (Springer, 1996), pp. 104–113
40. C.H. Lim, T. Kwon, in *Strong and Robust RFID Authentication Enabling Perfect Ownership Transfer*. 8th International Conference on Information and Communications Security (ICICS), Raleigh, NC, USA, 4–7 Dec 2006. Lecture Notes in Computer Science, vol. 4307 (Springer, 2006), pp. 1–20
41. S. Mangard, E. Oswald, T. Popp, *Power Analysis Attacks Revealing the Secrets of Smart Cards*. (Springer, Berlin, 2007)
42. D. Molnar, D. Wagner, in *Privacy and Security in Library RFID: Issues, Practices, and Architectures*. Proceedings of the 11th ACM Conference on Computer and Communications Security, Washington, DC, USA, 25–29 Oct 2004 (ACM Press, 2004), pp. 210–219
43. M. Neve, E. Peeters, D. Samyde, J.-J. Quisquater, in *Memories: A Survey of Their Secure Uses in Smart Cards*. Proceedings of the Second IEEE International Security in Storage Workshop, Washington, DC, USA, 31 Oct 2003 (IEEE Computer Society, 2003), pp. 62–72
44. C.Y. Ng, W. Susilo, Y. Mu, R. Safavi-Naini, in *New Privacy Results on Synchronized RFID Authentication Protocols Against Tag Tracing*. Proceedings of ESORICS, Saint Malo, France, 21–25 Sept 2009. Lecture Notes in Computer Science, vol. 5789 (Springer, 2009), pp. 321–336
45. C.Y. Ng, W. Susilo, Y. Mu, R. Safavi-Naini, RFID privacy models revisited. In Jajodia and Lopez [28], pp. 251–256
46. NXP Semiconductors. MIFARE Application Directory (MAD) — List of Registered Applications. http://www.nxp.com/acrobat/other/identification/mad_overview_042008.pdf, Apr 2008
47. NXP Semiconductors. MIFARE Smartcard ICs. http://www.mifare.net/products/smartcardics/, Sept 2008
48. Octopus Holdings. Web site of Octopus Holdings. http://www.octopus.com.hk/en/, June 2008
49. S. Micali, O. Goldreich, S. Goldwasser, How to construct random functions. J. ACM **33**(4), 792–807 (1986)
50. M. Ohkubo, K. Suzuki, S. Kinoshita, in *Cryptographic Approach to "Privacy-Friendly" Tags*. Presented at the RFID Privacy Workshop (MIT, Cambridge, MA, 15 Nov 2003); rfidprivacy.ex.com/2003/agenda.php
51. M. Ohkubo, K. Suzuki, S. Kinoshita, in *Efficient Hash-Chain Based RFID Privacy Protection Scheme*. International Conference on Ubiquitous Computing (UbiComp), Workshop Privacy: Current Status and Future Directions, Tokyo, Japan, 11–14 Sept 2005
52. K. Ouafi, R. Overbeck, S. Vaudenay, On the security of HB# against a man-in-the-middle attack. in *Advances in Cryptology — ASIACRYPT 2008*, ed. by J. Pieprzyk. 14th International Conference on the Theory and Application of Cryptology and Information Security, Melbourne, Australia, 7–11 Dec 2008, Proceedings. Lecture Notes in Computer Science, vol. 5350 (Springer, 2008), pp. 108–124
53. R.-I. Paise, S. Vaudenay, in *Mutual Authentication in RFID: Security and Privacy*. ASIACCS'08: Proceedings of the 2008 ACM Symposium on Information, Alexandria, VA, USA, 27–31 Oct 2008, Computer and Communications Security (ACM Press, 2008), pp. 292–299
54. D.C. Ranasinghe, D.W. Engels, P.H. Cole, in *Security and Privacy: Modest Proposals for Low-Cost RFID Systems*. Auto-ID Labs Research Workshop, Zurich, Switzerland, 23–24 Sept 2004

55. É. Levieil, P.-A. Fouque, in *An Improved LPN Algorithm*. Security and Cryptography for Networks, 5th International Conference, SCN 2006, Maiori, Italy, 6–8 Sept 2006, Proceedings. Lecture Notes in Computer Science, (Springer, 2006), pp. 348–359

56. A.-R. Sadeghi, I. Visconti, C. Wachsmann, in *User Privacy in Transport Systems Based on RFID E-tickets*. International Workshop on Privacy in Location-Based Applications (PiLBA), Malaga, Spain, 9 Oct 2008

57. A.-R. Sadeghi, I. Visconti, C. Wachsmann, in *Anonymizer-Enabled Security and Privacy for RFID*. The 8th International Conference in Cryptography and Network Security, 12–14 Dec 2009, Kanazawa, Ishikawa, Japan. Lecture Notes in Computer Science (Springer, 2009)

58. A.-R. Sadeghi, I. Visconti, C. Wachsmann, in *Location Privacy in RFID Applications*. Privacy in Location-Based Applications — Research Issues and Emerging Trends. Lecture Notes in Computer Science, vol. 5599 (Springer, Aug 2009), pp. 127–150

59. J. Saito, J.-C. Ryou, K. Sakurai, in *Enhancing Privacy of Universal Re-encryption Scheme for RFID Tags*. International Conference on Embedded and Ubiquitous Computing (EUC), Aizu-Wakamatsu City, Japan, Aug 2004, Proceedings. Lecture Notes in Computer Science, vol. 3207 (Springer, 2004), pp. 879–890

60. S.P. Skorobogatov, R.J. Anderson, in *Optical Fault Induction Attacks*. 4th International Workshop on Cryptographic Hardware and Embedded Systems (CHES 2002), Redwood Shores, CA, USA, 13–15 Aug 2002, Revised Papers. Lecture Notes in Computer Science, vol. 2523 (Springer Verlag, 2002), pp. 31–48

61. B. Song, C.J. Mitchell, *RFID Authentication Protocol for Low-Cost Tags*. Proceedings of the First ACM Conference on Wireless Network Security, Alexandria, VA, USA, 31 Mar-2 Apr 2008 (ACM Press, 2008), pp. 140–147

62. Sony Global. Web site of Sony FeliCa. http://www.sony.net/Products/felica/, June 2008

63. Spirtech. CALYPSO functional specification: Card application, version 1.3. http://calypso.spirtech.net/, Oct 2005

64. G. Tsudik, in *YA-TRAP: Yet Another Trivial RFID Authentication Protocol*. Proceedings of the 4th Annual IEEE International Conference on Pervasive Computing and Communications Workshops, Pisa, Italy, 13–17 Mar 2006 . Lecture Notes in Computer Science, vol. 2802 (IEEE Computer Society, 2006), pp. 640–643

65. P. Tuyls, L. Batina, in *RFID-Tags for Anti-counterfeiting*. The Cryptographers' Track at the RSA Conference, San Jose, CA, USA, 13–17 Feb 2006, Proceedings. Lecture Notes on Computer Science, vol. 3860 (Springer, 2006), pp. 115–131

66. P. Tuyls, B. Škorić, Tom Kevenaar (eds.), *Security with Noisy Data — On Private Biometrics, Secure Key Storage, and Anti-Counterfeiting* (Springer, New York, NY, 2007).

67. S. Vaudenay, in *On Privacy Models for RFID*. 13th International Conference on the Theory and Application of Cryptology and Information Security, Kuching, Sarawak, Malaysia, 2–6 Dec 2007 Proceedings. Lecture Notes in Computer Science, vol. 4833 (Springer, 2007), pp. 68–87

68. S.A. Weis, S.E. Sarma, R.L. Rivest, D.W. Engels, in *Security and Privacy Aspects of Low-Cost Radio Frequency Identification Systems*. 1st International Conference on Security in Pervasive Computing, Boppard, Germany, 12–14 Mar 2003 Revised Papers. Lecture Notes in Computer Science, vol. 2802 (Springer, 2003), pp. 50–59

Part VI
Hardware-Based Security Architectures and Applications

Authentication of Processor Hardware Leveraging Performance Limits in Detailed Simulations and Emulations

Daniel Y. Deng, Andrew H. Chan, and G. Edward Suh

1 Introduction

As the need for secure and trusted computation escalates, hardware architecture, in addition to traditional software techniques, is playing an increasingly important role in securing computer systems. Hardware serves as a foundation for trust in software; software security mechanisms can be compromised if hardware is insecure. Moreover, trusted hardware is starting to provide new security features. Recent Intel microprocessors are enhanced with Trusted eXecution Technology (TXT) [8] and many computing systems are already equipped with a trusted platform module (TPM) [22]. Researchers have also proposed "secure processor" solutions including XOM [12] and AEGIS [20] that provide hardware protection against physical tampering as well as software exploits.

In addition to enhancing the security of individual systems, secure processors also enable trusted applications over the network *if they can be authenticated and trusted by a remote system*. A system can be trusted to protect sensitive information and perform operations correctly based on trusted hardware even when its operator or environment cannot be trusted. As an example, imagine an Internet banking application where a user performs sensitive financial transactions on a terminal. In traditional systems, the banking server can only assume that a user uses a secure terminal even though the terminal may be compromised. If the terminal is equipped with a secure processor, the processor can attest software running on the terminal and guarantee an untampered execution with its security mechanisms. In a similar fashion, the secure processor can also enable certified distributed computation on the Internet, trusted peer-to-peer systems, trusted mobile agents, a strong form of digital rights management (DRM), etc.

To be trusted, secure hardware must convince other systems that they are indeed interacting with *authentic hardware of a trustworthy design*, not a software emulator or untrustworthy hardware. As an example, without authentication of TPM hardware, a virtual machine can simply pretend to be a TPM by emulating its external

D.Y. Deng (✉)
Cornell University, Ithaca, NY, USA
e-mail: dyd2@cornell.edu

A.-R. Sadeghi, D. Naccache (eds.), *Towards Hardware-Intrinsic Security,* Information
Security and Cryptography, DOI 10.1007/978-3-642-14452-3_14,
© Springer-Verlag Berlin Heidelberg 2010

behavior. Obviously, a simple identifier such as a serial number is insufficient for the authentication purpose because it is easy to forge. This chapter aims to enable bootstrapping of trust between remote systems by checking the authenticity of hardware.

Today's approach to authenticate an unknown system relies on the public key cryptography and a certificate authority (CA). In this approach, secure hardware contains private/public key pairs, where the private portion never leaves the chip. If an authentic public key is known, the corresponding hardware can be checked using a private key signature. For arbitrary computers on the Internet, however, determining whether a public key indeed belongs to trusted hardware is a difficult challenge and requires a certificate authority (CA) in today's approach. For example, to enable authentication of secure processors, a CA needs to obtain a public key of each processor at a trusted location where it knows the processor is authentic. Then, the CA creates a certificate by signing the processor's public key with its private key. In the field, a party who wants to authenticate an unknown system must first trust the CA, obtain the public key of the CA, and verify the processor's certificate to ensure that the public key belongs to a trusted secure processor.

Unfortunately, this certificate-based approach faces a number of limitations [6] when applied to hardware (processor). First, there is a serious concern for privacy; activities done with a particular processor can be linked together because one public key pair is used for authentication. Second, if a certificate is issued for a wrong public key or a private key becomes exposed, the authentication scheme is broken and an adversary can easily impersonate the corresponding secure processor. Finally, the centralized CA introduces cost and scalability issues. Today, for websites, companies such as Verisign charge website owners for their service as a trusted party. Requiring such costly services can severely limit the applications of secure processors.

This chapter proposes to directly check the authenticity of hardware based on its low-level implementation details instead of relying only on a certificate. More specifically, our approach exploits the fact that the microarchitecture of modern high-end processors, including speculative and out-of-order execution mechanisms, branch prediction, memory hierarchy is very complex and different for each processor model. In our approach, a secure processor is challenged to provide a checksum that depends on cycle-by-cycle activities of its internal micro-architectural mechanisms for a given code within a time limit. Even after years of research in developing fast simulation technologies for design space exploration and design validation, accurate simulation or emulation of processor microarchitecture is still extremely slow compared to real hardware. In fact, the gap between the hardware speed and the simulation speed is widening as the number of transistors on a chip increases exponentially. Moreover, producing a high-end microprocessor is becoming prohibitively expensive as processor complexity and fabrication costs increase. As a result, only an authentic secure processor hardware can compute a correct checksum fast enough.

This proposed technique complements the traditional certificate approach. There is no privacy concern because the authentication process only reveals that a processor is a particular model, but not which particular instance. The processor can easily have many independent key pairs. Also, impersonating a secure processor is difficult even if a legitimate private key or all of the microarchitecture details are known; an adversary still needs to compute the checksum as fast as real hardware. Finally, our approach enables secure processors to be introduced in a distributed fashion without relying on a few centralized certificate authorities.

Detailed microarchitecture simulations and an RTL implementation demonstrate that the proposed approach is indeed viable. The hardware extension to compute the microarchitecture checksum is simple and only requires negligible amounts of additional hardware resources. Moreover, even a small difference in the processor's microarchitecture results in significant deviation in the checksum, therefore an adversary will not be able to use other processors or a stripped down FPGA implementation to impersonate a secure processor. Also, even with recent advances in simulation technologies, simulations are still multiple orders of magnitude slower than real hardware; an adversary will not be able to use simulations to obtain valid checksums fast enough. We believe that the only viable attack on the proposed approach is to build a clone of the secure processor that has an identical microarchitecture and comparable performance but without the security features of an authentic secure processor, which is prohibitively difficult and expensive for most adversaries.

The rest of the chapter is organized as follows. Section 2 outlines our threat model and assumptions. Section 3 describes our proposed authentication approach. Sections 4 and 5 discuss the architectural mechanisms and challenge programs that are required to realize the approach. Section 6 evaluates the overhead and the security of our scheme through simulations and an RTL implementation. Section 7 compares related works and Sect. 8 concludes.

2 Threat Model

Figure 1 illustrates the security challenge that this chapter addresses. As shown in the figure, a verifier (V) interacts with a target secure processor (T) through an untrusted network. The target system (T) contains a private/public key pair (SK$_T$, PK$_T$), which is unique for each secure processor instance. We assume that a secure processor can internally generate private–public key pairs in a way that the private keys are only known to the processor hardware. For example, a processor can use a

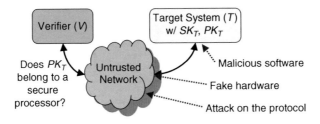

Fig. 1 The challenge and the threat model

hardware random number generator to produce a unique key as in TPMs [22] or use other methods such as physical unclonable functions [19].

The goal of the verifier is to ensure that the target system contains secure processor hardware of a particular model and obtain its public key (PK_T). Note that our authentication scheme only aims to associate a key with a particular processor model, but *not* to a particular instance. Once a public key is bootstrapped, the verifier can later use the public key and ensure that it is communicating with the same instance again. This chapter focuses only on the initial bootstrapping of trust and does not discuss additional security features that utilize the bootstrapped public key. Previous studies in secure architecture provide various mechanisms such as attestation, secure execution environments, encrypted storage, secure I/O to enable secure applications [8, 12, 20, 22].

Our threat model trusts manufacturer to properly implement security features and not to produce insecure clones. We assume that a secure processor chip is protected from physical tampering during execution so that its internal state and operation cannot be tampered with or observed directly by physical means. Therefore, an adversary cannot physically tamper with an on-chip authentication mechanism or directly extract private keys or authentic checksums in on-chip registers. We also assume that both hardware and software of the verifier is secure and do not consider attacks on the verifier. Additionally, this chapter mainly aims to handle common adversaries who may be reasonably well-funded hardware/software experts, but not government agencies with almost unlimited resources.

Given that an adversary cannot directly extract a private key from an authentic secure processor, a successful attack must fool the verifier to trust a public key that belongs to the adversary or an insecure processor. We assume that an adversary may control the network and may own an authentic secure processor chip, providing the adversary with three possible avenues to pursue. First, an adversary can monitor and possibly even redirect the communication between the verifier and the target system. Second, an adversary can install malicious software on the secure processor that it controls. However, we assume that the secure processor has an attestation mechanism as in TPMs and AEGIS so that the verifier can detect the malicious software once a legitimate public key is bootstrapped. Finally, an adversary can use various hardware, such as different processors, FPGAs, and custom chips, to impersonate an authentic secure processor.

3 Authentication Approach

The goal of a verifier (V) is to check that a public key PK_T from the target system (T) belongs to an authentic hardware, not a simulator or emulator, of a particular model. Here, we assume that an authentic secure processor has a private key that is only known to itself.

Our approach leverages the complex cycle-by-cycle low-level microarchitecture state of modern high-end processors. Effectively, a microprocessor pipeline is used as a complex function that maps a set of challenges to responses based on the implementation details. In our context, a challenge is a program or a test vector to run on the processor, and a response is a checksum computed based on cycle-by-cycle internal processor operations. We call this function, microarchitecture signature function (MSF). For security, the MSF must be different for each processor model and each challenge, be efficiently computable only by an authentic processor, and be difficult to duplicate or simulate quickly. For usability, the MSF must be deterministic and consistent across all instances of the same model.

Due to optimization techniques for instruction level parallelism such as speculation and reordering, the cycle-by-cycle state of a processor pipeline during a program execution will be different for each processor design. For example, an in-order pipeline is noticeably different from an out-of-order processor pipeline with speculation techniques even when both processors implement the same instruction set architecture. Therefore, the checksum computed for the internal processor functions will be different for each processor microarchitecture and each program and can serve as a fingerprint of a microprocessor model.

At the same time, this fingerprint (checksum) is difficult to forge. Only authentic secure processors that implement a particular microarchitecture can compute correct checksums quickly and efficiently. The microarchitecture of a modern high-performance processor is very complex and becoming even more complex as the number of transistors increases. Due to this complexity, accurate cycle-by-cycle simulations of internal processor operations are extremely slow and building a processor clone of a particular microarchitecture is very difficult and expensive.

To prevent an adversary from simply using an authentic secure processor to obtain a valid response and impersonate the processor, the interface to the MSF must be restricted to legitimate uses. In our design, a processor provides the following three interfaces to software layers:

- MSF_SIGN(challenge): Produce $H(\text{PK}_T \| response)$, the hash of the processor's public key and the response for a given challenge. This operation is used by the target system.
- MSF_VERIFY(challenge, signature, PK): Check if a given signature matches $H(PK \| response)$ for a given challenge. Returns TRUE or FALSE.
- MSF_CRP(): Return a Challenge–Response Pair (CRP) for a randomly chosen challenge.

Figure 2 shows our authentication process, with an example protocol. First, a verifier (V) initiates the process by sending an *AuthRequest* message to a target system (T), which replies with its public key PK_T and its hardware model, claiming to be a particular processor design. To validate the claim, the verifier sends a challenge to the target. The challenge could be a full program or simply a seed to a program (see Sect. 5). The target replies with a signature from MSF_SIGN: runs the challenge and hashes (cryptographic hash such as SHA256) the response with its public key to bind the public key with the response. The verifier checks the hash using either its built-in MSF_VERIFY function or a previously obtained challenge–response pair (CRP).

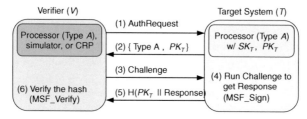

Fig. 2 The authentication protocol based on microarchitectural features

To prevent an adversary from using a simulator to obtain the response, the verifier sets a time limit on the response from the target. If the response is not received within the predetermined limit, the authentication fails. The time limit is set to be the time required for an authentic hardware to run the challenge plus the worst-case network latency. Note that this limit even with the network latency can be made much smaller than the time required for the fastest simulator to compute the correct response by increasing the execution time for a challenge. The verifier also changes the challenge for each authentication to ensure that an adversary cannot replay a recorded response.

To prevent denial of service attacks, the target system needs to limit the execution time of a challenge and the frequency of the authentication requests. The challenge programs must also be properly isolated so that they cannot perform sensitive system calls. Operating systems normally perform such isolation using software layers and virtual machine techniques.

Figure 3 compares our approach with a traditional certificate-based approach. In the certificate-based framework, a trusted certificate authority (CA) issues a certificate for each instance of a secure processor. The CA must somehow establish trust with each and every processor instance before it can issue a certificate. As a result, a few centralized CAs must provide service for millions of devices, each CA presents a single point of failure and this brings scalability and availability issues.

The proposed scheme enables processor authentication in a more distributed fashion as shown in Fig. 3b. First, a verifier can independently authenticate a target without a trusted third party if it has access to any authentic secure processors of the target's model (use MSF_VERIFY). Similarly, *if* the vendor releases an accurate simulator for a particular processor model, anyone can authenticate any instance of the model in a completely distributed fashion. Even if a verifier does not have an authentic processor of the target's model, the processor vendor or *any secure processor of the same model* can provide a randomly selected challenge–response pair (CRP) using the MSF_CRP primitive. Unlike a traditional CA, the trusted party here does not need to bootstrap each target instance. The verifier can save and use the CRP for an authentication process at any time. Note that this approach of distributing CRPs does not cause a security problem because obtaining a random CRP does not help an adversary in predicting the correct response for a particular challenge asked by a verifier.

In addition to its distributed nature, the proposed scheme preserves privacy because a single processor can easily use many different public keys. Also, the

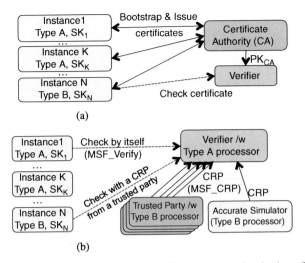

Fig. 3 The authentication infrastructures. (**a**) Certificate-based authentication, (**b**) MSF-based authentication

scheme enhances the security of traditional approaches by requiring authentic hardware. In traditional certificate-based systems, an adversary can easily impersonate an authentic processor if he or she can extract a private key of the processor even if the attack destroys the processor. However, in our approach, obtaining an authentic private key alone is not sufficient without also having a functional secure processor that can compute the checksum fast enough.

4 Hardware Design

This section describes hardware extensions to construct the MSF on modern high-end processors using internal pipeline checksums. For security, the checksums must reflect details of architectural features and be sensitive to even small differences in them. For reliability, the checksums must be consistent so that all processors of the same model can generate the same response.

4.1 Microarchitectural Features

Processor microarchitecture is defined by the processing core and the memory hierarchy. In modern superscalar processors, the core architecture is largely determined by the mechanisms for speculative and out-of-order execution and the memory hierarchy is mostly determined by cache architecture. We design our checksum mechanisms taking into consideration the internal operations that can effectively represent these key microarchitectural features.

Modern high-performance processors speculatively fetch and execute instructions by predicting the control flow and recovering from an incorrect prediction.

Because the branch prediction accuracy has a critical impact on the processor performance, the branch predictor is continuously improving and each architecture design uses a unique predictor in terms of its algorithm, size, and organization. For a difficult-to-predict stream of branches, different predictors are very likely to produce significantly different traces of correct and incorrect predictions and lead to different traces of fetched instructions. Therefore, a cycle-by-cycle record of fetched instructions exposes how support for speculative execution is implemented in a processor.

Modern processors also allow multiple instructions to be dispatched in each cycle in an out-of-order fashion in order to maximize the performance. Each processor architecture will differ in the exact mechanism to support out-of-order execution depending on the size of the buffer to store waiting instructions, how aggressively loads are speculatively reordered, etc. Hence, tracking the dispatched instructions exposes the provisions for out-of-order execution unique to each processor.

Caches improve performance by keeping recently used data close to the processing core so that they can be accessed more quickly. Cache design parameters have a significant impact on the latency of instruction and data accesses. Cache hit and miss traces will vary greatly depending on the size and organization of the cache. These variations are reflected in the trace of the fetched and issued instructions of the processor pipeline. Cache hits result in the fetch of new instructions and free space in the load/store queues while misses can cause stalling or idle slots in the pipeline.

4.2 Checksum Computation

To incorporate effects of various microarchitectural mechanisms, our checksum includes the fetched and dispatched instructions in each cycle. While the checksums can easily include more details in each pipeline stage, we found the fetched and dispatched instructions (augmented with cycle counts) capture almost all of the plausible microarchitecture discrepancies between any two processor designs. For example, a cycle-by-cycle record of fetched instructions reflect the speculation mechanisms such as branch predictors and instruction cache hits/misses. Similarly, the record of dispatched instructions shows the behavior of dynamic scheduling and data caches.

Figure 4a shows a modern superscalar processor augmented with checksum units to monitor fetched and dispatched instructions and a hash engine. For the MSF checksums, a processor is augmented with a checksum unit for each fetch and dispatch slot. For example, a 32-bit four-way superscalar processor will produce four 32-bit checksums for fetched instructions and four 32-bit checksums for dispatched instructions. The checksum units are zeroed and initiated at the start of a challenge program. At the end of the challenge, all checksums are combined by a cryptographic hash to generate the final response.

Figure 4b shows how each checksum is computed, which applies to both fetched and dispatched instructions. In each cycle, the checksum circuit either ADDs or XORs the incoming 32-bit value. The toggling between an ADD and an XOR allows

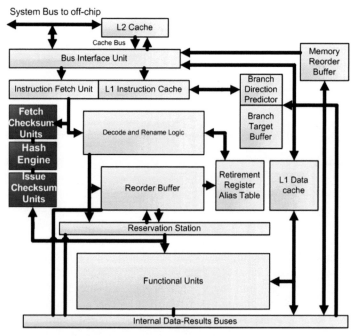

(a) A modern superscalar processor augmented
with hardware enhancements for secure authentication.

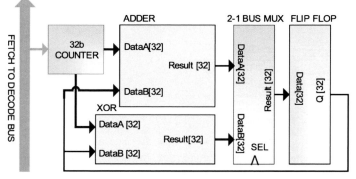

(b) A checksum circuit for monitoring fetched instructions.

Fig. 4 Processor extensions for MSF and an individual checksum unit. (**a**) A modern superscalar processor augmented with hardware enhancements for secure authentication. (**b**) A checksum circuit for monitoring fetched instructions

the results of the accumulation to be strongly ordered; deviation in the order of the incoming values would alter the checksum results. In the checksum, idle cycles in addition to real instructions are important because they may indicate interesting events such as a cache miss or branch misprediction. To account for these idle

cycles, our checksum units include the number of idle cycles encountered since
the start of a challenge on each idle cycle. Note that an adversary must know exact
cycle-by-cycle pipeline operations in order to compute a correct checksum. As a
result, checksums using simple additions and XORs are sufficient without expensive
cryptographic hash functions.

4.3 New Instructions

We introduce five new instructions summarized in Fig. 5 to support the three MSF
operations. First, the MSF_BEGIN and MSF_END instructions have the processor internally generate a response for a challenge program. The MSF_BEGIN instruction pre-
pares the processor for a deterministic execution and initiates the checksum computation. The instruction first disables interrupts, loads challenge code and data into
the L2 cache, and sets the TLB to preclude off-chip accesses. Then, the processor
is bootstrapped into a deterministic state by clearing any non-deterministic state
including the pipeline, L1 caches, cache controller buffers, and branch predictor
tables and starts the checksum units. The MSF_END stops the checksum units, compute a response by hashing checksums, stores the response in an internal register, and enables interrupts. Second, the MSF_SIGN and MSF_VERIFY instructions
use the stored response to provide MSF operations. The MSF_SIGN instruction
outputs the hash of the concatenation of its public key and the response. The
MSF_VERIFY instruction allows the hash from another processor in response to a
particular challenge to be checked. The instruction gets a public key (PK_T) and a
hash ($H(PK_T||\text{Response})$) as arguments, computes the hash with its own response,
and checks if the two hashes match. Finally, the MSF_CRP instruction runs a hard-
coded challenge code with a random seed and returns its response (see Sect. 5 for a
challenge code design).

4.4 Non-determinism

For reliable authentication, a processor must produce a consistent response to a
challenge every time the same challenge runs. Because a microprocessor is essen-

Instruction	Operands	
MSF_BEGIN	start & end addresses	Disable interrupts, put the processor in a deterministic state, and start the checksum units and the cycle counter
MSF_END	none	Stop the checksum units and the cycle counter, compute a response, and re-enable interrupts
MSF_SIGN	none	Outputs the hash of its public key and the response
MSF_VERIFY	public key, hash value	Check if the provided hash value matches the hash of the given public key with an internal response
MSF_CRP	none	Outputs a response for a challenge code with a randomly selected seed

Fig. 5 MSF instructions

tially a finite state machine, its cycle-by-cycle behavior will be deterministic if non-deterministic state and signals are eliminated. In fact, microprocessors often support short but deterministic execution for post-silicon validation. A previous study also demonstrated a longer deterministic execution for modern microprocessors [5].

The two sources for non-determinism in current microprocessors are inconsistent initial state and inconsistent arrival times of signals during an execution. In this work, to ensure consistent initial state, the processor sets up its state on MSF_BEGIN. During execution, there are two sources of non-determinism: off-chip memory accesses and interrupts. We eliminate the off-chip memory accesses by carefully designing the challenge program to only access on-chip memory and avoid accessing memory through off-chip buses that cross clock domains. Alternatively, we can also make off-chip accesses slow but deterministic. The processor disables maskable interrupts and ignores I/O for the duration of challenge execution. Extremely rare exceptions such as a power failure can still occur during challenge program execution. In this case, the processor simply aborts the checksum computation and the authentication needs to be retried later. Avoiding off-chip accesses also allows the processors with one microarchitecture design to produce the same response regardless of the operating frequency.

Besides the traditional sources of non-determinism discussed above, future advances in microprocessors may introduce new sources of non-determinism. We believe that all these issues can be effectively handled for a relatively short challenge execution. First, modern microprocessors often rely on dynamic voltage and frequency scaling technologies to reduce power and heat dissipation. For a short period of the checksum computation, these features can be "turned off" just like interrupts. Second, multi-core processors allow multiple applications to simultaneously share on-chip resources such as caches. To avoid non-deterministic interference, the hardware can ensure that only one core is active when executing a challenge. Finally, future microprocessors may use dynamic soft-error detection and recovery mechanisms, whose behavior depends on random soft errors. As these techniques are still in the development stage, we cannot provide detailed techniques to prevent non-determinism from soft errors here. As a broad overview, however, a checkpoint technique applied to recovering processor state on an error can be extended to checkpoint and deterministically restart the checksum units so that the checksum remains deterministic.

5 Challenge Program

The challenge program must satisfy a few properties to be effective in distinguishing between processor models. First, to generate unique responses, the challenge program must be able to exercise the functionality of a processor with high coverage so that even small differences in microarchitecture can be captured. Second, new challenges must be relatively easy to create so that the verifier can use a different challenge on each authentication attempt. Especially, processor hardware must be able to generate a random challenge for the MSF_CRP operation. Lastly, if hardware cannot support deterministic off-chip accesses, the challenge program must be small and self-contained to fit into an on-chip cache.

To differentiate microarchitectures, the challenge program should present many different patterns to a microarchitecture mechanism. At the same time, to make it difficult for an adversary to predict checksums, the patterns should result in high variance in microarchitectural behavior. In our design, we make judicious use of random patterns to achieve both objectives. For example, a random traversal of data can produce memory access patterns that result in unpredictable cache hits and misses depending on the cache configuration. Similarly, branches with random outcomes and targets of many different locations result in significantly varying instruction cache performance and varying predictions from different branch prediction algorithms.

In our design, the challenge program takes a seed for a pseudo-random number generator (PRNG) and performs repeated sequences of pseudo-random operations: memory accesses to pseudo-random locations to differentiate data memory organizations, pseudo-random branches to differentiate speculative execution mechanisms, jumps to pseudo-random locations to differentiate instruction memory organizations and BTBs, and long sequences of instructions with pseudo-random dependencies and operations to differentiate out-of-order execution schemes. In addition to differentiating microarchitectural features, the use of a seeded pseudo-random function has an added benefit that it is very easy to generate many different challenges by simply using a different seed. To support MSF_CRP, a processor can use fixed challenge code but with a random seed.

Figure 6 shows the pseudo-code of our challenge program. The challenge program takes a seed as an argument, which is used to seed the T-function [10], our pseudo-random number generator (PRNG). This PRNG is then used to initialize a small array, whose size can be set to control cache miss rates. In the main body, the challenge program generates a new random number x and reads the array element determined by the random number. Following the read, the program jumps to a random location in the program based on x (the switch statement). Then, a conditional branch based on a random condition executes. Finally, a number of unique instructions are placed in both branch paths. The instructions have random dependencies and operations to differentiate steering mechanisms in out-of-order processors. This read-jump-branch-compute sequence is then repeated a number of times to build a long sequence of operations and to sufficiently prolong the execution time to mask the worst-case network latency.

Note that the results of the challenge program, which are easy to compute on any processor, are unimportant in our context because the hash response is computed from the checksums that collect the internal pipeline operations. We also note that the challenge code design can leverage knowledge from processor validation and testing as they face similar goals of achieving high design coverage. Randomly generated test vectors are commonly used to elicit unexpected behaviors. Test programs can also be customized for a particular processor design [2].

```
void challenge(int seed) {
    MSF_BEGIN;              \\ prepare and start checksums
    x = seed;
    \\ create an a[] of size N
    do {
        x = x + ((x*x) | 5); \\ generate a new random number
        v = a[x & (N - 1)];   \\ load a random address
        switch(x & (M - 1)) {
        case 0:
            if((x & 32) == 0) { \\ branch on a random condition
            \\ Z random instructions with random dependencies
            } else {
            \\ Z random instructions with random dependencies
            } break;
        case 1:
            if((x & 32) == 0) { \\ branch on a random condition
            \\ Z random instructions with random dependencies
            } else {
            \\ Z random instructions with random dependencies
            } break;
            ...
        case M-1:
            if((x & 32) == 0) { \\ branch on a random condition
            \\ Z random instructions with random dependencies
            } else {
            \\ Z random instructions with random dependencies
            } break;
        }
    } while (j < N);    \\ do N times
    MSF_END;     \\ signals checksum units and cycle counter to stop
}
```

Fig. 6 The implemented challenge program. For our simulations, $Z=20$, $M=1,024$, $N=65,536$

6 Evaluation

6.1 Overheads

The proposed hardware extension has negligible area overheads, only adding checksum units, a cryptographic hash engine, and a cycle counter. Each checksum circuit requires a 32-bit adder, 32 XOR gates, a 32-bit 2-to-1 multiplexer, a latch, and a register. The adder is the largest and most complex element in the circuit and would incur the bulk of the area and power overheads. Prior work [23] showed that a 5 GHz, 32-bit adder in the 130 nm technology consumed about 0.03 mm^2. Therefore, eight adders (0.24 mm^2) represent less than 0.6% of the area of a recent four-way 130 nm Intel microprocessor (Pentium M) that had a core area of 39 mm^2 [4]. For the hash engine, recent ASIC implementations [15] show that a hardware SHA-256 engine requires in the order of 40–60 K transistors, an overhead that is negligible when compared to modern processors that total hundreds of millions of transistors. Further, the hash engine does not need to be high performance as the MSF instructions are only rarely used. In addition to the checksum and hash engine blocks, we also augment the execution core with the checksum units and the hash engine as well as the new instructions. However, this control logic should be fairly simple and should not result in any significant area overheads.

Performance of the processor will not be adversely affected by the checksum units added to the processor pipeline. The checksum units do not lie on the critical path of computation and operate in parallel. While monitoring fetched and dispatched instructions do represent an extra load capacitance that can potentially increase the minimum clock period, this problem can be addressed using low-input-capacitance buffers between the preceding pipeline stages and the checksum units to minimize the additional loading. Finally, the additional adders and hash engine consume static and dynamic power during a challenge execution. However, since the challenge–response authentication scheme is used infrequently and the added circuits can be aggressively clock-gated during normal system use to minimize the power overhead.

6.2 Effectiveness

To study its effectiveness in differentiating microarchitectures, the proposed scheme was implemented in superscalar processors built on a simulator (SESC) [14] and an RTL model (IVM) [24]. For each processor, we compared differences in cycle-by-cycle behavior when a single architectural parameter is changed.

Table 1 shows the percentage of cycles that a different instruction was observed by each checksum unit when a single microarchitecture parameter was varied for a four-way superscalar processor modeled in SESC. Fetch Avg and Fetch Min show

Table 1 Differences (%) in cycle-by-cycle fetched and dispatched instructions when varying a single microarchitecture parameter

Varied parameters		Fetch		Dispatch	
('+': change from the baseline)		Avg	Min	Avg	Min
L1 I-cache					
- latency (clks)	1, 2, 3, 4	22.9	22.4	23.3	22.8
- size (KB)	4, 8, 16, 32, 64	22.4	21.2	22.7	21.6
- associativity	1, 2, 4, 8, 16	21.9	21.1	22.3	21.4
L1 D-cache					
- latency (clks)	1, 2, 3, 4	22.9	22.6	23.3	22.9
- size (KB)	4, 8, 16, 32, 64	21.0	20.7	21.3	21.1
- associativity	1, 2, 4, 8, 16	21.2	21.1	21.6	21.5
L2 latency(clks)	6, 8, 10, 12, 14	22.7	21.4	23.0	21.7
BTB					
- size (entries)	512, 1K, 2K, 4K	21.6	21.1	21.2	20.8
- associativity	1, 2, 4	21.0	20.8	21.1	20.9
Branch predictor	YAGS, gskew, gshare, hybrid	23.7	23.3	23.4	23.0
Fetch widths	2, 3, 4, 5, 6	19.3	16.5	22.4	20.1
Retire widths	2, 3, 4, 5, 6	22.8	21.1	23.1	21.4
Issue widths	2, 3, 4, 5, 6	23.0	21.5	29.8	22.1
Int FU latency	+0, +1, +2, +3	20.9	20.0	21.2	20.3
FP FU latency	+0, +1, +2, +3	22.1	21.5	22.4	21.8
ROB size	80, 96, 112, 128, 144, 160	23.0	22.3	23.4	22.9

the average and the minimum differences for the fetched instructions, respectively, and Dispatch Avg and Dispatch Min show the average and the minimum for the dispatched instructions, respectively. Percentages appear low because checksum units often do not have valid instructions to process. The difference exceeds 90% if we only count cycles with valid instructions.

To further validate the effectiveness of our approach, we also implemented the checksum units in the Illinois Verilog Model (IVM) [24]. The IVM is a Verilog model of a superscalar, dynamically scheduled pipeline that is designed to be similar to the Alpha 21264 and the AMD Athlon. The RTL experiments varied processor parameters such as scheduler size, BTB size, choice predictor size. As shown in Table 2, the results confirm that each configuration generates a significantly different response even for the same challenge code. Also, the RTL implementation showed that inputs to checksum units indeed depend on the entire processor; it is difficult to extract a smaller circuit just to compute the checksums. Both simulation and RTL experiments demonstrate that each processor behaves very differently at the microarchitectural level even when only one configuration is changed.

Our authentication scheme also requires that each challenge for the same processor elicit unique responses. We ran 50,000 different challenge seeds on the same configuration and a different response was observed for each challenge. This shows that the responses will be unique for each challenge and difficult to predict even if responses for the other challenges were observed.

6.3 Deterministic Execution

To confirm that the microarchitecture checksums can be reliably reproduced, we studied deterministic execution of a challenge on the Verilog model (IVM) and Pentium 4. Our RTL model implemented mechanisms described in Sect. 4.4; before computing checksums, the processor clears its state such as branch history tables, caches, and the processor pipeline, and pre-loads on-chip caches and TLBs to avoid off-chip accesses. In IVM, running a challenge multiple times, with arbitrary programs in between, produced an identical response. To further verify the feasibility

Table 2 Percentage of differences in cycle-by-cycle issued instructions in IVM

Varied parameters	Min (%)	Avg (%)
Fetch queue size	17	25
Scheduler size	29	29
BTB size	14	26
Branch order buffer size	19	25
Local branch history length	15	29
Memory dependent prediction size	26	26
Load/store queue size	28	28
Global history register size	24	28
Choice predictor table size	15	20

of getting deterministic responses, we executed our challenge code on a Pentium 4 machine running Linux and measured the cycle count for each execution of a challenge program using a built-in performance counter. For deterministic execution, we disabled interrupts (CLI instruction), cleared the pipeline (CPUID), and initialized caches and TLB entries with multiple pre-runs before running the challenge. The experiment confirmed that a short execution[1] can be deterministic. Repeatedly running the challenge with a fixed seed returned consistent cycle counts and different seeds returned different cycle counts.

6.4 Security Discussion

The simulation study and the RTL implementation demonstrated that the response based on internal architecture checksum is different for each microarchitecture and each challenge. Therefore, a response uniquely *identifies* a processor model. Here, we discuss the security of the proposed scheme in *authenticating* the processor model. Given the threat model in Sect. 2, an adversary can try to attack our scheme in three different ways: tamper with the network protocol, misuse an authentic processor, and create a hardware/software clone. In the following discussion, note that an adversary's goal in our context is to fool a verifier to accept a public key PK_{Fake} that does not belong to an authentic secure processor. Therefore, to be successful, the adversary must be able to present a valid hash $H(PK_{Fake}||Response)$ for a given challenge within the time limit. Here, we discuss potential ways that an adversary can try to obtain the valid response for a given challenge.

Our threat model assumes that the adversary controls the communication between the verifier and the target. Therefore, the adversary can easily make an authentic processor fail to be authenticated by changing messages. The proposed mechanism does not prevent denial-of-service attacks. The adversary can also try man-in-the-middle attack to simply relay messages and make the verifier accept a public key of another authentic processor. However, this attack is not an issue because higher level protocols can check if the target system indeed owns the corresponding private key. Finally, an adversary may try to predict the challenge that a verifier will use based on previous history. If the guess is correct, the adversary can violate the time limit and possibly pre-compute the response with a simulator. However, this attack is infeasible because the verifier randomly selects a challenge.

Given there is no plausible attack on the protocol, an adversary can try to use an authentic processor to extract a response. However, none of the three MSF instructions allows an adversary to impersonate the authentic processor. The MSF_SIGN instruction gives out a hashed response, $H(PK_T||Response)$. However, an adversary cannot replace PK_T with PK_{Fake} or obtain Response from the hash given a

---

[1] The challenge code ran from 20 ms to 1 s.

non-malleable one-way hash function. The `MSF_VERIFY` instruction only reveals whether a hashed response is correct or not. Finally, the `MSF_CRP` does not let an adversary choose a particular challenge that is asked by the verifier. Note that different processor models cannot be used either because they generate different checksums.

Without exploiting existing authentic processors, an adversary must compute the correct response directly. However, this approach requires an adversary to reverse engineer the exact microarchitecture of a target processor, which is quite difficult and expensive. Even if the detailed microarchitecture is known, the adversary still needs to compute the exact cycle-by-cycle pipeline operations *as fast as* the authentic processor. Years of research in performance modeling and processor verification suggest that this proposition is practically infeasible.

Due to the complexity of modern processors, accurate simulations of internal microprocessor operations are extremely slow. Even low-fidelity academic simulators run three to four orders of magnitude slower than the actual processor being modeled. As an example, the SESC simulator used in our evaluations resulted in a 2200X slowdown in comparison to executing the challenge on actual hardware. More accurate performance simulators used in industry often are about a million times slower than the target [13]. Moreover, even the industry performance simulators are not sufficient to obtain accurate cycle-by-cycle checksums because they often ignore some details as their goal is to *explore* the design space. Faithful RTL simulations that are used in functional verification will be accurate enough, but are even slower (only run at tens of Hertz [21]). Also, our RTL studies showed that inputs to checksum units depend on the entire RTL model; therefore it will be very difficult to compute the checksums using a simpler sub-circuit of the processor.

Recent advances in simulation methodology propose to use field programmable gate arrays (FPGAs) for timing models [3]. This approach has the promise of increasing simulation speeds by upward of 100 times over that of pure software simulators. An adversary can also achieve additional speed ups by running simulations on future microprocessors and FPGAs. However, for high-performance microprocessors built in the state-of-the-art fabrication technologies, accurate simulations will still be orders of magnitude slower than authentic hardware during the processor's life cycle. For example, if a processor is used for 10 years, industry-level performance simulators will still be at least 1,000 times slower at the end of the processor's lifespan.

As a result, to match the performance of actual hardware, the adversary must build a custom chip that is essentially a clone of the target microprocessor but without the security features. However, such an attack will be simply too costly for most applications and adversaries as high-end microprocessors cost enormous amounts of time and money to design and produce. The mask and fabrication costs will only increase for future process generations, and the design and verification of a microprocessor will cost even more.

7 Related Work

Public key infrastructure (PKI) based on a certificate authority (CA) is the most widely used framework for remote authentication of unknown parties. In this framework, a certificate authority (CA) signs a certificate that binds an identifier to a public key. For example, Verisign can issue a certificate that endorses the validity of the binding between a bank's public key and its website address. Trusted platform modules (TPMs) [22] adopt the traditional PKI for authenticating trusted devices. This approach, however, faces difficult challenges as discussed in Sect. 1. The proposed scheme complements the weaknesses of PKI such as privacy, security, and cost.

A simple approach to authenticate trusted hardware is to embed the same private key in all trusted devices. This approach greatly simplifies binding of a public key to trusted hardware because there is only one public key that is common to all devices. However, this approach suffers from serious security issues; if a single device is compromised and the private key is leaked, all devices are broken. As an example, Microsoft Xbox embedded one common secret key in all devices, which was extracted by a person and now any Xbox console can be reprogrammed to run any code [7].

Seshadri et al. proposed time-bound verification functions to authenticate software executing on a remote system [16, 17], which exploits the fact that malicious software will slow down an execution. However, such time bounds can only be defined for known hardware configurations. This chapter aims to verify that the hardware is a trusted model so as to enable additional software-based techniques such as theirs.

Genuinity [9] tracks side-effects of program execution to determine whether a remote system is genuine hardware or a virtual machine. Their proposal takes a software-only approach that relies on inaccurate hardware performance counters [11] and was shown to be insecure because malicious software can execute and alter the processor state without affecting the side-effects [18]. Our proposed authentication approach is secure because the checksum units are embedded *in hardware* and computed for detailed cycle-by-cycle operations; only genuine hardware can access checksums and each instruction affects the checksums.

As discussed in the chapters in Part I, physical unclonable functions (PUFs) enable secure authentication of each IC instance by mapping a set of challenges to a set of responses based on an intractably complex physical system. While both use a challenge–response protocol, PUFs address a different problem compared to the proposed scheme; PUFs enable secure authentication of a particular IC instance *once it is trusted and registered* whereas the proposed scheme addresses the question of how to trust an unknown system initially.

Research by Agrawal et al. proposed to use side-channel information to detect malicious circuits in an IC [1]. This is related to our approach as this side-channel approach also tries to determine if an unknown IC can be trusted by checking implementation details. The differences are in capabilities and applications. The side-channel approach requires physical access to an IC to measure power consumption and addresses detecting malicious circuits inserted by manufacturers. On the other hand, the proposed approach is applicable *remotely* and assumes that the authentic processors are manufactured by a trusted party.

8 Conclusion

In this chapter, we presented a new challenge–response scheme to check the authenticity of a processor based on the performance gap between authentic hardware and simulations or emulations. We augmented a modern processor pipeline with modest hardware enhancements to capture internal processor state during challenge execution to produce a unique response for each processor model. The challenge program is designed to take advantage of pseudo-random functions to produce random-looking memory accesses, branches, and various instruction sequences so that processors with different microarchitecture designs produce different internal behaviors. Experimental results demonstrate that the results from the checksum mechanisms are different for each challenge to the same microarchitecture and for the same challenge to microarchitectures varying in a single configuration parameter. These experimental results suggest that the scheme is effective in authenticating different processor models.

This chapter mainly focuses on authentication of a single high-performance processor. In the future, we plan to extend the current scheme to multi-core architectures and embedded systems. For multi-core processors, our scheme can be extended to exploit the interconnect architecture in addition to the core microarchitecture and the caches. For this purpose, a challenge can be designed as a parallel program. In the embedded systems, whose microarchitecture may be simple enough for fast simulations, we believe that the proposed approach can be extended to incorporate even lower level circuit details such as layouts.

References

1. D. Agrawal, S. Baktir, D. Karakoyunlu, P. Rohatgi, B. Sunar, in *Trojan Detection Using IC Fingerprinting*. Proceedings of IEEE 2007 Symposium on Security and Privacy, Oakland, CA, USA, 20–23 May 2007, pp. 296–310
2. A. Aharon, D. Goodman, M. Levinger, Y. Lichtenstein, Y. Malka, C. Metzger, M. Molcho, G. Shurek, in *Test Program Generation for Functional Verification of Powerpc Processors in IBM*. Design Automation Conference,1995, pp. 279–285
3. D. Chiou, H. Sunjeliwala, D. Sunwoo, J. Xu, N. Patil, in *FPGA-Based Fast, Cycle-Accurate, Full-System Simulators*. Proceedings of the 2nd Annual Workshop on Architecture Research using FPGA Platforms, Austin, TX, USA, 12 February 2006
4. H. de Vries, http://www.chip-architect.com/
5. B. Greskamp, S.R. Sarangi, J. Torrelas, in *Designing Hardware that Supports Cycle-Accurate Deterministic Replay.* Proceedings of 2006 Workshop on Complexity-Effective Design, Boston, MA, USA, 18 June 2006, 22–25

6. P. Gutmann, PKI: It's not dead, just resting. Computer **35**, 41–40 (Aug 2002)

7. A. Huang, in *Keeping Secrets in Hardware: The Microsoft Xbox Case Study*. Cryptographic Hardware and Embedded Systems - CHES Redwood City, CA, USA, 13–15 Aug 2002. Lecture Notes in Computer Science, vol. 2523 (Springer, Heidelberg, 2002)

8. Intel. Intel Trusted Execution Technology. http://www.intel.com/technology/security/, 2007

9. R. Kennel, L.H. Jamieson, in *Establishing the Genuinity of Remote Computer Systems*. Proceedings of the 12th Annual USENIX Security Symposium, Washington, DC, USA, 4–8 Aug 2003, pp. 295–308

10. A. Klimov, A. Shamir, in *A New Class of Invertible Mappings*. 4th International Workshop on Cryptographic Hardware and Embedded Systems - Revised Papers, Redwood City, CA, USA, 13–15 Aug 2002, pp. 471–484

11. W. Korn, P.J. Teller, G. Castillo, in *Just How Accurate Are Performance Counters?* Proceedings of the 2001 IEEE International Conference on Performance, Computing, and Communications, Phoenix, AZ, USA, 4–6 Apr 2001, pp. 303–310

12. D. Lie, C. Thekkath, M. Mitchell, P. Lincoln, D. Boneh, J. Mitchell, M. Horowitz, in *Architectural Support for Copy and Tamper Resistant Software*. Proceedings of the 9th International Conference on Architectural Support for Programming Languages and Operating Systems (ASPLOS-IX), Cambridge, MA, USA, 12–15 Nov 2000, pp. 168–177

13. M. Pellauer, J. Emer, A. Thiagarajan: Implementing a Partitioned Performance Model on an FPGA. http://publications.csail.mit.edu/abstracts/abstracts07/pellauer-abstract/hasim.html

14. J. Renau, B. Fraguela, J. Tuck, W. Liu, M. Prvulovic, L. Ceze, S. Sarangi, P. Sack, K. Strauss, P. Montesinos, SESC simulator, Jan 2005. http://sesc.sourceforge.net

15. A. Satoh, T. Inoue, in *ASIC-Hardware-Focused Comparison for Hash Functions MD5, RIPEMD-160, and SHS*. Proceedings of the International Conference on Information Technology: Coding and Computing (ITCC'05), Las Vegas, NV, USA, 4–6 Apr 2005

16. A. Seshadri, M. Luk, E. Shi, A. Perrig, L. van Doorn, P. Khosla, in *Pioneer: Verifying Code Integrity and Enforcing Untampered Code Execution on Legacy Systems*. Proceedings of the 20th Annual Symposium on Operating Systems Principles, Brighton, UK, 23–26 Oct 2005, pp. 148–162

17. A. Seshadri, A. Perrig, L. van Doorn, P. Khosla, in *SWATT: SoftWare-Based ATTestation for Embedded Devices*. 2004 IEEE Symposium on Security and Privacy, Oakland, CA, USA, 9–12 May 2004

18. U. Shankar, M. Chew, J.D. Tygar, in *Side Effects Are not Sufficient to Authenticate Software*. Proceedings of the 13th USENIX Security Symposium, San Diego, CA, USA, 9–13 Aug 2004, pp. 89–101

19. G.E. Suh, S. Devadas, in *Physical Unclonable Functions for Device Authentication and Secret Key Generation*. Proceedings of the 44th Design Automation Conference, San Diego, CA, USA, 4–8 June 2007

20. G.E. Suh, C.W. O'Donnell, I. Sachdev, S. Devadas, in *Design and Implementation of the* AEGIS *Single-Chip Secure Processor Using Physical Random Functions*. Proceedings of the 32nd Annual International Symposium on Computer Architecture, Madison, WI, USA, 4–8 June 2005

21. H. Tago, K. Hashimotof, N. Ikumi, M. Nagamatsu, M. Suzuoki, Y. Yamamoto, in *Importance of CAD Tools and Methodologies in High Speed CPU Design*. Proceedings of the Asia and South Pacific Design Automation Conferences (ASP-DAC 2000), Yokohama, Japan, 25–28 Jan 2000, pp. 631–633

22. Trusted Computing Group. TCG Specification Architecture Overview. http://www.trustedcomputinggroup.com/

23. S. Vangal, M.A. Anders, N. Borkar, E. Seligman, V. Govindarajulu, V. Erraguntla, H. Wilson, A. Pangal, V. Veeramachaneni, J.W. Tschanz, Y. Ye, D. Somasekhar, B.A. Bloechel, G.E. Dermer, R.K. Krishnamurthy, K. Soumyanath, S. Mathew, S.G. Narendra, M.R. Stan,

S. Thompson, V. De, S. Borkar, 5-GHz 32-bit integer execution core in 130-nm dual-V/sub T/CMOS. IEEE J. Solid-State Circuits **37**, 1421–1432 (Nov 2002)

24. N.J. Wang, J. Quek, T.M. Rafacz, S.J. Patel, in *Characterizing the Effects of Transient Faults on a High-Performance Processor Pipeline*. Proceedings of the 2004 International Conference on Dependable Systems and Networks, Florence, Italy, 28 June–1 July 2004

Signal Authentication in Trusted Satellite Navigation Receivers

Markus G. Kuhn

1 Introduction

Physical location can be an important security parameter, whether for location-based access control or to audit the whereabouts of goods and people. In outdoor applications, location is often most easily determined with a global navigation satellite system (GNSS) receiver. This means today primarily GPS [9, 11], but the list is growing (GLONASS, Galileo, Beidou/Compass, etc.). Each of these operates a constellation of Earth-orbiting satellites that broadcast a high-precision time signal, along with a low bit rate data stream (50–1,000 bit/s) that carries orbital position (ephemeris) predictions and calibration data. Receivers measure the time-of-arrival differences of at least four satellite signals and then solve a system of equations to determine both their current location and time, with accuracies of a few meters and tens of nanoseconds. Even higher accuracies can be achieved by using nearby reference receivers for calibration.

GNSS receivers may be integrated with tamper-resistant cryptographic modules, for security-critical applications where the person in possession of the device has an interest in it misreporting its location. The purpose of such devices may be to attest their current location to a remote observer, via an authentication protocol. It may also attest the device's recent location and velocity history via an authenticated recording or enable some functionality based on its location. Potential application examples include

- anti-theft tracking systems for vehicles and transport containers, which automatically alert the owner if a vehicle no longer follows its expected route;
- prisoner-tagging systems that permit probation officers to remotely monitor curfews and probation conditions;

M.G. Kuhn (✉)
Computer Laboratory, University of Cambridge, Cambridge CB3 0FD, UK
e-mail: Markus.Kuhn@cl.cam.ac.uk

A.-R. Sadeghi, D. Naccache (eds.), *Towards Hardware-Intrinsic Security,* Information
Security and Cryptography, DOI 10.1007/978-3-642-14452-3_15,
© Springer-Verlag Berlin Heidelberg 2010

- road tax and insurance fees for motor vehicles calculated in on-board units based on actual usage, with algorithms that incorporate information about speed, route, and travel times in order to take into account externalities and risk;
- road speed limits enforced electronically using on-board navigation systems that determine the current location, look up the local speed limit, and communicate that to both driver and engine controller.

Some of these applications are already deployed, others may well evolve from existing road usage or congestion charging systems, or the tachographs or speed limiters found already in many commercial vehicles.

The design of tamper-resistant embedded computers is already on its way to become a well-understood engineering discipline [2, 14], supported by a range of commercially available components, such as intrusion sensors [3], battery-backed RAM key storage with emergency zeroization mechanisms [8], shielding against compromising emanations, and other side-channel countermeasures.

Therefore, the focus is here on the other main vulnerability of tamper-resistant GNSS receivers: that their antenna input could be fed with a simulated signal rather than from the satellites. A specialized portable signal generator could synthesize a GNSS antenna signal that causes the receiver to report an alternative position, velocity or time to the connected cryptographic module. A simple example would be a device that records the route taken by a lorry and then replays the coordinates slower to a GNSS signal simulator that the driver has installed to replace the satellite reception antenna, such that a speed-limiter function is not triggered, but the tachograph still shows a realistic-looking record of the driven route. (Police have already uncovered similar manipulations of speed-sensor signals in existing tachographs [1].)

What measures could a GNSS receiver implement to assess the authenticity of a received GNSS signal and the resulting navigation solution? It is important to note that the notion of authenticity of a GNSS signal goes beyond the usual meaning of message authenticity in cryptographic protocols. We have to protect not only the authenticity of the transmitted navigation data, but also that of the relative arrival times (the *pseudo-ranges*) of the transmitted spread-spectrum waveforms, within better than a microsecond. Both together form the basis for calculating the navigation solution.

1.1 Environmental Assumptions

This discussion focuses in particular on tamper-resistant GNSS receivers (also called trusted receivers) that are assumed to be in the hands of the adversary. We make the following assumptions:

The trusted receiver consists of an antenna, a circuit for demodulating and tracking the GNSS signal, and a secure microcontroller. The microcontroller stores cryptographic secret keys and uses these to attest (e.g., by time stamps, digital signatures, or similar cryptographic protocols) to a remote party the current location (or the recent location history).

The receiver's RF front end, signal-processing circuitry, local oscillator, and the secure microcontroller are all enclosed in a tamper-responding shield that the adversary is unable to penetrate without destroying the secret keys stored inside.

This tamper-responding enclosure is equally securely attached to the object whose location is ultimately of interest (car, laptop, etc.). (This attachment could be secured, for example, by strong mechanical bonding, by detachment sensors, or by some cryptographic distance-bounding protocol to another tamper-resistant CPU in the monitored object.)

The adversary has full control over the RF signal received by the device and, in particular, may disconnect the antenna and connect the receiver's RF input instead to a signal generator programmed to emulate GNSS broadcast signals, with the aim to cause the secure microcontroller to process fake position information. Alternatively, where the antenna is not easily detachable, the adversary may also place the tamper-resistant receiver's antenna inside a shielded enclosure, along with transmission antennas connected to a signal generator.

The first commercially available GNSS simulators have been very expensive and specialized devices. However, during the past decade, numerous low-cost components (high-speed DACs, FPGAs, DSPs) and standardized platforms for building software-defined radio applications (GNU Radio project, Ettus USRP, various DSP processor/FPGA evaluation boards, etc.) have become available. This makes it practical now to design high-quality GNSS signal simulator prototypes with a hardware budget in the region of 1–2 k$ [7]. The result of such design efforts can easily be shared as open-source software, which will substantially increase the number of people able to understand, implement, and customize such devices. With the increased availability of GNSS simulation capabilities, attacks involving GNSS signal simulators should be expected as soon as attractive targets emerge, namely mass market applications that involve remote-attestation GNSS receivers (e.g., location-based access control, pay-as-you-drive road charging systems) where the holder of the trusted receiver has an incentive to spoof its input signal.

1.2 Related Technologies

Much of the existing literature on GNSS signal spoofing and jamming has focused on a remote attacker scenario, where the receiver is believed to be in the hands of a user who is interested in it finding a correct navigation solution (e.g., a soldier) and where the antenna is still exposed to genuine GNSS broadcast signals. A remote attacker can only add additional signals to the receiver's RF environment from a distance. Anti-jamming and anti-spoofing countermeasures aim to suppress these, to preserve the availability of the (also present) genuine signals. Examples for such countermeasures include

- the use of directional antennas (beam-forming networks) to suppress unusually strong GNSS signals, which are unlikely to be coming from a genuine satellite;
- an adaptive filter for suppressing interfering narrowband signals;

- the combination of two tracking circuits, where the job of the first is to track the spoofed (often stronger) signal, such that it can be subtracted from the input in order to allow the second tracking circuit to follow the remaining, weaker genuine signal.

In contrast, we assume here a local attacker who can easily suppress any trace of the genuine GNSS signal at the RF input during an attack, where the entire antenna signal may be fake. Rather than looking for traces of a weak genuine signal in the presence of a stronger spoof signal, the signal authenticity mechanisms discussed in Sect. 2 focus on discovering signal characteristics that help to distinguish between a genuine and a simulated GNSS antenna input.

1.3 Goals

Any practical GNSS signal simulator will produce an idealized signal that lacks some of the subtle characteristics found in a genuine signal. Ultimately, any mechanism for assessing signal authenticity can only be effective if the receiver's designer uses a more accurate model of a genuine signal than the signal simulator's designer. With enough effort and resources, any of the methods discussed in Sect. 2 can be circumvented, either by carefully emulating all the tested characteristics or by appropriately modifying a genuine signal. However, such a simulator may not in practice be an attractive means of defeating a given security application (operating cost, mobility, physical dimensions, etc.). Also, if it were openly sold, it would have to provide capabilities substantially beyond the type of simulators normally used legitimately for the development, testing, and maintenance of GNSS receivers. Therefore, it could be identified as having been specifically designed to circumvent the proposed security mechanisms for assessing signal authenticity, and its sale might be illegal under existing cybercrime legislation.

We can distinguish between two broad categories of methods for assessing signal authenticity:

- *Instant methods* assess signal authenticity almost as soon as a navigation solution has been found and should not extend the duration of the normal lock-on process by more than a few seconds. They are of particular interest where an action (such as a network login with location-based access control) has to be blocked instantly if there are substantial doubts regarding the authenticity of the navigation solution.
- *Cumulative methods* monitor the GNSS signal over many hours or days and report in the end whether there has been substantial evidence of a fake signal during this period. Such methods may be applicable in accounting applications (e.g., road charging), where the damage that a successful attacker can cause is proportional to the time that the fake signal is accepted by the receiver. Where cumulative methods can be used, a wider range of verification techniques is available.

Applications that can rely on cumulative methods have another security advantage: the detection of a simulated signal need not be made known to the user of the receiver immediately and therefore deprives an adversary from the rapid feedback that helps optimizing a signal generator. Instead, the signal-authenticity assessment can just be silently recorded, helping the operator of the protected application to estimate the level and nature of attacks taking place and to focus investigation and countermeasures appropriately.

2 Techniques

2.1 Secret Spreading Sequences

GPS and planned GNSSs use direct-sequence spread-spectrum modulation in their broadcast signals. The low bit rate (<1 kbit/s) data signal is XORed with a high bit rate (> 1 Mbit/s) pseudo-random spreading sequence, before the result is used to modulate the phase of a carrier sine wave.

A range of possible techniques rely on the fact that GPS broadcasts both its civilian (C/A) and military (Y) signals at a power-spectral density substantially below the background noise level. Receivers with omnidirectional antennas are therefore unable to decode the individual "chip" symbols of the spreading sequences and can only detect a cross-correlation with a known sequence of at least a few hundred chips. In addition, the GPS Y signal is, due to its encryption, not predictable by non-military users and therefore difficult to reproduce in a simulator. Galileo is foreseen to broadcast similar weak broadband signals and to provide a similarly encrypted (and therefore for most users unpredictable) signal in its public-regulated service (PRS), although the details have yet to be finalized and published.

2.1.1 Conditional Access

One possibility for assuring signal authenticity is, of course, to keep the spreading sequence (ranging codes) used secret and non-repeating. A conditional-access system, similar to those already widely implemented in the direct-broadcast satellite pay-TV industry, has to ensure that the cryptographic keys needed to predict the spreading sequence for the near future are distributed to tamper-resistant modules handed out to authorized subscribers. Such modules then determine, based on received entitlement management messages, to which level of service the user of each module is entitled and then extract from also received entitlement control messages the necessary cryptographic keys for accessing these services. The US military uses already a form of conditional access for the encrypted Y signal, and subscriber modules appear to be planned for the Galileo commercial, safety-of-life, and military services.

The tamper-resistant subscriber modules have to be very carefully designed such that they cannot be abused by a spoofing attacker as a component of a signal generator that can predict the secret sequences about to be broadcast. This would typically involve performing the correlation and tracking operation inside the module, such that the keys used to generate the next parts of the spreading sequence never leave the tamper-resistant envelope. Another challenge, which has already been studied in detail over the past 20 years in the context of pay-TV conditional access systems, is to design a broadcast-encryption and traitor-tracing key management system that can recover its security after a small number of subscriber modules have been broken [5]. This is not an easy task if the available broadcast data channel has only a low bit rate.

If these aspects can be secured, the main option remaining to an attacker interested in simulating a conditional-access signal is to use tracking high-gain antennas. These could improve the signal-to-noise ratio (SNR) to a level that allows reliably detecting the individual chip symbols in the broadcast spreading sequences in real time. The attacker can then slightly delay and remix them in the signal generator (selective delay attack with high-gain antennas, see [10]) to simulate how they appear relative to each other at the pretended location. Such attacks can be made more cumbersome in two ways:

- Keep the broadcast power density well below the background noise level, in order to maximize the physical antenna dimensions required (e.g., large parabolic dish or long helical antennas).
- Keep the symbol rate high, in order to make it more difficult for an attacker to forward the signals received at a stationary set of directional antennas to a mobile signal generator.

2.1.2 Delayed Release of Spreading Sequences

A method to achieve similar signal-integrity assurance as a conditional-access system can provide, but without the overhead and risk of compromise of a tamper-resistant subscriber module and associated key distribution infrastructure, was proposed independently by Scott [13] and Kuhn [10]. The idea is that the spreading sequences used are secret at the time of their broadcast, but information to reconstruct them is broadcast with a delay of a few seconds. This allows tamper-resistant receivers to discover, with a short delay, the genuine broadcast signals, using FFT-based cross-correlation on recorded segments of the entire transmission band. At the same time, this forces the designer of a signal generator to delay the signal also by a few seconds, which an independently synchronized UTC clock in the receiver can easily detect.

The delayed release of the spreading sequence remains the most practical and resilient single-integrity assurance method currently known:

- It does not require support from a network of reference stations.
- It does not rely on the security of a subscriber key distribution infrastructure.

Such a scheme could be piggybacked on top of an existing service that broadcasts using a secret spreading sequence. The latter would have to be generated by a pseudo-random bit sequence (PRBS) generator that is seeded with a new secret start value in regular intervals. The satellites then simply would have to occasionally broadcast a subset of the PRBS seeds that have been used, but with some delay. For example, the encryption scheme for the Galileo PRS signal could be designed such that keys that generate only short intervals of the spreading sequence can be released without affecting the security of its conditional-access users. The key K obtained by the conditional-access modules would in such a system not be applied directly to generate the spreading sequence. Instead, it would be used to encrypt a timestamp t that identifies a short time interval (e.g., 1 s) in order to obtain a short-term intermediate key $K_t = E_K(t)$ which is then used to seed the PRBS generator that generates the actual pseudo-random noise (PRN) spreading sequence, 1 s at a time. E is some suitable keyed pseudo-random function, e.g., a cipher, message authentication, or secure-hash function. A small subset of the short-term intermediate keys K_t is then released with a short delay. The interval length (e.g., 1 s), the subset of the released intermediate short-term keys K_t (e.g., one K_t every 20 s), and the delay (e.g., 10 s) have to be chosen such that their publication does not enable practical spoofing of regular receivers of the conditional-access service, whose tracking loops would have to be designed to be immune to regular but brief bursts of old spreading sequence.

2.1.3 Permanently Secret Sequences

Can we adapt the basic idea from the preceding Sect. 2.1.2 if we have users who never get access to the spreading sequence used by the satellites? This is the case, for example, for civilian users regarding the GPS Y code. A reference station still can record the spreading sequence, but has to use high-gain antennas that lift the SNR sufficiently to allow it to receive and detect the spreading sequence directly, convert it into a bit stream (10.23 Mbit/s for the GPS Y code), and arrange for that to be forwarded to the tamper-resistant receiver, who will correlate it with a pre-agreed brief concurrent recording of the full transmission band. This method, discussed in more detail by Psiaki [12], works similarly as the one outlined in the preceding Sect. 2.1.2, but is more expensive to implement:

- It requires a reference station with large tracking antennas (ideally at least four, fewer if only probabilistic verification is required).
- It requires a higher bandwidth secure communication link to the tamper-resistant receiver. Entire spreading sequences received during the pre-agreed time window will have to be provided to the receiver for delayed cross-correlation there, rather than just K_t seed values that generate them. (Transmitting the received and signed raw spectrum during remote attestation from the tamper-resistant receiver is another option, but requires an even higher bit rate communication channel.)

2.2 Individual Receiver Antenna Characteristics

2.2.1 Directional Characteristics

If the receiver antenna is installed at a fixed location, or mounted on a car, the receiver might be able to observe the directional variation of amplitude (and perhaps even phase?) as the satellites move along the sky with known azimuth and elevation. Mounted on a car, the orientation of the antenna will normally only vary in azimuth (yaw), due to curves, and to a limited degree in elevation (pitch), due to hills. Both angles can there be inferred from the velocity vector determined by the receiver, assuming that roll movements are very limited and temporary. In particular, the yaw motions of a car will cause the satellites to quickly scan a substantial part of the directional characteristic of the antenna, which can be monitored for changes.

A receiver could characterize the directional characteristic of its antenna from the received signal strength, in particular if data about absolute signal strength from the automatic gain controller and the correlator are available. The designer of the trusted receiver could choose an antenna type specifically for its structurally rich directional pattern, for instance, a fractal antenna rather than a simple dipole and could even individually vary the exact antenna shape and encase it in opaque resin, in order to increase the effort needed by an attacker to recognize and model its characteristic. While an attacker could measure the individual antenna pattern of each replaced antenna and program the signal simulator accordingly, this adds substantially to the effort needed to implement an attack, ideally beyond being economically attractive for a mass-market fraud device.

2.2.2 Impedance Test

Where a custom RF front end is being designed for a tamper-resistant receiver, this opens the possibility to add circuitry that characterizes the frequency-dependent impedance of the connected antenna occasionally, raising an alarm if that changes substantially. Possible techniques include time-domain reflectometry, VSWR measurement (if there is a transmission line), or vector network analysis. Especially if the antenna has been produced deliberately with characteristic invisible manufacturing variations, the need to keep the antenna attached (i.e., use a shielded enclosure around it) or to emulate the antenna impedance represents a substantial complication in the appropriate connection of a signal generator, possibly one that makes mass-market sale of signal generators far less feasible.

2.3 Consistency with Reference Receivers

One group of signal authenticity measures compares characteristics of the received GNSS signals with the same characteristics measured at the same time by a network of trusted reference receivers. These reference receivers can either be dedicated stations, secured by traditional anti-jamming measures (e.g., distance, directional antennas), or they can be obtained by assuming that the majority of the signal characteristics reported by a fleet of trusted receivers is genuine, allowing outlier detection.

2.3.1 Time

All existing or planned GNSSs broadcast coordinated universal time (UTC) with an accuracy better than a microsecond. More accurate signal simulation techniques often involve incorporating data from reference stations, and this usually requires delaying the generated signal. Therefore, a trusted GNSS receiver should first of all verify the UTC received by GNSS with an independent authenticated source of UTC. This can be accomplished by operating a local UTC clock, independent from any received GNSS signal. This clock should be synchronized regularly via an authenticated challenge–response time protocol, like NTP. Such network time protocols can, depending on the communication link, achieve UTC accuracies of a few tens of milliseconds or better. The resulting clock accuracy is mostly a function of how frequent these phase and frequency adjustments can be made compared to the undisciplined frequency stability of the local oscillator. If the received UTC(GNSS) differs from the UTC(NTP) in the local clock by substantially more than the latter's uncertainty (e.g., a few tens of milliseconds), a clear indication has been found that either the GNSS signal or the independent source of UTC has been manipulated.

2.3.2 Navigation Data

The independently synchronized UTC clock in a trusted receiver can also be used to timestamp a revision history of the navigation messages received from individual satellites. This revision history, which records (with a resolution of a few tens of milliseconds) when which bit in the navigation message has been observed to change, can then be compared with the corresponding revision history collected by the reference receivers.

Bypassing this measure would require an attacker to either be able to anticipate the content of navigation messages that are newly uploaded into satellites or implement in the signal generator a specialized receiver that provides a real-time feed of the navigation signal. The proprietary binary protocols of typical existing GPS receiver chipsets output changes to navigation messages only with significant delay, usually awaiting the completion of frames and parity checks. If a signal simulator is merely fed with such delayed navigation-message updates, its use would be detected by this measure.

This test relies on the satellite operator not publishing all updates to the broadcast data in advance. The more frequent the navigation data changes in unpredictable ways the more effective the test is. For this reason, designers of future GNSS signals could add to navigation messages unpredictable random bits, such as time-dependent message authentication codes, or hash chains.

2.3.3 Pseudo-Ranges

If more than four satellites are in view simultaneously, an over-determined system of equations will lead to the navigation solution. Satellite clock and ephemeris errors, as well as atmospheric path delays, will then cause inconsistencies, usually of several meters. A tamper-resistant receiver with access to raw pseudo-range measurements could compare these inconsistencies with those observed by a nearby reference receiver. Inconsistencies caused by the atmosphere will vary geographically and therefore would force the adversary to have access to a reference receiver in the vicinity of the emulated location. (Experience with differential GPS suggests that pseudo-range inconsistencies show a substantial loss of correlation at distances larger than a few tens of kilometers.)

There are regional networks of differential GPS stations that publish pseudo-range inconsistencies[1] that both the trusted receiver and the adversary could refer to. However, as long as they publish their information only with a delay larger than the auto-correlation width of the data, they could be used for verifying pseudo-range inaccuracies without enabling a signal generator to simulate them in real time.

2.4 Receiver-Internal Plausibility Tests

Beyond the minimally necessary processing needed to achieve a navigation solution, receivers can implement additional consistency checks without requiring a connection to a network of reference receivers. A number of such tests have been proposed and are often referred to in the literature as receiver autonomous integrity measures (RAIM). They were originally aimed primarily at detecting accidental malfunctions in the GNSS, such as one of the satellites suffering from a phase jump or frequency deviation in its local oscillator, or the broadcast of incorrect or out-of-date navigation messages. They have also been proposed to detect very simple types of GNSS signal simulation [15] and would force the attacker to use a more complete simulation model, including realistic and up to date navigation data and parameters.

2.4.1 Elevation Limit

A very simple check involves verifying that each satellite from which a signal is received actually claims to be above the horizon at the moment. This test was proposed by Wen et al. [15] to detect if a very simple type of signal simulator is used that always transmits a fixed number of satellite signals (e.g., 10), even if their simulated position is well below the horizon. This test can be implemented with many consumer receivers, which output the azimuth and elevation of all tracked satellites. Some receivers may already search only during a cold start for the

---

[1] For example, the OS Net RINEX data server available on http://gps.ordnance-survey.co.uk/ for the British Isles or (continent wide at currently much lower station density) the data from augmentation services such as EGNOS/SISNeT.

spreading sequences of satellites below the horizon. This test is obviously also very easy to circumvent by the designer of a signal simulator, which simply has to gradually attenuate signals as the simulated satellite's elevation reaches the horizon.

2.4.2 Power Limits

With typical satellite altitudes of more than 20,000 km, the receiver–satellite distance, and therefore the best-case received signal strength, varies relatively little with elevation. It is guaranteed by the GPS specification to never exceed -150 dBW [6, 6.3.1]. A substantially stronger signal would indicate a manipulation. Power can be measured at different levels: (a) across the entire band, in form of the automatic gain control (AGC) signal, and (b) for a single satellite, in form of the correlation value reported by the prompt correlator in the code-tracking DLL. The across-the-band GPS L1 power level is largely dominated by thermal and receiver noise and therefore varies only little in normal operation. While a small amount of excess power beyond that, per satellite, can be explained by constructive multipath interference, anything stronger must be considered suspicious. On the other hand, there is no lower bound, as line-of-sight obstacles can always explain a lack of signal. Unlike a remote adversary, a local spoofing attacker should not find it difficult to adjust the power of the signal realistically, making this test less of a hurdle.

2.4.3 Doppler-Shift Verification

Many GNSS receivers track the phase of the received carrier signal, or more often that of a down-converted intermediate-frequency (IF) equivalent, after they have removed the ranging code, by implementing a Costas loop [4]. When such a loop has locked on, the input of its numerically controlled oscillator (NCO) is a function of the relative speed of both the transmitter and receiver antenna in an inertial coordinate system (Doppler shift, ± 10 kHz) as well as the frequency error of the local oscillator that is used to both down-convert and sample the incoming signal (typically a few parts per million). When the receiver tracks several satellites simultaneously, the frequency error of the local oscillator cancels out in the difference between the respective NCO inputs, and what remains (apart from tracking noise) is only the difference in the Doppler shifts between the satellites. A receiver can predict the Doppler shift of each satellite from the received ephemeris data and its own location and velocity and compare these predictions with the observed Doppler-shift differences. The elimination of the local-oscillator error allows the application of tight tolerances in such checks, limited mainly by the uncertainty of the speeds involved and tracking noise. Such a test will require the designer of a simulator to accurately emulate the Doppler shift and will detect some comparatively simple simulators that do not.

Regular GNSS receivers will also estimate the Doppler shift in order to speed up initial signal acquisition, but may not apply any checks on the frequency once they are tracking a signal. They will try all reasonable Doppler shifts during a cold start.

When connected to a signal simulator without accurate Doppler-shift generation, such receivers may take longer to acquire a signal but may otherwise not complain.

However, building a simulator that accurately reproduces Doppler shift is not that difficult. In a complex number baseband representation of a quadrature amplitude-modulated signal, Doppler shift Δf can be applied by multiplying the signal with $e^{2\pi i \Delta f t}$, thereby rotating the complex (or IQ) coordinate system with an angular velocity proportional to the Doppler shift. After several simulated individual satellite baseband signals have been frequency shifted this way, they can be added together before being fed into a single transmitter (with IQ input) that up-converts the signal to the carrier frequency. This is much cheaper than the individually tuned per-satellite transmitter claimed to be necessary in a Doppler-accurate simulator by Wen et al. [15].

2.4.4 Code–Carrier Phase Comparison

The signal generators implemented in the satellites synthesize all aspects of the broadcast signal from a single atomic clock. As a result, the phases of all the emitted carriers and the pseudo-random-noise (PRN) code sequences and data signals modulated on top are strictly phase locked, i.e., there is a constant number of carrier periods per PRN chip and a constant number of PRN chips per data bit. Nevertheless, most receivers implement two independent tracking loops, a Costas loop for tracking the carrier and a PLL with early-late discriminator for tracking the PRN spreading sequence. This is because most receivers first down-convert the microwave carrier band to an intermediate frequency of much less than 100 MHz. This frequency down-conversion introduces the frequency of the receiver's local oscillator as an additional variable and thereby destroys the fixed code–carrier phase relationship, making two tracking loops necessary for initial acquisition. Once both loops have locked on and the receiver switches from acquisition into tracking mode, many receivers use the feedback of the (less noisy) carrier-tracking loop to aid the PRN code-tracking loop [9, Chap. 5].

A signal simulator based on standard software-defined radio platforms (e.g., USRP) will digitally synthesize an IQ or IF signal that is then up-converted into the GNSS transmission band. Unless the synthesis of all the frequencies in this process is carefully phase locked and matched, the IF up-conversion process can easily break the fixed code–carrier phase relationship of a genuine signal. Regular receivers will not notice this during acquisition and may not be disturbed by it much either during tracking, unless they do accurate phase accounting. Receivers that merely report a Doppler-shift frequency that crudely indicates the feedback signal in the carrier-tracking Costas loop are unlikely to help detect such deviations. What is needed instead is a register in each tracking loop that accurately integrates the frequency corrections that both tracking loops apply, in order to show the accumulated phase correction achieved (e.g., in meters). If this phase correction then starts to differ substantially between the carrier and code-tracking loop, this would be a strong indication that the signal emerged from a simulator whose designer did not worry too much about that phase relationship. Most normal GPS receivers do not accurately integrate the frequency correction onto a phase correction; however, special carrier-based differential GPS receivers, used in some geodetic and robotic applications, may collect the raw data necessary to verify the code–carrier phase relationship.

2.4.5 Multi-band Reception

A receiver that covers all the GNSS bands on which a satellite broadcasts (e.g., GPS L1 = 1.5754 GHz and L2 = 1.2276 GHz) can impose rather more substantial requirements on a signal simulator. In a genuine signal, the different carrier bands

- will be attenuated in nearly (but due to diffraction not exactly) the same way by line-of-sight obstacles;
- will show phase shifts caused by atmospheric diffraction, but remain phase locked.

A signal simulator might transmit only the signals in a single band (e.g., only GPS L1). If it broadcasts in multiple bands, it might lack the phase lock, phase shift, and close but imperfect power-level relationship typical of concurrently observed different carrier frequencies from the same satellite. Even if one of the carriers is modulated only with an unknown encrypted signal (e.g., Y on GPS L2), it can still be correlated against the same encrypted signal on any other carrier, in order to measure phase-shift and compare attenuation.

2.4.6 Ephemeris Data Check

The orbital position (ephemeris) data broadcast by each satellite should preferably be verified by comparing it with what is received at a secure reference receiver or by verifying any cryptographic authenticity features included (digital signatures, message authentication codes, hash chains, etc.). GPS currently lacks the latter, but future systems might support cryptographic authenticity checks of ephemeris data.

Where neither of these options are feasible, a plausibility check against long-term invariants of the orbital data remains a possibility. Each satellite has a limited amount of fuel onboard, in order to change orbit, resulting in a maximum velocity change $||\Delta v||$ achievable during its lifetime. This fuel can be used not only for station keeping, but also to reconfigure the orbital constellation, e.g., after satellite failures.[2] Likewise, satellite engines have limited thrust (especially ion engines), limiting the acceleration $||\Delta v||/\Delta t$. If these limits and the rate of natural orbital perturbations are available, along with an algorithm that estimates a lower bound for the $||\Delta v||$ needed to move the orbit of an satellite in a given time interval Δt from known past ephemeris data to the currently broadcast ones, these can be compared as a broad plausibility test.

---

[2] The GPS satellites are rumored to even be able to change the inclination of their orbits somewhat to achieve better polar coverage, should the need arise.

However, the security gains achieved this way are limited: there appears to be no advantage to our local attacker from substantially deviating in the navigation data from the orbits of the satellites currently in space. False ephemeris data might be more useful in remote attacks, where the attacker wants to minimize the likelihood that the receiver reacquires the genuine signal, whereas we assume here the genuine signal to be easily suppressed.

2.4.7 Jump Detection

Another commonly proposed type of spoofing detector looks for discontinuities in the received signals, e.g., the pseudo-ranges or the resulting solutions for the location and local clock error, or bounds such changes with independent sensors (inertial navigation, odometer, dead reckoning, etc.). It is certainly prudent and practical to monitor the continuity of GNSS time against an independent, battery-backed local clock (see also Sect. 2.3.1). Such techniques also make sense to protect against remote attackers who start to spoof the signal after the receiver had already locked onto the genuine one. However, the applicability of such techniques against a local attacker seems rather limited, as the latter can replace the antenna with a signal generator while the receiver and alternative sensors are switched off. It also is not a practical instant check in situations where the GNSS receiver is only briefly switched on for an attestation operation, never running long enough to monitor the long-term continuity of satellite signals.

2.4.8 Quality Metrics

Several quality metrics have been proposed in the literature for GNSS signals. If the quality of the received signal is substantially better than anything the receiver ever has seen with its real antenna attached, this might indicate the use of a signal generator. Examples of quality metrics include

- the residual error in the navigation solution (which solves an over-determined system of equations if more than four satellites are in sight);
- the deviation of the actual cross-correlation result from the ideal (e.g., triangular) auto-correlation function of the PRN signal.

2.5 Some Other Ideas

2.5.1 Individual Transmitter Characteristics

Signal analysis techniques have been developed that identify individual radio transmitters based on the influence that electronic component tolerances have on the exact shape of the emitted RF waveform. Parameters measured for transmitter fingerprinting include in particular

- carrier-frequency deviation;
- transients occurring when the carrier is switched on and off;

- amplitude and phase roll-off of the band-pass filters used to shape the output spectrum (which affect the shape of the eye pattern in digital modulation).

Normal GNSS signal generators are likely to use exactly the same mathematical function to synthesize the waveform for each satellite, adjusted only by obvious parameters such as Doppler shift, range phase shift, range attenuation, and spreading sequence. Real-world satellites may have additional other characteristics (hopefully within the tolerances allowed by the RF interface definition). However, carrier-frequency deviation is already carefully calibrated in GNSS signals, and as the signals are broadcast continuously, there is no opportunity to observe on/off transients. This leaves filter roll-off, which is difficult to measure directly given the very low signal-to-noise ratios typical of GNSS systems, especially where the spreading sequence is unknown to the receiver (e.g., GPS Y signal). It may show up, however, as satellite individual and receiver bandwidth-dependent variations in the exact shape of the cross-correlation function.

2.5.2 Spectrum Analysis

The RF input should normally see an expected minimum noise level not only within the transmission band (e.g., 20 MHz wide), but across the entire radio spectrum, along with evidence of other, non-GNSS transmitters in adjacent bands. Substantial reduction of this out-of-spectrum noise level could indicate the use of a signal generator. This would require a more widely tunable receiver to measure. An attacker who wants to fake this wider input spectrum would either have to use a substantially more wideband signal generator (more expensive, more power required) or would have to mix the synthesized GNSS spectral-band content with real background noise from an antenna (possibly with the GNSS band attenuated by a band-stop filter or using spectrum frequency shifted from a different band).

2.5.3 Extended Search for GNSS Signals

A regular GNSS receiver will lock onto a correlation peak with a particular spreading sequence as soon as one is found or may search for a local maximum or the earliest peak among several nearby ones, in the interest of robust multipath behavior. A signal authenticity verifying receiver could, in addition, continue to scan combinations of correlation delay and Doppler shift and warn about the presence of more satellite signals than can be expected from the genuine transmitter constellation (e.g., the same spreading sequence at two range delays or Doppler shifts). This test is particularly useful if a local attacker mixes the simulated signal with background spectrum from an antenna to evade the test outlined in the previous Sect. 2.5.2.

3 Comparison

The receiver technology required in order to implement the measures discussed in the preceding Sect. 2 differs substantially from method to method. Some require substantial extensions, or even alternative receiver architectures, compared to what is commonly implemented in existing civilian receivers. Commercial low-cost GPS chipsets receive only L1 C/A code. Most chips merely output time, location, and the identity and claimed azimuth and elevation of tracked satellites, using the very limited, but standardized, NMEA 0183 "sentences" ASCII format. Some GPS chips can also be switched into an additional, vendor-specific, binary communication protocol that gives access to additional data, such as the ephemeris, almanac, and health information received from individual satellites. A very small number of GPS receiver chipsets provide even access to "raw" tracking data for each tracked satellite, such as pseudo-range, Doppler shift, carrier–noise ratio, as well as internal receiver variables such as AGC gain setting, and local-oscillator error from the navigation solution.

For many of the proposed methods, the only practical prototype implementation method involves a software-defined radio approach, where the 2–40 wide MHz GNSS band of interest is down-converted into an IQ baseband representation, loaded block-by-block into RAM, and then all tracking and analysis algorithms are implemented in software [4].

Table 1 attempts to give an overview of the requirements and properties of each proposed method. The "Access" column describes at what level the measure needs to access the receiver's processing pipeline and thereby gives an indication what existing GPS receiver chips could support such a measure: "RF" means that support has to be integrated in the RF front end, "IQ" means that a software-defined receiver that receives down-converted IQ samples and then implements all further processing in software could implement the measure, "Raw" means that the proprietary binary protocols of some existing GPS receiver chips provide enough data, and "NMEA" means that the standard NMEA output of most existing GPS chips will suffice. The

Table 1 Overview of the presented authenticity verification methods

Method	Section	Access	Ref.	Extra requirements	Type
Conditional access	2.1.1	IQ		Signal support, SIM	Instant
Delayed release	2.1.2	IQ		Signal support, NTP	Instant
Permanently secret	2.1.3	IQ	Y	NTP	Instant
Directional char.	2.2.1	Raw			Cumulative
Impedance test	2.2.2	RF		TDR, etc.	Instant
Time	2.3.1	NMEA		NTP, battery clock	Instant
Navigation data	2.3.2	Raw	Y	NTP	Both
Pseudo-ranges	2.3.3	Raw	Y	NTP	Cumulative
Elevation limit	2.4.1	NMEA			Instant
Power limits	2.4.2	Raw			Instant
Doppler	2.4.3	Raw			Instant
Code–carrier phase	2.4.4	IQ		Or tracking-loop integrators	Instant
Multiple bands	2.4.5	IQ		Multiple down-converters	Both
Ephemeris	2.4.6	Raw			Instant
Jump	2.4.7	NMEA		Battery-backed clock	Cumulative
Quality metrics	2.4.8	IQ			Both
Transmitter character	2.5.1	IQ			Both
Spectrum analysis	2.5.2	IQ		Tunable down-converter	Both
Extended search	2.5.3	IQ			Both

"Ref" column indicates whether communication with a separate, secure reference receiver station is required.

4 Conclusions

There clearly exist circumvention techniques for all the authenticity verification methods outlined in this survey. The mechanisms available today for protecting GNSS signals against tampering by local attackers still can at best offer a level of security comparable to most other types of tamper-resistant hardware. They all fall well short of the ambition behind the Kerckhoffs' principle so popular in cryptology: detailed knowledge of the protection mechanisms used may still substantially aid in their circumvention. Nevertheless, some of the presented mechanisms (e.g., secret spreading sequences, individual antenna characteristics) have the potential to prevent easy-to-use mass-market circumvention products. Others at least force the designer of a circumvention tool to add rather specialized functions, whose obvious purpose would be to circumvent these checks. The latter may help to enforce legal restrictions on their commercial availability. Some may be most useful as intrusion-detection tools that report suspicious signals for further investigation, rather than to automatically decide on their authenticity. In combination, they provide a formidable toolkit for managing the risk of local attackers on trusted GNSS receivers in many potential applications.

Acknowledgments This work was supported by the European Commission under FP7 grant 228443 (TIGER project).

References

1. R.J. Anderson, *On the Security of Digital Tachographs*. In European Symposium on Research in Computer Security (ESORICS), Louvain-la-Neuve, Belgium, 16–18 Sept 1998. Lecture Notes in Computer Science, vol. 1485 (Springer, 1998), pp. 111–125
2. R.J. Anderson, M.G. Kuhn, *Tamper Resistance — A Cautionary Note*. In The Second USENIX Workshop on Electronic Commerce Proceedings, Oakland, CA, USA, 18–21 Nov 1996 (USENIX Association, 1996) pp. 1–11
3. Anti-tamper physical security for electronic hardware. GORE
   http://www.gore.com/en_xx/products/electronic/specialty/antitamper.html
4. K. Borre, D.M. Akos, N. Bertelsen, P. Rinder, S.H. Jensen, *A Software-Defined GPS and Galileo Receiver* (Birkhäuser, Boston, MA, 2007)
5. J.A. Garay, J. Staddon, A. Wool, *Long-Lived Broadcast Encryption*. In Advances in Cryptology (CRYPTO), Santa Barbara, CA, USA, 20–24 Aug 2000. Lecture Notes in Computer Science, vol. 1880 (Springer, 2000), pp. 333–352
6. GPS Interface Control Document, ICD-GPS-200C, 2003-01-14
7. T.E. Humphreys, B.M. Ledvina, M.L. Psiaki, B.W. O'Hanlon, P.M. Kintner, Assessing the spoofing threat. GPS World **20**(1), 28–38 (Jan 2009)
8. S. Joshi, Addressing the physical security of encryption keys. Maxim Eng. J. **62**, 7–11 (2008)
   http://pdfserv.maxim-ic.com/en/ej/EJ62.pdf
9. E.D. Kaplan, C.J. Hegarty, *Understanding GPS: Principles and Applications*, 2nd edn. (Artech House, Norwood, MA 2006)

10. M.G. Kuhn *An Asymmetric Security Mechanism for Navigation Signals*. In 6th Information Hiding Workshop, Toronto, Canada, 23–25 May 2004. Lecture Notes in Computer Science, vol. 3200 (Springer, 2004), pp. 239–252
11. B.W. Parkinson, J.J. Spilker Jr., *Global Positioning System: Theory and Applications – Volume I. Progress in Astronautics and Aeronautics*, Vol. 163 (American Institute of Aeronautics and Astronautics, Washington, DC, 1996), ISBN 1-56347-106-X
12. M.L. Psiaki, Spoofing Detection for Civilian GNSS Signals via Aiding from Encrypted Signals. Proceedings of ION GPS/GNSS (Savannah, GA, USA, 22–25 Sept 2009)
13. L. Scott, *Anti-Spoofing & Authenticated Signal Architectures for Civil Navigation Systems*. In Proceedings of ION GPS/GNSS (Institute of Navigation, Portland, OR, USA, 9–12 Sept 2003), pp. 1543–1552
14. S.H. Weingart, *Physical Security Devices for Computer Subsystems: A Survey of Attacks and Defenses*. Cryptographic Hardware and Embedded Systems (CHES), Worcester, MA, USA, 17–18 Aug 2000. Lecture Notes in Computer Science, vol. 1965 (Springer, 2000), pp. 45–68
15. H. Wen, P.Y.-R. Huang, J. Dyer, J. Archinal, J. Fagan, Countermeasures for GPS Signal Spoofing. Proceedings of ION GPS/GNSS (Long Beach, CA, USA, 13–16 Sept 2005

On the Limits of Hypervisor- and Virtual Machine Monitor-Based Isolation

Loic Duflot, Olivier Grumelard, Olivier Levillain, and Benjamin Morin

1 Introduction

In the past few years, there has been a lot of different attempts to build trusted platforms allowing users to access sensitive and non-sensitive data in a compartmentalized way, i.e., such that applications dealing with sensitive data are fully isolated from those dealing only with public data. Such systems are often called compartmented systems as they allow the user to access simultaneously data of different levels of sensitivity. Isolation is often provided by some hardware functionalities [2, 22] and some kind of hardware [36, 38] or software [18, 24] virtualization layers. Of course, confidence in the abstraction layer is mandatory for the user to be confident in applications isolation. Such confidence can be obtained through formal method-based development [20] or static and dynamic analysis. Platforms such as those presented in [16, 31] are examples of such compartmented systems. In France, a challenge has been launched in October 2008 by the French National Research Agency that brings three teams in a competition for the design of a compartmented system usable by Internet users [17].

These initiatives generally assume that most of the hardware components (for instance, the CPU, the chipset, the keyboard) of the platform can be trusted. In this chapter we study the exact level of security-compartmented solutions based on hardware abstraction layers provide in case such assumptions prove to be wrong. One of the main contributions of this chapter is to address new threats such as DIMM (dual inline memory modules used for RAM storage on modern platforms) backdoors and to describe the impact of such backdoors on compartmented systems. In Sect. 2, we describe traditional compartmented systems and the attacker model we consider. In Sect. 3, we present a taxonomy of attacks against hypervisor- or virtual machine monitor-based systems. In Sect. 4 we study DIMM backdoors and in Sect. 5 we describe how they can be used by attackers to retrieve sensitive information or as

L. Duflot (✉)
French Network and Information Security Agency (ANSSI), Paris, France
e-mail: loic.duflot@ssi.gouv.fr

A.-R. Sadeghi, D. Naccache (eds.), *Towards Hardware-Intrinsic Security,* Information
Security and Cryptography, DOI 10.1007/978-3-642-14452-3_16,
© Springer-Verlag Berlin Heidelberg 2010

a means for privilege escalation. Section 6 presents potential countermeasures and Sect.art 7 concludes the chapter.

2 Compartmented Systems

Compartmented (or multilevel) systems are systems dealing with information associated with different sensitivity levels. For instance, the system could be manipulating public data as well as sensitive information (such as personal data, banking- or health-related documents). Obviously isolation between the public world and the sensitive world is fundamental. For the sake of simplicity, the examples we provide are generally based on a x86 (or x86-64) CPU. However, the analysis should hold true on a variety of different platforms.

2.1 Traditional Architectures and Definition of a Trusted Computing Base

The trusted computing base (TCB) is the set of components that need to be trustworthy so as to ensure trust in the overall platform. A typical example of a trusted computing base is composed of the CPU, the chipset, the BIOS, and the operating system.

In our case, the system could be composed of an hypervisor (such as Xen [36]) running two different guest operating systems in parallel (see Fig. 1a), one of them running applications using sensitive data and the other being connected to a non-sensitive network (for instance, Internet). Alternatively (see Fig. 1b), the system could be composed of a minimalist Linux operating system running two different virtual machine monitors (VMware Workstation [38] or Qemu [4], for instance) in parallel, each running an operating system, one of them dealing with sensitive data and the other dealing with non-sensitive data.

In this chapter we consider both cases. For systems such as the one depicted on Fig. 1a, the trusted computing base would typically be composed of (part of) the hardware, possibly the BIOS[1] and the hypervisor. For systems such as the one described on Fig. 1b the trusted computing base would be composed of (part of) the hardware, possibly the BIOS, the minimalist Linux, and the virtual machine monitors.

These models are believed to provide strong isolation. In this chapter, we study the actual limits of these architectures assuming that there is no software implementation flaw in the virtual machine monitor layer. This assumption may seem very strong but many different projects aim at designing secure by design microkernels using formal methods (e.g., OKL4/seL4 [20]).

---

[1] Technologies like Intel® TxT and AMD SVM/skinit aim at excluding the BIOS from the trusted computing base.

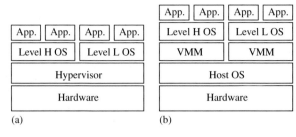

Fig. 1 Compartmented systems based on (**a**) an hypervisor and (**b**) virtual machine monitors

2.2 Attacker Model

In this chapter we assume that the attacker is initially only able to

- either run arbitrary code within the context of a non-privileged (non-root on Linux/UNIX systems for instance) application in the domain with the lowest sensitivity (we call attacks using such a vector "local attacks");
- or send packets to the network adapter associated with this domain (we call attacks based on this vector "remote attacks").

This model corresponds to a compartmented system whose lowest level is connected to the Internet. The user himself is not necessarily an attacker but could connect to non-trusted Web sites (and run scripts or flash applications controlled by an attacker) or open non-trusted documents. This attack model describes a very weak attacker that does not have any local access to the machine and no administration privileges.

3 Attack Paths

3.1 Taxonomy of Attack Vectors

In order to find what the relevant attacks against such a system are, we consider that parts of the software and hardware components of the platform can be trusted. Our definition of "trust" is that the components work according to their specifications (i.e., they are not flawed, there are no implementation bugs, undocumented functions, or backdoors), and they are used in the best possible way (their security model can be understood by the upper layer designers). This assumption is, of course, very strong and unrealistic in practice but provides us with an exhaustive view of attack vectors against the considered system.

If we suppose that the entire platform can be trusted, i.e. the hardware (including CPU, chipset, network controller, keyboard, screen) and the software (BIOS, OS, and virtual machine monitors), this leaves no attack vector for the attacker we just presented. In the present section, we explore the impact of relaxing the trust assumptions about the different hardware and software components, one after the other. Each non-trustworthy component can lead to new attack vectors for the attackers.

Of course, more complex and sophisticated schemes can be elaborated to bypass the isolation mechanisms and corrupt the target platform, by exploiting a combination of the following flaws at different levels.

3.1.1 Cooperative Attacks Between Corrupted Guest OSes

We first consider that the operating systems running on top of the virtual machines cannot be trusted. It is then very likely that an attacker will be able to use side or even covert channels to recover sensitive data (an encryption key, for instance) from the higher level.

Attacks exploiting such channels have been described extensively in the literature. Such attacks may use caches [6, 7, 28, 35] or branch prediction [26] as a channel between different applications or virtual machines.

3.1.2 Attacks on the Virtual Machine Monitor or Hypervisor

This category includes different kinds of potential attacks against hypervisors or virtual machine monitors, which have been much discussed in the literature [1, 32]:

- local attacks exploiting a flawed security model of the virtual machine monitor. Indeed, it may be possible for an invited OS to request the hypervisor for an access to privileged memory areas through DMA accesses;
- local attacks exploiting vulnerabilities of the virtual machine monitors, allowing an attacker to escape the guest environment to tamper with the host operating system;
- remote attacks exploiting vulnerabilities of the virtual machine monitors (IP stack, drivers).

Many vulnerabilities have been found during the past 2 years in major virtualization products.

3.1.3 Local Exploitation of the BIOS

Still considering the software level, an attacker can tamper with the BIOS and its associated functions to introduce a backdoor that may be invisible to the OS or the VM monitor.

For example, the BIOS update mechanism has been studied as an attack vector [33, 37], as well as the System Management Mode [9, 13, 15, 39] and the ACPI tables [14, 19].

3.1.4 Attacks Using a Flaw in the Chipset or the CPU

If the chipset or the CPU[2] of the platform cannot be trusted, backdoors or bugs inside them may be used either by a remote attacker or by a local one.

In the past 2 years, exploiting or detecting CPU bugs and backdoors received much attention in the security community. For instance, in [25], Kapersky discusses the impact of bugs in Intel CPUs, some of which give attackers full control over the machine; in [27], King et al. present the design and implementation of flawed hardware that supports general purpose attacks and in [8], Shamir discusses how a CPU arithmetic unit bug might be used by attackers to carry out attacks against cryptographic algorithms. David et al. [10] introduce hardware-supported concealment techniques for rootkits. Agrawal et al. [3] as well as Duflot [12] investigate the consequences of backdoor introduction in hardware.

3.1.5 Local Attacks Using HID

We now consider that human interaction devices (HID) such as keyboards, mice, and display screens cannot be considered trustworthy. The attacker could launch local attacks exploiting a bug or using a backdoor in any of these devices as a means for privilege escalation.

Possible use of user interface devices by attackers has already been addressed from a side-channel perspective [5, 40]; however, the possible uses of keyboard and screen bugs or backdoors by attackers do not seem to have been studied yet. For instance, an attacker could modify a screen to export the display over the air, possibly using Wifi.[3] Another possibility would be to trap a keyboard to record the keystrokes and replay them in an encoded form to a process controlled by the attacker in the non-sensitive environment.

3.1.6 Local Attacks Using RAM DIMM

Finally we assume that the RAM DIMMs cannot be trusted any more because of a bug or backdoor that has been introduced during the manufacturing process. Therefore, an attacker may try to exploit such a backdoor to gain more privileges.

To the best of our knowledge, this type of assumption has not been considered before in the literature. The main contribution of this chapter is to study the implications of this kind of vulnerability and illustrate its feasibility. We will show what an attacker may achieve should such a backdoor exist inside DIMM modules.

---

[2] We would also consider here any other controller that would not be embedded in the chipset such as a network controller.

[3] On many laptops, the Wifi antenna is indeed wrapped around the screen.

4 Design of a DIMM Backdoor

We have seen in the previous section that most of the potential attack mechanisms were not relevant if we assume that the CPU, the chipset, and the virtual machine monitor are trustworthy. The only attack vectors left to the attacker are then multilevel peripherals, namely keyboard, mouse, screen, and DIMM, that are shared between domains.

In the remainder of the chapter, we study the threats associated with DIMM bugs or backdoors. Of course, a backdoor with features similar to those presented in the following sections could also be implemented inside the chipset. However, from now on, we assume that the target machine is a compartmented system composed of a trustworthy CPU and a trustworthy chipset (i.e., a trustworthy motherboard). The only hardware component we do not consider trustworthy is the DIMM. To the best of our knowledge, this chapter is the first one to address this issue.

The lower software layers are also supposed to be trusted, but the attacker is still able to run code on the machine on the very precise conditions described in Sect. 2.2 (see Fig. 2).

4.1 Overview of DDR DIMM

RAM-integrated storage devices are called DIMM (dual inline memory modules). They are typically composed of different RAM circuits and embed a memory controller that manages those circuits. On modern computers, the chipset can communicate with the DIMMs of the computer using the DDR/DDR2 interface (DDR SDRAM stands for Double Data Rate Synchronous Dynamic RAM). The overall principle of the DDR [23] protocol is that DDR DIMMs are organized in banks (or circuits). It is expected that each DIMM corresponds to one bank. The chipset can only address one bank at a time using a "chip select." Banks are in turn composed of ranks. For each bank, only one rank can be selected at a time. The ranks typically correspond to the different RAM circuits of the DIMM but in principle, the physical

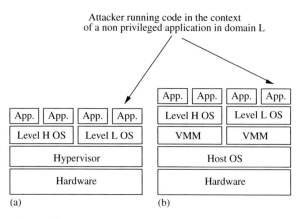

Fig. 2 Local attacker model

layout of the DRAM chips on the DIMM does not have to reflect the rank layout. Ranks correspond to an array of data where data can be accessed using a row and column index.

Physical addresses management at the chipset level is very simple. The chipset enumerates all the different banks on the DDR bus and associates the first bank (Bank0) with addresses ranging from 0 to Bank0_Max_capacity-1 of the bank, the second one (Bank1) with addresses ranging from Bank0_Max_capacity to Bank0_-Max_capacity+Bank1_Max_capacity-1, and so on. Correspondence between banks and addresses is thus straightforward.

Of course, physical addresses are translated into bank, rank, rows, and columns addresses by the chipset and sent on the DDR interface. Contrary to the chipset, the DIMM itself only deals with ranks, rows, and columns and does not know their correspondence with physical addresses; banks are implicit, as they physically correspond to a selection of the correct DIMM. However, the only information the DIMM is lacking to reconstruct physical addresses is the physical base address for the bank. In a single DIMM system, this base address will likely be 0. On a multiple DIMM system, because DIMM sizes are standardized, the number of possibilities is very limited.

4.2 Principle of the Backdoor

As the attacker is able to interfere with the DIMM manufacturing process and modify the software that will be running on the embedded controller of the DIMMs, she could include hidden functions or backdoors inside the DIMM and later try to bypass the isolation enforced by a virtual machine monitor by invoking these functions from a non-privileged application running in the least-sensitive domain L (see Fig. 2).

The goal of the attacker could be either to get access to confidential data or to fully compromise the higher clearance domain. In any case, the attacker has to be able to locate a target structure (either the confidential data itself of the operating system structure she is aiming at modifying).

According to our threat model, the attacker is not able to determine the physical memory mapping used by both domains, but she is able to allocate memory pages. The backdoor has to be simple but should be usable in such a context.

For the sake of simplicity we will also first assume that the system only uses one DIMM (in that case physical addresses and rank, row, column tuples are equivalent). We consider the case of systems using multiple DIMMs in Sect. 6.

4.3 Proof of Concept Implementation

Even if the idea of a RAM-based backdoor can seem far fetched and not really realistic, it has proven to be very easy to implement on an emulated hardware platform and almost trivial to use in practice.

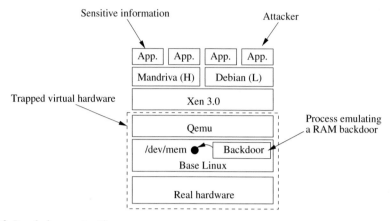

Fig. 3 Proof of concept setting

Figure 3 presents the architecture of the proof-of-concept implementation. The hardware platform is emulated by Qemu [4], which is running on a minimal Linux system, called Base Linux.[4] Qemu emulates a complete hardware platform; so, in theory, the hidden functions should be implemented inside Qemu. However, Qemu uses instruction translation mechanisms, the goal of which is to execute instructions whose effects are entirely contained within the elements managed by the control program (such as memory locations and arithmetic registers) directly on the hardware. As a consequence, accesses to the memory pages allocated by Qemu are not intercepted by Qemu, nor by the underlying Base Linux, and prevent us from implementing the backdoor inside Qemu. Nonetheless, from the Qemu perspective, we can consider that the Base Linux lies somewhere between the chipset emulated by Qemu and the main system memory (see Fig. 3). This is why we have chosen to implement our proof-of-concept backdoor as a process running at the application level of the Base Linux system. This way, the backdoor acts as if it was implemented inside the DIMMs themselves. Memory accesses by the malware are simulated by means of the /dev/mem pseudodevice exposed by the Base Linux kernel.

Our basic assumption is that the attacker has perfect knowledge of the DIMM hidden functions, but she is not able to determine the physical addresses of memory buffers that she allocates, neither does she know the memory layout of the target system and applications.

Therefore, the attacker has to design a backdoor that can be used whatever the target system is. The attacker also has to create a communication channel between herself and the backdoor allowing her to somehow send parametrized commands to the backdoor and to read back status information and responses. As the attacker

---

[4] The Base Linux should not be confused with the hypervisor or the host OS shown in Fig. 1.

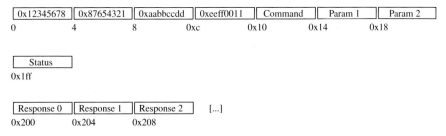

Fig. 4 Command layout

does not know the physical address of the buffer she is allocating, the easiest way to send commands to the backdoor is to decide on a fixed pattern that will be required to precede any command (for instance, a 128-bit pattern such as 0x12345678, 0x87654321, 0xaabbccdd, 0xeeff0011). The backdoor will permanently look for this pattern in memory and interpret the subsequent memory words as a command and its parameters if it finds the pattern. The layout of such a command is presented in Fig. 4. The assumption is of course that no legitimate application would write such a pattern to memory. This assumption seems realistic if the pattern is long enough and reasonably well chosen (i.e., the pattern does not correspond to any usable code, ASCII string). Anyway, should an application use the pattern legitimately, the command following it in memory is likely to be meaningless from the backdoor's point of view (and thus would not be interpreted), or its execution would result in a crash of the application. To reduce significantly the probability of such a mistake, we could add a CRC pattern ending each command.

The easiest way for the trojan function to send data to the attacker is to use the memory that immediately follows the command and its parameters. Indeed, allocated memory buffers are composed of several pages (4 kB) of memory on x86 platforms. Contiguous virtual memory is not required to correspond to contiguous physical memory (see Fig. 5), but a correctly aligned virtual memory page will always correspond to 4 kB of contiguous memory zones. Therefore, if the attacker uses a correctly aligned 4 kB buffer to store the pattern and the command, then the backdoor knows that the next 4 kB starting at the pattern address are actually allocated to the attacker and that data to be sent back to the attacker can safely be written here. Moreover, this strategy really speeds up the pattern-matching process as the backdoor only has to look for the pattern on 4 kB boundaries, which will be far more efficient than to browse the whole memory space.

Commands will thus use a 4 kB buffer with a layout such as the one presented on Fig. 4.

The next step is for the attacker to determine which commands are necessary for her to blindly carry out an attack without any indication of the physical memory layout of the target system. At least, the attacker needs the two following commands:

- a `read(A)` command allows the attacker to read 32 bits of date stored at address A;
- a `write(A,V)` command allows the attacker to write 32-bit data V at address A;

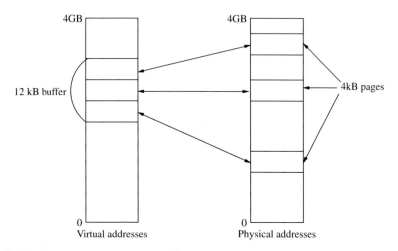

Fig. 5 Virtual memory management on x86 systems

As we will see later, sending these commands and getting an answer can be a long process. That is why we add a third command, needed for a practical attack:

- a `find_pattern(P,N,S)` command looks for a pattern P in memory, starting from address S, and returns the address A of the pattern and optionally dump N bytes of RAM content immediately following the pattern.

The attacker could also craft more complex commands in the controller (`read` or `write` commands with a size parameter, a `find_pattern_write` command that would be essentially the same as the `find_pattern` command but would allow the attacker to modify those data that follow the pattern) but these commands are not necessary for the attacks we present in the following section. Figure 6 shows a possible layout for those three commands. The status bit (offset 0x1ff) is set by the backdoor when the command has successfully been executed by the backdoor. In our proof of concept, we also included a text log output to the console that is used to describe the operations the backdoor has carried out.

Because of cache-related limitations which will be studied more in depth in Sect. 5.1, the attacker must make sure that the command line, the status bit, and the data returned by the backdoor are not stored in the same cache line (see Fig. 6). Because of this, the amount of data that can be dumped is limited to 0xdfb bytes (limit of the command buffer (0xfff) minus the offset of the returned data (0x204)).

For the sake of readability of the proof-of-concept implementation, we will consider that addresses exchanged between the attacker and the hardware backdoor are physical addresses. Of course, a real implementation would require the addresses to be encoded in rank, rows, column, but the proof-of-concept schemes (see Sect. 4.3) would work exactly in the same way.

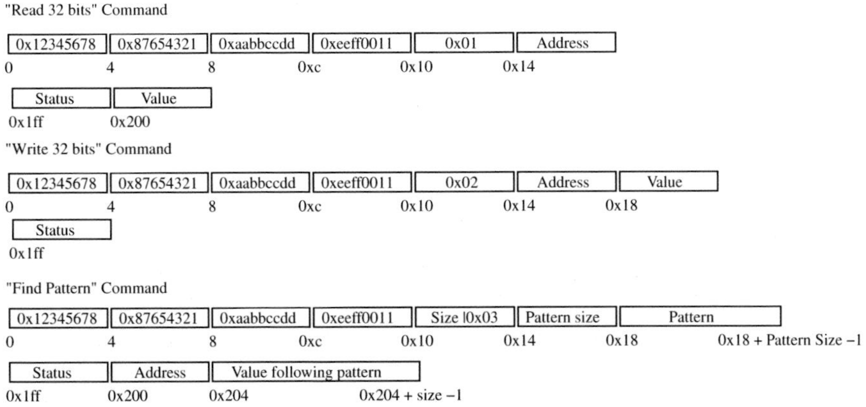

"Read 32 bits" Command

0x12345678	0x87654321	0xaabbccdd	0xeeff0011	0x01	Address
0	4	8	0xc	0x10	0x14

Status	Value
0x1ff	0x200

"Write 32 bits" Command

0x12345678	0x87654321	0xaabbccdd	0xeeff0011	0x02	Address	Value
0	4	8	0xc	0x10	0x14	0x18

Status
0x1ff

"Find Pattern" Command

0x12345678	0x87654321	0xaabbccdd	0xeeff0011	Size |0x03	Pattern size	Pattern
0	4	8	0xc	0x10	0x14	0x18	0x18 + Pattern Size −1

Status	Address	Value following pattern
0x1ff	0x200	0x204	0x204 + size −1

Fig. 6 Layout of the three commands

5 Exploitation

5.1 Difficulties

In this section, we try to analyze the technical difficulties that the attacker will have to overcome when trying to use the backdoor.

The first difficulty is to actually locate her target structure. In case the attacker is willing to access data from domain H (which denotes a high-sensitivity domain; in contrast low-sensitivity domain is called domain L), she first has to determine the correct address in memory of this information. In order to do so, the attacker has to use the backdoor to determine the address of a pattern she knows will be present at a fixed offset from the data she is willing to access. However, the fact that the pattern and the sought information are close in the virtual address space does not mean that they actually are in the real address space. Indeed, if the pattern and the information are stored in different pages of memory, the attacker will not know the correspondence between the pattern address and the information address unless she has knowledge of the page tables and directories for domain H. If the attacker cannot find such a pattern or if the address of the pattern cannot be used because of domain H virtual memory management strategy, then the attacker will have to dump the whole memory space using the "read" command. In case the attacker is willing to modify the target structure, she will also have to locate it using similar means.

Another major difficulty is related to memory caching. In our model, the attacker can only send commands to the backdoor and read status and return values returned by the backdoor through buffers allocated in user space. If domain L uses a "Write Back" cache management strategy [21], write operations are performed in cache and are not propagated to main memory before the corresponding cache line is evicted or the operating system decides to flush the caches. As a consequence, the attacker does not know when the commands she sent will actually make their way to the main memory, where the backdoor can find them. Furthermore, read operation might also only occur in cache. For instance, if the attacker tries to read the status bit and finds it cleared, all subsequent read of the status bit may only be performed in cache. The attacker will continue to read a cleared status flag even though the DIMM actually modified the status bit in the meantime. Cache coherency will not be enforced as the DIMM is not supposed to modify data stored in physical memory on its own initiative. For the attack to succeed, the attacker will typically have to wait a few seconds in order to be sure that the DIMM did receive the command and have time to send the response back. The need for such a delay between a request and an answer justifies using more complex commands allowing for dumping longer memory areas, instead of just a word, so as to reduce the number of requests.

Alternatively, an attacker with sufficient knowledge of the platform cache mechanisms could try and evict the cache line manually by filling large data arrays.

It is interesting to notice that, even if our experiment implements the backdoor in a process running on the Linux system, instead of the DIMM module, we face the same cache difficulties. The physical pages affected by the backdoor are indeed mapped at two different virtual addresses: in the Base Linux running Qemu, and in domain L; however, the cache strategy in domain L is "Write Back" whereas the page is not cached at all by the /dev/mem pseudo device, so a modification of the physical memory page by the backdoor process is not cached, and does not trigger a cache update for the second mapping. The mechanism is thus exactly similar to what would have happened if the malware had been installed in the DIMM.

Another potential difficulty will also be that physical memory pages may be moved by the operating system (or even swapped to the disk) as the system is trying to organize physical memory in the best way (usually, unused pages are swapped to the disk and brought back to physical memory later at a different address). However, this shall not be too much of a problem for the attacker as target structures are not likely to be swapped (it is often recommended that sensitive data and kernel structures are not swapped to ensure either confidentiality or good system performance), and the buffers allocated by the attackers would not be swapped if the attacker takes care to access them frequently.

5.2 Use of the Hidden Functions to Access Sensitive Data

We have been able to use our proof-of-concept implementation of a DIMM backdoor in the context presented on Fig. 3. The attacker is running code in the context of a non-privileged application running in domain L (Debian domain) and is able to either recover sensitive data from domain H or modify the kernel of the operating system.

For the first proof of concept, we considered that the user was running an application in domain H storing its online banking (OLB) password and that the attacker is willing to get access to this passphrase. We assume that the attacker knows what the application is and for instance knows another string (St) that will be stored with

the password (for instance the full user name or a prefix such as "OLB password:").
The attack steps are as follows:

1. send the `find_pattern(St, Size_max, 0)` command where St is the "OLB password:" string, Size_max is the maximum size that the backdoor will dump. The command must be stored at the beginning of a correctly aligned 4 kB buffer allocated by the attacker;
2. wait for a few seconds for the backdoor to find the pattern in memory and read the status bit. If the status bit is set, read the data sent back by the backdoor at offset 0x204;
3. if the data is not what the attacker is looking for (i.e., a false positive[5]), the first two steps should be repeated after updating the starting address of the `find_pattern` command.

Below is a description of what the attacker has to do.

```
#####(Written by the attacker)
#define PAGE_SIZE 0x1000
#define N_INT (PAGE_SIZE / sizeof(int))
//create the buffer where the command will be written
int * command_buff = malloc(PAGE_SIZE*2);
//make sure the buffer is aligned correctly
command_buff = (command_buff + N_INT) & 0xfffff000;
//define the command
int * attack_command = build_find_pattern_command
         (pattern, size_max, size_to_dump);
//send command
memcpy(command_buff, attack_command, command_size)

######(Backdoor log output when executed)
Received Find Pattern Command
Command success
Pattern Found at address 0x2e185418

#####(Written by the attacker)
//read the address of the pattern
address = command_buff[0x80];
//read what follows the pattern
//(sent back by the backdoor)
memcpy(value_read , command_buff + 0x81, size_max);
//display what follows the pattern
```

---

[5] The attacker shall keep in mind of certain false positives that are inherent to this scheme. Depending on the start address given to the `find_pattern` command, the pattern she is looking for will indeed be found in the command itself and might also be found in the process launched on domain L.

```
//(should be a printable  account number)
printf("Account number: %s\n", (char*) value_read);
```

5.3 Use of the Backdoor as a means for Privilege Escalation

In this second experiment, we show how the attacker running code in the context of a non-privileged application of domain L could get full control of domain H by modifying the kernel of the Mandriva operating system running within domain H. For the proof of concept, we show how the attacker can modify the setuid() system call of the Mandriva operating system so that any process running the setuid() system call is granted superuser ("root") privileges. We assume that the attacker knows that the target operating system actually is a regular Mandriva operating system beforehand. In that case, the attacker knows the structure of the setuid system call code and which part of it must be modified. The code of the setuid system call (or part of the system call) can be used as the pattern to search. In our example, we look for the first bytes (Syscall_pattern) of the system call. We know that once the address of this pattern is known, the attacker has to replace the code that could be found 0xf4 bytes further by the following code (New_code). Indeed, the original code located at setuid+0xf4 corresponds to the place where the user effective id is set (mov [%esi], 0x16c(%eax) - 6-byte instruction – where %esi points to the requested user id, and 0x16c(%eax) is the current user effective id. This instruction is followed by another 6-byte instruction (mov [%esi], 0x174(%eax)) that is useless in our context. Those two instructions should be replaced by the instruction mov $0, 0x16c(%eax) (10-byte instruction) followed by two nop instructions to ensure that the control flow of the system call is not modified. Indeed, $0 is the superuser id, so mov $0, 0x16c(%eax) grants any caller of the setuid() system call with superuser privileges.

In practice, the attacker has to

1. send a find_pattern(Syscall_pattern,Size_max,0) command where Size_max is the maximum size that the backdoor will dump. Again, the attacker will have to use a correctly aligned 4 kB buffer to do so;
2. wait for a few seconds for the backdoor to find the pattern in memory and read the status bit. If the status bit is set, read the data sent back by the backdoor at offset 0x204, and the address A of the pattern at offset 0x200 within the buffer;
3. if the data does not correspond to what she is looking for (false positive), the first two steps must be repeated after updating the starting address of the find_pattern command;
4. send three write commands, respectively, at addresses A+0xf4, A+0xf8, A+0xfc to overwrite the part of the system call where the user effective ID is set.

```
// mov $0, 0x16c(%eax) correspond to
//bytecode 0x00000000016c80c7
// nop to 0x90
```

```
write(A+0xf4,0x016c80c7)
write(A+0xf8, 0)
write(A+0xfc, 0x90900000)
```

Once this is done, any process running in domain H will be able to escalate to "root" privilege by running the setuid() system call using their own identity. For instance, users login will be granted the "root" user ID.

```
/* Regular behaviour      */    /* After use of the backdoor*/
Mandriva Linux Release 2008.0   Mandriva Linux Release 2008.0
Kernel 2.6 on an i686 / tty1    Kernel 2.6 on an i686 / tty1
Login: user                     Login: user
Password:                       Password:

#id                             -bash$id
uid=50(user)[...]euid=50(user)  uid=50(user)[...]euid=0(root)
#whoami                         -bash$whoami
user                            root
```

If the attacker does not know what the target operating system is, she has to use the backdoor "read" command to analyze the layout of the target operating system to identify the target structure before carrying the attack scheme.

6 Countermeasures

In the previous section, we have seen that unconditional isolation cannot be achieved if the system used a single DIMM that cannot be trusted. However, if we consider using multiple DIMMs (each of which is possibly trapped), the virtual machine monitor can enforce a memory allocation strategy such that each DIMM is used by only one domain. This way, the exploitation of the backdoor on each DIMM is still possible, but the attacker is unable to either retrieve data or carry out any privilege escalation scheme. As a consequence, it seems fair to recommend to use a different DIMM for each sensitivity level (and one devoted to the hypervisor) whenever possible.

An alternative countermeasure could be to use DIMMs from different manufacturers in order to mitigate the risk of the attacker having knowledge of how to exploit a backdoor in every DIMM module the system is using. If the system uses several DIMMs but does not assign one of them to the higher level and to the virtual machine monitor or hypervisor, the attacker may still carry out the scheme using the backdoor of the shared DIMM.

Lastly, RAM encryption may also be an efficient countermeasure, as it protects the confidentiality of the data stored in DIMMs, thereby preventing the backdoor from accessing sensitive information. Several projects aim at designing these kinds of architectures, such as Xom [29], Aegis [34], Hide [41], and CryptoPage [11]. These approaches basically assume that no component shall be trusted apart from the CPU, the role of which is to encrypt data that are written to memory and decrypt those that are read from memory. Memory encryption would prevent the DIMM backdoor from finding the patterns requested by the attacker, as it does not have access to the decryption key. In addition to confidentiality, some of these approaches also provide integrity protection (e.g., by means of a Merkle tree [30]) and prevent blind modifications of guest kernel data structures. As these techniques only assume that the CPU is trustworthy, they are efficient countermeasures against the attack we present.

7 Conclusion and Future Work

This chapter presents an original attack against compartmented systems, based on the assumption that a bug or backdoor is present in RAM (DIMM) modules. We show how an attacker can take advantage of these hidden functions to escalate her privileges or retrieve sensitive data from a protected domain, even if the software, CPU, chipset, and peripherals are trusted. We also illustrate how such a backdoor can be implemented and exploited.

Future work will include analysis of the impact of a bug or a backdoor in keyboards and display screens on the overall security of a virtual machine monitor- and hypervisor-based system.

For example, a keyboard could be trapped to record all the keystrokes in an internal buffer and then replay them in an encoded form to a process controlled by the attacker. A way to encode the keystrokes would be to simulate user pressing shift and control keys rapidly, for instance, or to use undefined scancode numbers. Furthermore, we could also imagine that the non-sensitive guest OS is compromised, and give some feedback to the rogue keyboard, to trigger a buffer emission, or acknowledge the data received, by means of a covert channel from the OS to the keyboard (e.g., through the virtualized keyboard LEDs that can be activated on the OS's own initiative).

References

1. L. Absil, L. Duflot, in *Programmed I/O Accesses: A Threat to Virtual Machine Monitors.* Pacific Security Conference PacSec07, Tokyo, Japan, 29–30 Nov 2007
2. Advanced Micro Devices (AMD). *AMD Virtualisation Solutions*, 2007. http://www.amd.com/virtualization/
3. D. Agrawal, S. Baktir, D. Karakoyunlu, P. Rohatgi, B. Sunar, in *Trojan Detection Using IC Fingerprinting.* Proceedings of the IEEE Symposium on Security and Privacy, Oakland, CA, USA, 20–23 May 2007, pp. 296–310
4. F. Bellard, *QEMU Open Source Processor Emulator*, 2007 http://wiki.qemu.org/
5. Y. Berger, A. Wool, A. Yeredor, in *Dictionary Attacks Using Keyboard Acoustic Emanations.* CCS'06: Proceedings of the 13th ACM Conference on Computer and Communications Security, Alexandria, VA, USA, 30 Oct–3 Nov, 2006 (ACM Press, New York, NY, 2006), pp. 245–254

6. D. J. Bernstein, *Cache Timing Attacks on AES*. Technical Report, The University of Illinois at Chicago, 2005.
7. G. Bertoni, V. Zaccaria, L. Breveglieri, M. Monchiero, in *AES Power Attack Based on Induced Cache Miss and Countermeasure*. ITCC'05: Proceedings of the International Conference on Information Technology: Coding and Computing, Las Vegas, NV, USA, 4–6 Apr 2005
8. E. Biham, Y. Carmeli, A. Shamir, in *Bug Attacks*. CRYPTO, Santa Barbara, CA, USA, 17–21 Aug 2008
9. BSDDaemon, coideloko, and D0nAnd0n, System Management Mode Hack: Using SMM for Other Purposes. *Phrack Magazine*, 2008. http://www.phrack.org/
10. F. David, E. Chan, J. Carlyle, R. Campbell, in *Cloaker: Hardware Supported Rootkit Concealment*. Proceedings of the IEEE Symposium on Security and Privacy, Oakland, CA, USA, 18–21 May 2008
11. G. Duc, R. Keryell, Cryptopage: An efficient secure architecture with memory encryption, integrity and information leakage protection. Ann. Comput. Secur. Appl. Conf., 483–492 (Shanghai, China, 6–8 Sept 2006)
12. L. Duflot, in *CPU Bugs, CPU Backdoors and Consequences on Security*. ESORICS 2008: Proceedings of the 13th European Symposium on Research Computer Security, Malaga, Spain, 6–8 Oct 2008
13. L. Duflot, O. Grumelard, O. Levillain, B. Morin, in *Getting into the SMRAM: SMM Reloaded*. CanSecWest Applied Security Conference 2009, Vancouver, Canada, 18–20 Mar 2009
14. L. Duflot, O. Levillain, B. Morin, in *ACPI: Design Principles and Concerns*. Trust 2009, Oxford, UK, 6–8 Apr 2009
15. S. Embleton, S. Sparks, in *The System Management Mode (SMM) Rootkit*. Black Hat Briefings, Washington, DC, USA, 18–21 Feb 2008
16. EMSCB Consortium. Turaya EMSCB, 2005. http://www.emscb.com/content/pages/emscb.turaya.htm
17. French National Research Agency. Secure and isolated operating system challenge, 2008. http://secsi.adullact.net/
18. GNU. *Linux VServer*, 2007. http://linux-vserver.org
19. J. Heasman, in *Implementing and Detecting an ACPI BIOS Rootkit*. Blackhat Federal 2006, Washington, DC, USA, 23–26 Jan 2006
20. G. Heiser, K. Elphinstone, I. Kuz, G. Klein, S. Petters, Towards trustworthy computing systems: Taking microkernels to the next level. ACM SIGOPS Oper. Syst. Rev. **41**(4), 3–11 (July, 2007)
21. Intel Corp. *Intel 64 and IA 32 architectures software developer's manual volume 3A: system programming guide part 1*, 2007
22. Intel Corp. *Intel 64 and IA 32 architectures software developer's manual volume 3A: system programming guide part 2*, 2007
23. JEDEC. *DDR2 specification*, Nov 2009
24. P.-H. Kamp, R.N.M. Watson, in *Jails: Confining the Omnipotent Root*. Proceedings of the 2nd International SANE Conference, Maastricht, The Netherlands, 22–25 May 2000
25. K. Kaspersky, in *Remote Code Execution Through Intel CPU bugs*. Hack In The Box Security Conference, Kuala Lumpur, Malaysia, 27–30 Oct 2008
26. O. Kaya, J.-P. Seifert, On the Power of Simple Branch Prediction Analysis. Cryptology ePrint Archive, 2006. http://eprint.iacr.org/2006/351.pdf
27. S. King, J. Tucek, A. Cozzie, C. Grier, W. Jiang, Y. Zhou, in *Designing and Implementing Malicious Hardware*. Proceedings of the First USENIX Workshop on Large Scale Exploits and Emergent Threats, LEET'08, San Francisco, CA, USA, 15 Apr 2008
28. C. Lauradoux, in *Collision Attacks on Processors with Cache and Countermeasures*. WeWorC '05: Western European Workshop on Research in Cryptology, Leuven, Belgium, 5–7 July 2005
29. D. Lie, C. Thekkath, M. Mitchell, P. Lincoln, D. Boneh, J. Mitchell, M. Horowitz, Architectural support for copy and tamper resistant software. ACM SIGPLAN Not. **35**(11), 168–177 (2000)

30. R. Merkle, *Secrecy, Authentication and Public Key Systems – A Certified Digital Signature*. Ph.D. thesis, Department of Electrical Engineering, Stanford University, 1979
31. National Security Agency. NetTop, 2009. http://www.nsa.gov/research/tech_transfer/fact_sheets\discretionary{-}{}{}/nettop.shtml
32. J. Rutkowska, R. Wojtczuk, in *Preventing and Detecting Xen Hypervisor Subversions*. Blackhat Briefings, Washington, DC, USA, 18–21 Feb 2008
33. A. Sacco A. Ortega, in *Persistent BIOS Infection*. CanSecWest Conference, Vancouver, Canada, 18–20 Mar 2009
34. G.E. Suh, D. Clarke, B. Gassend, M. van Dijk, S. Devadas, in *Aegis: Architecture for Tamper-Evident and Tamper-Resistant Processing*. ICS '03: Proceedings of the 17th Annual International Conference on Supercomputing, San Francisco, CA, USA, 23–26 June 2003 (ACM, New York, NY, 2003), pp. 160–171
35. Y. Tsunoo, T. Saito, T. Suzaki, M. Shigeri, H. Miyauchi, in *Cryptanalysis of DES Implemented on Computers with Cache*. CHES '03: Proceedings of the 4th Workshop on Cryptographic Hardware and Embedded Software, Cologne, Germany, 7–10 Sept 2003
36. University of Cambridge. *Xen Virtual Machine Monitor*, 2007. http://www.cl.cam.ac.uk/research/srg/netos/xen/
37. J. Vanegue, in *Hacking PXE Without Reboot*. BA-Con Argentina, 2008
38. VMware Inc. *VMware Virtualisation Software*, 2007
39. R. Wojtczuk, J. Rutkowska, in *Attacking Intel Trusted Execution Technology*. Blackhat Federal 2009, 2009
40. L. Zhuang, F. Zhou, J.D. Tygar, in *Keyboard Acoustic Emanations Revisited*. CCS '05: Proceedings of the 12th ACM Conference on Computer and Communications Security Alexandria, VA, USA, 7–11 Nov 2005 (ACM Press, New York, NY, 2005), pp. 373–382
41. X. Zhuang, T. Zhang, S. Pande, Hide: An infrastructure for efficiently protecting information leakage on the address bus. ACM SIGOPS Oper. Syst. Rev. **38**(5), 72–84 (2004)

Efficient Secure Two-Party Computation with Untrusted Hardware Tokens (Full Version)*

Kimmo Järvinen, Vladimir Kolesnikov, Ahmad-Reza Sadeghi, and Thomas Schneider

1 Introduction

Secure and efficient evaluation of arbitrary functions on private inputs has been subject of cryptographic research for decades. In particular, the following scenario appears in a variety of practical applications: a service provider (server S) and user (client C) wish to compute a function f on their respective private data, without incurring the expense of a trusted third party. This can be solved interactively using Secure Function Evaluation (SFE) protocols, for example, using the very efficient garbled circuit (GC) approach [23, 36]. However, GC protocols potentially require a large amount of data to be transferred between S and C. This is because f needs to be encrypted (garbled) as $\tilde{f}$ and transferred from S to C. In fact, the communication complexity of GC-based SFE protocols is dominated by the size of the GC, which can reach megabytes or gigabytes even for relatively small and simple functions (e.g., the GC for AES has size 0.5 MB [30]). Further, if security against more powerful adversaries is required, the use of the standard cut-and-choose technique implies transfer of multiple GCs. (For covert adversaries, the transfer of only one GC is sufficient [12].)

While transmission of this large amount of data is possible for exceptional occurrences, in most cases, the network will not be able to sustain the resulting traffic. This holds especially for larger scale deployment of secure computations, e.g., by banks or service providers, with a large number of customers. Additional obstacles include energy consumption required to transmit/receive the data, and the resulting reduced battery life in mobile clients, such as smartphones.[1]

T. Schneider (✉)
Horst Görtz Institute for IT-Security, Ruhr-University Bochum, Germany
e-mail: thomas.schneider@trust.rub.de

*A short version of this chapter appears at FC'10 [18].

[1] In some cases, the impact can be mitigated by creating and transferring GCs in the precomputation phase. However, this is not fully satisfactory. First, even more data needs to be transferred since demand cannot be perfectly predicted. Further, this creates other problems, such as requiring large long-term storage on client devices.

A.-R. Sadeghi, D. Naccache (eds.), *Towards Hardware-Intrinsic Security,* Information Security and Cryptography, DOI 10.1007/978-3-642-14452-3_17,
© Springer-Verlag Berlin Heidelberg 2010

Further, computational load on S (computing $\widetilde{f}$) is also a significant problem, especially in the case of large-scale deployment of SFE.

1.1 Our Setting, Goals, and Approach

Motivated by the possibility of large-scale and decentralized SFE deployment we aim to remove the expensive communication requirement and to shift some of S's computation to C. To this end, we note that in SFE protocols, and, in particular, in GC, the role of the server can be split between two entities, with the introduction of a new entity – secure token T, which is placed in client C's possession, but executes S's code thus offloading S. Further, it is possible to eliminate most of the communication between C and S and replace this with local communication between C and T. A number of technical issues arises in this setting, which we address in this work.

More specifically, we discuss and analyze hardware-supported SFE, where the service provider S issues a *secure* (see below) hardware token T to C. C communicates locally with T and remotely with S. There is no direct channel between T and S, but of course C can pass (and potentially interfere with) messages between T and S. T is created by S, so S trusts T; however, as C does not trust S, she also does not trust the token T to behave honestly.[2]

Attack model. We consider all three standard types of adversaries: semi-honest (follows protocol but analyzes the transcript), malicious (arbitrary behavior, cheating is always caught), and covert (cheating is caught with a certain deterrence probability, e.g., $1/2$).

Hardware assumption. We assume T is tamper proof or tamper resistant. We argue that this assumption is reasonable. Indeed, while every token can likely be broken into given sufficient resources, we are motivated by the scenarios where the payoff of the break is far below the cost of the break. This holds for relatively low-value transactions such as cell phone or TV service, where the potential benefit of the attack (e.g., free TV for one user) is not worth the investment of thousands or tens of thousands of dollars to break into the card. For higher value applications one could raise the cost of the attack by using a high-end token T, e.g., a smart card certified at FIPS 140-2, level 3 or 4.

Hardware restrictions. As we assume the token to be produced in large quantities, we try to minimize its costs (e.g., chip surface) and make the assumptions on it as weak as possible. In particular our token requires only restricted computational capabilities (no public-key operations) and small constant secure RAM. We consider T with and without small constant secure non-volatile storage.

---

[2] Note, if C in fact trusts T to behave honestly, then there exists a trivial solution, where C would let T compute the function on her inputs [16].

Further, we envision smart phones playing the role of the client (and a SIM card the role of the token). Therefore we aim to limit client's computational power and storage ability.

1.2 Envisioned Applications

As mentioned above, we aim to bring SFE closer to a large-scale deployment. The need to minimize communication forces us to rely on tokens, the issuance of which requires certain logistical capabilities. Therefore, we believe client–server applications are most likely to be the early adopters of SFE. Further, the natural trust model (semi-honest or covert server and malicious client) allow for efficient GC protocols. Finally, many client–server application scenarios naturally include financial and other transactions which involve sensitive, in particular privacy-sensitive, information.

Today, many service providers already distribute trusted tokens to their users. Examples include SIM cards in users' cell phones and smart cards in users' TV set-top boxes. Bank- and credit cards often contain embedded secure chips. Special purpose (e.g., diagnostic) medical devices, which need to be monitored and controlled, are often issued to users at home. In these settings, it is natural to use the existing infrastructure to enhance the security and privacy guarantees of the offered products and to offer new products previously impossible due to privacy violation concerns. We consider the following examples in more detail.

Privacy protection in targeted advertisement and content delivery. Cable TV (phone, Internet) providers gain a large part of their revenue from advertisements, so it is desired to increase their effectiveness by considering individual user preferences, purchase history, and other collected personal information. On the other hand, privacy guidelines and laws severely limit the kinds of information that can be collected and how it can be used. Further, even legal use of personal information by the service provider may be viewed as privacy violation and cause negative perception of the company. At the same time, using SFE to select advertisements guarantees customers' privacy and, moreover, ensures that the company cannot breach their privacy policies even by accident.

Of course, other content (songs, movies, TV shows) can be target delivered as rewards or incentives, while preserving complete privacy of the user. Discount coupons and certificates, powerful spending incentives well liked by users, bring much more value to both the issuer and the user, if their delivery is based on sensitive personal and location information.

Privacy preserving remote medical diagnostics. Health care is moving faster than ever toward technologies that offer personalized online self-service, medical error reduction, consumer data mining, and more (e.g., [11]). Such technologies have the potential of revolutionizing the way medical data is stored, processed, delivered, and made available in an ubiquitous and seamless way to millions of users all over the world. Here service provider S usually owns the diagnostic software and/or hardware that operates on C's data and outputs classification/diagnostic results. A concrete example in this context is the classification of electrocardiogram (ECG) data. A privacy-enhanced version for remote ECG diagnostics requires to transfer GCs of size $\approx$ 63 MB [2]. With our token-based protocol this can be reduced substantially to approximately 100 kB as no garbled circuits need to be transferred.

Other applications concern *financial transactions* such as monetary transfers, bidding, or betting, and *biometric authentication*.

1.3 Our Contributions and Outline

Our main contribution is architecture design, implementation, a number of optimizations, and detailed analysis of two-party SFE aided by a server-issued low-cost tamper-proof token. The communication complexity of our protocols is linear in the size of the input and is independent of the size of the evaluated functionality. Further, most of the work of S can be offloaded to T.

We use GC techniques of [20] and offer no-cost XOR gates. We rely on cheap hardware – the token T only executes symmetric-key operations (e.g., SHA and AES). T has small constant-size RAM (much smaller than the size of the circuit), but we do not resort to implementing expensive secure external RAM. We also show how to optimize for low-power, low-memory client C.

We provide two solutions; in one, T keeps state in secure non-volatile storage (a monotonic counter), while in the other, T maintains no long-term state.

We consider semi-honest, covert [12], and malicious [22] adversaries; our corresponding communication improvements are shown in Table 1.

1.3.1 Outline

We start with outlining our model and architecture in Sect. 3. We describe the protocols for both stateful and stateless T and state the security claim in Sect. 4. In Sect. 5 we give further optimizations to speed up computation of T by caching and to optimize for low-power, low-memory C. In Sect. 6, we discuss technical details of our FPGA prototype implementation, present timings, and measurements and show practicality of our solution.

1.4 Related Work

Related work on using tokens for secure computations falls in the following three categories, summarized in Table 2.

Table 1 Communication between server S and client C for secure evaluation of function f with n inputs, statistical security parameter s, and deterrence probability $1 - 1/r$

Security	Previous work		This work
Semi-honest	[36]	$\mathcal{O}(|f| + n)$	$\mathcal{O}(n)$
Covert	[12]	$\mathcal{O}(|f| + sn + r)$	$\mathcal{O}(sn + r)$
Malicious	[22]	$\mathcal{O}(s|f| + s^2 n)$	$\mathcal{O}(s^2 n)$

Table 2 Secure protocols using hardware tokens. Columns denote the number of tokens, who trusts the token(s), if token(s) are stateful or stateless, and perform public-key operations. Properties more desired for practical applications in bold font

Type	References	Functionality	# Tokens	Trusted by	Stateful	PK ops
(A)	[15]	UC commitment	2	Both	Yes	Yes
	[8, 19]	UC commitment	2	**Issuer**	Yes	Yes
	[7]	UC commitment	2	**Issuer**	**No**	Yes
	[26]	UC commitment	1	**Issuer**	Yes	**No**
(B)	[14]	Set intersection, ODBS	1	Both	Yes	**No**
	[13]	Non-interact. OT	1	Both	Yes	Yes
	[34]	Verif. Enc., Fair Exch.	1	Both	Yes	Yes
(C)	[10]	**SFE**	2	Both	Yes	Yes
	[16]	**SFE**	1	Both	Yes	Yes
	This work	**SFE**	1	**Issuer**	**Yes/No**	**No**

(A) Setup assumptions for the universal composability (UC) framework. As shown in [4], UC SFE protocols can be constructed from UC commitments. In turn, UC commitments can be constructed from signature cards trusted by both parties [15], or from tamper-proof tokens created and trusted only by the issuing party [7, 8, 19, 26]. Here, [7] consider stateless tokens and [26] require only one party to issue a token. This line of research mainly addresses the feasibility of UC computation based on tamper-proof hardware and relies on expensive primitives such as generic zero-knowledge proofs. Our protocols are far more practical.

(B) Efficiency improvements for specific functionalities. Efficient protocols with a tamper-proof token trusted by both players have been proposed for specific functionalities such as set intersection and oblivious database search (ODBS) [14], non-interactive oblivious transfer (OT) [13], and verifiable encryption and fair exchange [34]. In contrast, we solve the general SFE problem.

(C) Efficiency improvements for arbitrary functionalities. Clearly, SFE is efficient if aided by a trusted third party (TTP), who simply computes the function. SFE aided by hardware TTP was considered, e.g., in [10, 16]. In contrast, we do not use TTP; our token is only trusted by its issuer.

2 Preliminaries

Notation. We denote symmetric security parameter by t (e.g., $t = 128$), and pseudo-random function (PRF) keyed with k and evaluated on x by $\mathsf{PRF}_k(x)$. PRF can be instantiated with a block cipher, e.g., AES, or a cryptographic hash function H, e.g., SHA-256, which we model as a Random Oracle (RO). AES is preferable if PRF is run repeatedly with same k as AES's key schedule amortizes. Message authentication code (MAC) keyed with k and evaluated on message m is denoted by $\mathsf{MAC}_k(m)$. We use an MAC that does not need to store the entire message but can operate "online" on small blocks, e.g., AES-CMAC [32] or HMAC [21].

2.1 Garbled Circuits (GC)

Yao's Garbled Circuit approach [36], excellently presented in [23], is the most efficient method for secure evaluation of a boolean circuit C. We summarize its ideas in
the following. The circuit *constructor* (server S) creates a *garbled circuit* $\widetilde{C}$: for each
wire w_i of the circuit, he randomly chooses two garblings $\widetilde{w}_i^0, \widetilde{w}_i^1$, where $\widetilde{w}_i^j$ is the
garbled value of w_i's value j. (Note: $\widetilde{w}_i^j$ does not reveal j.) Further, for each gate
G_i, S creates a *garbled table* $\widetilde{T}_i$ with the following property: given a set of garbled
values of G_i's inputs, $\widetilde{T}_i$ allows to recover the garbled value of the corresponding
G_i's output, but nothing else. S sends these garbled tables, called *garbled circuit*
$\widetilde{C}$ to the *evaluator* (client C). Additionally, C obliviously obtains the *garbled inputs*
$\widetilde{w}_i$ corresponding to inputs of both parties: the garbled inputs $\widetilde{y}$ corresponding to
the inputs y of S are sent directly and $\widetilde{x}$ are obtained with a parallel 1-out-of-2
oblivious transfer (OT) protocol [1, 27]. Now, C can evaluate the garbled circuit $\widetilde{C}$
on the garbled inputs to obtain the *garbled outputs* by evaluating $\widetilde{C}$ gate by gate,
using the garbled tables $\widetilde{T}_i$. Finally, C determines the plain values corresponding to
the obtained garbled output values using an output translation table received by S.
Correctness of GC follows from the way garbled tables $\widetilde{T}_i$ are constructed.

Improved Garbled Circuit with free XOR [20]. An efficient method for creating garbled circuits which allows "free" evaluation of XOR gates was presented
in [20]. More specifically, a garbled XOR gate has no garbled table (*no communication*) and its evaluation consists of XORing its garbled input values (*negligible
computation*) – details below. The other gates, called *non-XOR gates*, are evaluated
as in Yao's GC construction [36] with a *point-and-permute technique* (as used in
[25]): The garbled values $\widetilde{w}_i = \langle k_i, \pi_i \rangle \in \{0, 1\}^{t'}$ consist of a symmetric key
$k_i \in \{0, 1\}^t$ and a random permutation bit $\pi_i \in \{0, 1\}$ (recall, t is the symmetric
security parameter). The entries of the garbled table are permuted such that the
permutation bits π_i of a gate's garbled input wires can be used as index into the
garbled table to directly point to the entry to be decrypted. After decrypting this
entry using the garbled input wires' t-bit keys k_i, evaluator obtains the garbled
output value of the gate. The encryption is done with the symmetric encryption
function $\mathsf{Enc}^s_{k_1,\dots,k_d}(m)$, where d is the number of inputs of the gate and s is a
unique identifier used once. Enc can be instantiated with $m \oplus \mathsf{H}(k_1 || \dots || k_d || s)$,
where H is a RO. This requires 2^d invocations of H for creating and 1 invocation for
evaluating a garbled non-XOR gate. To avoid random oracles, Enc can be instan-
tiated with $m \oplus H(k_1 || s) \oplus .. \oplus H(k_d || s)$ instead, where H is a correlation robust
hash function (cf. [30] for details). This needs $d \cdot 2^d$ invocations of H for creating
and d invocations for evaluating a non-XOR gate. In practice, H can be chosen
from the SHA-2 family. The main observation of [20] is that the constructor S
chooses a global key difference $\Delta \in_R \{0, 1\}^t$ which remains unknown to evaluator C and relates the garbled values as $k_i^0 = k_i^1 \oplus \Delta$. Clearly, the usage of such
garbled values allows for *free evaluation of XOR gates* with input wires w_1, w_2 and
output wire w_3 by computing $\widetilde{w}_3 = \widetilde{w}_1 \oplus \widetilde{w}_2$ (no communication and negligible
computation).

3 Architecture, System, and Trust Model

We present in detail our setting, players, and hardware and trust assumptions.

As shown in Fig. 1, there are three parties – client $\mathcal{C}$, server $\mathcal{S}$, and tamper-resistant token $\mathcal{T}$, issued and trusted by $\mathcal{S}$. Our goal is to let $\mathcal{C}$ and $\mathcal{S}$ securely evaluate a public function f on their respective private inputs x and y.

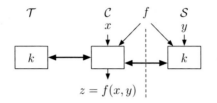

Fig. 1 Model overview

Communication. $\mathcal{C} \leftrightarrow \mathcal{S}$: We view this as an expensive channel. Communication $\mathcal{C} \leftrightarrow \mathcal{S}$ flows over the Internet and may include a wireless or cellular link. This implies small link bandwidth and power consumption concerns of mobile devices. We wish to minimize the utilization of this channel.

$\mathcal{T} \leftrightarrow \mathcal{C}$: As $\mathcal{T}$ is held locally by $\mathcal{C}$, this is a cheap channel (in terms of both bandwidth and power consumption), suitable for transmission of data linear in the size of f, or even greater.

$\mathcal{T} \leftrightarrow \mathcal{S}$: There is no direct channel between $\mathcal{T}$ and $\mathcal{S}$, but, of course, $\mathcal{C}$ can pass (and potentially interfere with) messages between $\mathcal{T}$ and $\mathcal{S}$.

Trust. $\mathcal{C} \leftrightarrow \mathcal{S}$: As in the standard SFE scenario, $\mathcal{C}$ and $\mathcal{S}$ do not trust each other. We address semi-honest, covert, and malicious $\mathcal{C}$ and $\mathcal{S}$.

$\mathcal{S} \leftrightarrow \mathcal{T}$: $\mathcal{T}$ is fully trusted by $\mathcal{S}$, since $\mathcal{T}$ is tamper resistant. $\mathcal{S}$ and $\mathcal{T}$ share a secret key k, used to establish a secure channel and to derive joint randomness.

$\mathcal{T} \leftrightarrow \mathcal{C}$: $\mathcal{C}$ does not trust $\mathcal{T}$, as $\mathcal{T}$ is the agent of $\mathcal{S}$ and may communicate with $\mathcal{S}$ through covert channels.

Storage, computation, and execution. $\mathcal{C}$ and $\mathcal{S}$ are computationally strong devices which can perform both symmetric- and asymmetric-key operations.[3] Both have sufficient memory, linear in the size of f. We also address the setting where $\mathcal{C}$ is a weak mobile device with restricted memory in Sect. 5.1. $\mathcal{C}$ has control over $\mathcal{T}$ and can reset it, e.g., by interrupting its power supply. As justified in Sect. 1.1, $\mathcal{T}$ is a cheap special purpose hardware with minimum chip surface: $\mathcal{T}$ has circuitry only for evaluating symmetric-key primitives in hardware (no public-key or true random number generator) and has a small secure RAM. It may (Sect. 4.3) or may not (Sect. 4.4) have small non-volatile secure storage[4], unaffected by the resets by $\mathcal{C}$.

---

[3] If needed, $\mathcal{C}$'s capabilities may be enhanced by using a trusted hardware accelerator.

[4] $\mathcal{T}$'s key k is a fixed part of its circuit and is kept even without non-volatile storage.

4 Token-Assisted Garbled Circuit Protocols

In our presentation, we assume reader's familiarity with the GC technique, including free XORs of [20] (cf. Sect. 2.1), and concentrate on the aspects specific to the token setting. We start with a high-level description of our protocol. Then, in Sects. 4.2, 4.3 and 4.4, we present the technical details of our construction – efficient circuit representation, and GC generation by stateful and stateless tokens.

4.1 Protocols Overview and Security

Our constructions are a natural (but technically involved) modification of standard GC protocols, so as to split the actions of the server into two parts – now executed by S and T – while maintaining provable security. We offload most of the work (notably, GC generation and output) to T, thus achieving important communication savings and partially offloading S's computation to T.

We start our discussion with the solution in the semi-honest model. However, our modification of the basic GC is secure against malicious actions, and our protocols are easily and efficiently extendible to covert and malicious settings.

At the high level, our protocols work as shown in Fig. 2: C obtains the garbled inputs $\widetilde{x}, \widetilde{y}$ from S, and the garbled circuit $\widetilde{f}$ corresponding to the function f from T. Then, C evaluates $\widetilde{f}$ on $\widetilde{x}, \widetilde{y}$ and obtains the result $z = f(x, y)$.

It is easy to see that the introduction of T and offloading to it some of the computation does not strengthen S and thus does not bring security concerns for C (as compared to standard two-party GC). On the other hand, separating the states of S and T, placing C in control of their communication, and C's ability to reset T introduces attack opportunities for C. We show how to address these issues with the proper synchronization and checks performed by S and T.

Our main tool is the use of a unique session id sid for each GC evaluation. From sid and the shared secret key k, S and T securely derive a session key K, which is then used to derive the randomness used in GC generation. Jumping ahead (details in Sect. 4.3), we note that sid uniqueness is easily achieved if T is stateful simply by

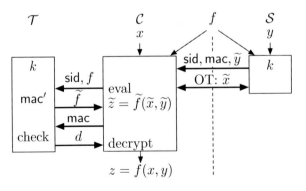

Fig. 2 Protocols overview

setting sid equal to the value of the strictly monotonic session counter ctr maintained by T. However, if T is stateless, C can always replay S's messages. In Sect. 4.4 we show how to ensure that replays do not help C.

Since S and T derive the same randomness for each session, the (same) garbled circuit $\widetilde{f}$ can be generated by T. Unfortunately, the weak T cannot store the entire f. Instead, C provides the circuit corresponding to function f gate by gate to T and obtains the corresponding garbled gate of $\widetilde{f}$. The garbled gate can immediately be evaluated by C and needs not to be stored. C is prevented from providing a wrong f to T, as follows. First, S issues a MAC of f, e.g., $\mathsf{mac} = \mathrm{MAC}_k(\mathsf{sid}, f)$, where f is the agreed circuit representation of the evaluated function (cf. Sect. 4.2). Further, T computes its version of the above MAC, mac', as it answers C's queries in computing $\widetilde{f}$. Finally, T reveals the decryption information d that allows C to decrypt the output wires only if C provides the matching mac.

4.1.1 Garbled Inputs

The garbled input $\widetilde{y}$ of S can be computed by S and sent to C, requiring $|y| \cdot t$ bits communication, where t is the security parameter. Alternatively, if T is stateful, S can establish a secure channel with T, e.g., based on session key K, send y over the channel and have T output $\widetilde{y}$ to C. This achieves the optimal communication between S and C of $|y|$ bits.

The garbling $\widetilde{x}$ of C's input can be transferred from S to C with a parallel OT protocol, which requires $\mathcal{O}(|x|t)$ bits of communication. Alternatively, the efficient OT extension of [17], which reduces many OTs to a small number of OTs (depending on security parameter t), can be adopted to our token-based scenario as described next. This reduces the communication between S and C to $\mathcal{O}(t^2)$, which is independent of the size of the input x.

4.1.2 Extending OT Efficiently with Token

The efficient OT extension of [17] for *semi-honest receiver* C can be used to replace the possibly huge number of $|x|$ parallel OTs with a substantially smaller number of t OTs only, where t is a security parameter. As the token T is computationally weak and cannot perform public-key operations, the t real OTs in the protocol of [17] are performed between the computationally strong devices C and S. Afterward, S sends its state, i.e., the values obtained in the parallel OT protocol to T over a secure channel. This requires $\mathcal{O}(t^2)$ communication from S to T (forwarded by C). Finally, T completes the protocol with C, which requires several invocations of a correlation robust hash function (e.g., SHA-256) only, but no more public-key operations.

The fully secure OT extension protocol of [17] that is secure against *malicious receiver* C requires additional cut-and-choose, which results in correspondingly increased communication between C and S. Nevertheless, the communication between C and S remains independent of the number of inputs $|x|$.

4.1.3 Extension to Covert and Malicious Parties

Standard GC protocols for covert [12] or malicious [22] adversaries rely on the following cut-and-choose technique. S creates multiple GCs $\widetilde{C}_i$, deterministically derived from random seeds s_i, and commits to each, e.g., by sending $\widetilde{C}_i$ or $H(\widetilde{C}_i)$ to C. In covert case, C asks S to open all but one garbled circuit I by revealing the corresponding $s_{i \neq I}$. For all opened circuits, C computes $\widetilde{C}_i$ and checks that they match the commitments. The malicious case is similar, but C asks S to open half of the circuits, evaluates the remaining ones, and chooses the majority of their results.

These protocols similarly benefit from our token-based separation of the server into S and T. As in the semi-honest protocol, the GC generation can be naturally offloaded to T, achieving corresponding computation and communication relief on the server and network resources. GC correctness verification is achieved by requesting S to reveal the generator seeds $s_{i \neq I}$. (Of course, these "opened" circuits are not evaluated.) Note that requirements on T are the same as in the semi-honest setting. Further, in both covert and malicious cases, the communication between C and S is independent of the size of f. The resulting communication complexity of these protocols is summarized in Table 1.

4.1.4 Security Claim

For the lack of space, in this work we present our protocols implicitly, by describing the modifications to the base protocols of [20]. We informally argue the security of the modifications as they are described. Formal proofs can be naturally built from proofs of [20] and our security arguments. At the very high level, security against S/T follows from the underlying GC protocols, since S is not stronger here than in the two-party SFE setting. The additional power of C to control the channel between S and stateful T is negated by establishing a secure channel (Sect. 4.3). C's power to reset stateless T is addressed by ensuring that by replaying old messages C gets either what he already knows or completely unrelated data (Sect. 4.4).

Theorem 1 *Assuming T is tamper-proof, protocols described throughout this section are secure in the semi-honest, covert, and malicious models respectively.*

4.2 Circuit Representation

We now describe our circuit representation format. Our criteria are compactness, the ability to accommodate free XOR gates of [20], and ability of T to process the encoding "online", i.e., with small constant memory. Recall, our T operates in request-response fashion. C incrementally, gate by gate, "feeds" the circuit description to T, which responds with the corresponding garbled tables.

We consider circuits with two-input boolean gates. We note that our techniques can be naturally generalized to general circuits.

Our format is derived from standard representations, such as that of Fairplay [25], with the necessary changes to support our requirements. For readability, we

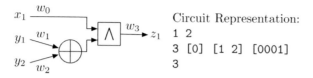

Fig. 3 Example for circuit representation

describe the format using a simple example circuit shown in Fig. 3. This circuit computes $z_1 = x_1 \wedge (y_1 \oplus y_2)$, where x_1 is the input bit of C and y_1, y_2 are two input bits of S. The corresponding circuit representation shown on the right is composed from the description of the inputs, gates, and outputs as follows.

Inputs and wires: The wires w_i of the circuit are labeled with their index $i = \{0, 1, ...\}$ (wires with a larger fan-out are viewed as a single wire). The first X wires are associated with the input of C, the following Y wires are associated with the input of S, and the internal wires are labeled in topological order starting from index $X + Y$ (output wires of XOR gates are not labeled, as XOR gates are incorporated into their successor gates as described in the next paragraph). The first line of the circuit description specifies X and Y (Fig. 3: $X = 1, Y = 2$).

Gates are labeled with the index of their outgoing wire; each gate description specifies its input wires. XOR gates do not have gate tables and are omitted from the description. Rather, non-XOR gates, instead of pointing to two input wires, include two input wire *lists*. If the input list contains more than one wire, these wire values are to be XORed to obtain the corresponding gate input. Gate's description concludes with its truth table. In Fig. 3, the second line describes the AND gate, which has index 3, and inputs w_0 and $w_1 \oplus w_2$.

Outputs: The circuit description concludes with Z lines, which contain the indices of the Z output wires (Fig. 3: the only ($Z = 1$) output wire is w_3).

Large XOR sub-circuits. In this representation, XOR gates with fan-out > 1 occur multiple times in the description of their successor gates. In the worst case, this results in a quadratic increase of the circuit description. To avoid this cost, we insert an identity gate after each XOR gate with a large fan-out.

4.2.1 Transformation of Circuit into Token Format

To convert a given circuit with n-input gates into the format described before, the gates are first decomposed into two-input gates and afterward the XOR gates are grouped together:

Decomposition into 2-input gates. Decomposing the n-input gates of a circuit into multiple 2-input gates can be done in a straightforward way using Shannon's expansion theorem [31] or the QuineMcCluskey algorithm, which results in smaller circuits [35]. For small n (e.g., for the common case of $n = 3$), the optimal replacement can be found via brute-force enumeration of all possibilities [30].

Grouping of XOR gates. The XOR gates can be grouped together as follows: To each input wire and each output wire of a non-XOR gate i we assign the set $\{i\}$.

Afterward we transfer the gates of the circuit in topological order and annotate to the output wire of each XOR gate the following set which is computed from the sets of its input wires S_1, S_2 as $S = S_1 \oplus S_2 := (S_1 \cup S_2) \setminus (S_1 \cap S_2)$. Finally, the remaining non-XOR gates are output in the token format using the sets associated to the input wires, which contain those wires that need to be XORed together for the specific input. As merging two sets of size at most $|C|$ entries each can be done in $\mathcal{O}(|C|)$ operations, the overall complexity of this algorithm is in $\mathcal{O}(|C|^2)$.

4.3 GC Creation with Stateful Token (Secure Counter)

The main idea of our small-RAM-footprint GC generation is having $\mathcal{T}$ generate garbled tables "on the fly." This is possible, since each garbled table can be generated only given the garblings of input and output wires. In our implementation, we pseudorandomly derive the wire garbling from the session key and wire index. The rest of this section contains relevant details.

Session Initialization. SFE proceeds in sessions, where one session is used to securely evaluate a function once. $\mathcal{T}$ has a secure monotonic session counter ctr which is (irreversibly) incremented at the beginning of each session. The session id sid is set to the incremented state of ctr. (We omit the discussion of synchronization of ctr between $\mathcal{T}$ and S, which may happen due to communication and other errors.) Then, the session key is computed by S and $\mathcal{T}$ as $K = \mathsf{PRF}_k(\mathsf{sid})$ and subsequently used to provide fresh randomness to create the GC.

As required by the construction of [20] (cf. Sect. 2.1), the two garbled values of the same wire differ by a global difference offset Δ. This offset is derived from K at session initialization and kept in RAM throughout the session.

Subsequently, garbled wire values w_i are derived on the fly from K as

$$\widetilde{w}_i^0 = \mathsf{PRF}_K(i), \qquad \widetilde{w}_i^1 = \widetilde{w}_i^0 \oplus \Delta. \tag{1}$$

Garbled Gates. $\mathcal{T}$ receives the description of the circuit, line by line, in the format described in Sect. 4.2 and generates and outputs to C corresponding garbled gates, using only small constant memory. $\mathcal{T}$ first verifies that the gate with the same label had not been processed before. (Otherwise, by submitting multiple gate tables for the same gate, C may learn the real wire values.) This is achieved by keeping the monotonically increasing processed gate counter gctr, verifying that gate's label glabel > gctr, and setting gctr = glabel. $\mathcal{T}$ then derives and stores garblings of the gate's input and output wires according to (1). (For input lists, the wire's garbling $\widetilde{w}^0$ is computed as the XOR of garblings of the listed wires, and $\widetilde{w}^1$ is set to $\widetilde{w}^0 \oplus \Delta$. Note that this requires constant RAM.) Finally, based on these garblings, gate's garbled table is computed and output to C.

Garbled Outputs. Recall, $\mathcal{T}$ must verify circuit correctness by checking mac generated by S. Thus, $\mathcal{T}$ does not release the output decryption tables to C until after the successful check. At the same time, the check is not complete until the entire circuit had been fed to $\mathcal{T}$. To avoid having $\mathcal{T}$ store the output decryption tables or involving S at this stage, T simply encrypts the output tables using a fresh key K' and outputs the key only upon a successful MAC verification.

4.4 GC Creation with Stateless Token (No Counter)

As discussed above, while non-volatile secure storage (the counter ctr) is essential in our protocol of Sect. 4.3, in some cases, it may be desired to avoid its cost. We now discuss the protocol amendments required to maintain security of SFE with the support of a token whose state can be reset by, e.g., a power interruption.

First, we observe that S is still able to maintain state and choose unique counters. However, T can no longer be assured that sid claimed by C is indeed fresh. Further, T does not have a source of independent randomness and thus cannot establish a secure channel with S, e.g., by running a key exchange.

We begin with briefly describing a replay vulnerability of our protocol of Sect. 4.3, when T is executed with same sid. First, C properly executes SFE. Second time he runs T with the same sid, but feeds T an incorrect circuit, receiving valid garbled tables for each of the gates, generated for the *same* wire garblings. Now, even though T will not accept mac and will not decrypt the output wires, C had already received them in the first execution. It is easy to see that C "wins."

Our solution is to ensure that C does not benefit from replaying T with the same sid. To achieve this, we require that each wire garblings are derived from the (hash of the) entire gate description (i.e., id, truth table, and list of inputs), as described below. If C replays and gives a different gate description, she will not be able to relate the produced garbled table with a previous output of T.

We associate with each wire w_i a (revealed to C) hash value h_i. For input wires, h_i is the empty string. For each other wire i, h_i is derived (e.g., via Random Oracle) from the description of the gate i (which includes index, truth table, and list of inputs; cf. Sect. 4.2) that emits that wire: $h_i = \mathsf{H}(\langle gate\_description \rangle)$. The garbled value of wire w_i now depends on its hash value h_i: $\widetilde{w}_i^0 = \mathsf{PRF}_K(h_i)$ and $\widetilde{w}_i^1 = \widetilde{w}_i^0 \oplus \Delta$. Finally, to enable the computation of the garbled tables, C must feed back to T the hashes h_i of the input wires and receive from T and keep for future use the hash of the output wire. As noted above, C's attempts to feed incorrect values result in the output of garbled tables that are unrelated to previous outputs of T and thus do not help C.

5 Further Optimizations

In order to allow C to evaluate the garbled circuit on the fly without caching the garbled gates, the gates of the circuit must be given in topologic order, i.e., all gates on which a gate depends have to occur before. As for most circuits many topologic orders exist, one can choose a specific topologic order according to further optimizations. We give two examples in the following – optimize C's memory requirements

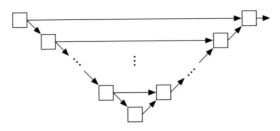

Fig. 4 Circuit with k gates that requires $\Theta(k)$ memory for evaluation

for GC evaluation (Sect. 5.1) or reduce $\mathcal{T}$'s runtime by caching intermediate values (Sect. 5.2).

5.1 Optimizing Memory of Client

For evaluating the garbled circuit, $\mathcal{C}$ starts with the garbled values of the inputs. Then, $\mathcal{C}$ obtains the garbled gates one by one from the token and evaluates them. The obtained garbled output for the gate needs to be stored in $\mathcal{C}$'s memory until it is used the last time as input into a garbled gate.

This allows to find a good topologic order of the circuit, which requires less memory for storing the intermediate results.

As pointed out in [3], the problem of finding a topologic order of a circuit that minimizes the needed amount of memory to store intermediate values is equivalent to the register allocation problem, which is well studied in the context of compilers. In fact, algorithms for register allocation [6, 33] can be used to find an optimal, respectively, good topologic order of the circuit, which reduces the amount of memory needed for its evaluation.

In the worst case, the memory needed to evaluate a circuit C is linear in the circuit size as one might need to keep garbled values of many wires. Figure 4 shows an (artificially constructed) example circuit with k gates, which requires $\Theta(k)$ memory for evaluation.

5.2 Optimizing Runtime of Token by Caching

During creation of the garbled circuit, the token $\mathcal{T}$ derives the garbled values corresponding to the gates' inputs and outputs with a PRF. If the garbled value has already been derived before and stored in a cache, the garbled value does not need to be derived again. When the garbled values are stored into or loaded from the cache can be encoded in the description of the circuit provided by $\mathcal{C}$.

5.2.1 Performance Improvements by Caching

At a first glance it seems that the cache provides the possibility for a time memory tradeoff. This is clearly the case if T allows sequential execution of instructions only (e.g., in a single-threaded software implementation on a smart card) as each cache hit results in reduced computation time. However, if T allows parallel execution (e.g., on FPGAs, ASICs, or CPUs with multiple cores), the garbled values can be derived in parallel to creating the garbled tables. As demonstrated in our FPGA prototype implementation in Sect. 6, the derivation of the garbled values using AES (10 cycles in our implementation) is substantially faster than creating the garbled tables using SHA (70 cycles in our implementation). In this setting, only gates with many inputs would benefit from caching.

5.2.2 Cache Sizes

If each garbled value should be derived only once and taken from the cache every time it is used afterward, the cache needs to have size $\mathcal{O}(|f|)$. In practice, the size of the cache is, however, bounded to a constant number of cache entries.

Cache with constant size. To determine a good ordering of the gates which maximizes the number of cache hits one can use standard algorithms for instruction scheduling and register allocation (instructions correspond to gates and registers to cache entries) [33].

Cache with size 1. In many cases, a cache with only one cache entry is sufficient already. This is due to the fact that many commonly used functionalities (e.g., addition or comparison) are composed from a line of gates in which the output of a gate directly serves as the input into the next gate.

The following efficient algorithm (time and space complexity in $\mathcal{O}(|f|)$) determines a topological order with many hits for the one entry cache: Insert – starting from the output gates – an edge to each of the gates whose outputs are input into the current gate. Afterward run a depth-first search on the outputs which marks a gate as visited and visits all successor gates which have not been visited yet. After having visited all successor gates, the gate is output as next in the topologic order. This algorithm eliminates "dead" gates and results in many cache hits for the one entry cache.

6 Proof-of-Concept Implementation

We have designed a proof-of-concept implementation to show the practicability of our token-assisted GC protocols of Sect. 4. In the following we describe our architecture for the stateful token case of Sect. 4.3. Extension to the stateless case is straightforward. We instantiate PRF with AES-128, H and H with SHA-256, and MAC with AES-CMAC.

6.1 Architecture

Figure 5 depicts the high-level architecture of our design consisting of a two-stage pipeline and an MAC core. Stage 1 of the pipeline creates the garbled input and output values of a gate using an AES core, and stage 2 computes the garbled table with a SHA core. The two-stage pipeline increases performance as two gates can be processed concurrently. The MAC core computes the authentication message mac' of the circuit provided by $\mathcal{C}$ (cf. Sect. 4.1).

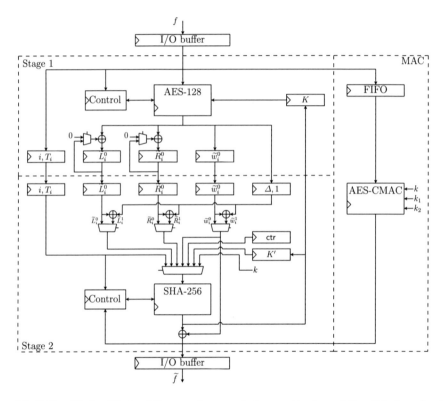

Fig. 5 Simplified architectural diagram of our proof-of-concept implementation. Selectors of multiplexers and write enables of registers are set by additional control logics

Design Principle. To achieve maximum speed with minimum hardware resources, we followed a general guideline exploiting parallelism as long as it can be done without using several instances of the same algorithm. For example, we opted to compute the four entries of a garbled table with a single SHA core instead of using four parallel SHA cores, which would have increased performance, but only with a significant increase in area. As the only exception, we included a separate MAC core rather than reusing the AES core of stage 1 because it would have severely complicated the control of the pipeline.

Description of Operation. In the session initialization (cf. Sect. 4.4), the SHA core in stage 2 derives session key K and output encryption key K' from key k and current counter value $\mathsf{ctr} = \mathsf{sid}$, which is used as key for the AES core. Then, the key difference Δ is derived with the AES core and stored in a register. The circuit is provided gate by gate into the input buffer in the format described in Sect. 4.2 (gate table denoted by T_i in Fig. 5). Stage 1 starts to process a gate by deriving the garbled output value $\widetilde{w}_i^0$. Then, the two garbled inputs of the gate ($\widetilde{L}_i^0$, $\widetilde{R}_i^1$ in Fig. 5) are derived by XORing the garblings listed in the input wire lists one by one (see Sect. 4.3). When all garblings are derived they are forwarded to stage 2 and stage 1 processes the next gate. Stage 2 computes the garbled table and the encrypted output decryption tables and writes them into the output buffer. The MAC core operates independently from the two-stage pipeline.

6.2 Prototype Implementation

We implemented our architecture in VHDL.

Implementation Details. For the AES core we chose an iterative design of AES-128 with a latency of 10 clock cycles per encryption. The core includes an online key scheduling unit. The S-boxes are implemented as suggested in [5]; otherwise, the core is a straightforward implementation of the standard [28]. The SHA core implements SHA-256 with a latency of 67 clock cycles per 512-bit block. The core is a straightforward iterative design of the standard [29]. The MAC core includes an AES core implemented as above; otherwise the core is a straightforward implementation of AES-CMAC [32]. As the subkeys, k_1 and k_2, depend only on the key k they were precomputed and hardwired in the design.

FPGAs. We compiled the VHDL code for a low-end FPGA, the Altera Cyclone II EP2C20F484C7 FPGA, with Quartus II, version 8.1 (2008). We emphasize that this FPGA is for prototyping only, as it lacks secure embedded non-volatile memory for storing ctr (e.g., the Xilinx Spartan-3AN FPGAs has integrated Flash memory for this). The resulting area and memory requirements are listed in Table 3. The design occupies 60% of logic cells and 21% of memory blocks available on the device and runs at 66 MHz (the critical path for clock frequency is in the AES core). These results show that the design is, indeed, feasible for a low-cost implementation, for example, with low-cost FPGAs which, in turn, is mandatory for the practicability of the token-based scheme.

Smart Cards. In particular, we note that the requirements are sufficiently low also for contemporary smart card technologies, because AES-128 and SHA-256 require

Table 3 Results on an Altera Cyclone II FPGA (the hierarchy is as shown in Fig. 5)

Entity	Stage 1	Stage 2	MAC	IO	Total
Area (Logic cells)	3,317 (30%)	4,539 (40%)	3,059 (27%)	263 (3%)	1,1231 (60%)
Memory (M4K)	0 (0%)	8 (73%)	1 (9%)	2 (18%)	11 (21%)

only about 3,400 and 11,000 gates, respectively [9]. As our protocol requires no public-key operations on the token, small smart cards are sufficient.

6.2.1 Performance

We determined the latency of our implementation with ModelSim, version 6.3g (2008). Overall, the latency is given as #clock_cycles $= 158G_1 + 312G_2 + 154O + 150$, where G_1, G_2 is the number of 1-input gates, respectively, 2-input gates and O is the number of outputs, assuming that each gate has at most 21 inputs in its input lists (if more, stage 2 needs to wait for stage 1) and that data I/O does not introduce additional delays. If we use a correlation robust hash function instead of RO, the coefficient for G_2 would roughly double for up to 42 inputs (cf. Sect. 2.1).

Example 1 Our implementation generates a GC for 16-bit comparison ($G_1 = 0$, $G_2 = 16$, $O = 1$) in 5,296 clock cycles ($\approx 80\,\mu s$ with 66 MHz clock). In Software, this takes roughly 0.5 s on an Intel Core 2 6,420 at 2.13 GHz [24].

Example 2 Generating a GC for AES-128 encryption ($G_1 = 12,614$, $G_2 = 11,334$, $O = 128$) takes 5,549,082 clock cycles ($\approx 84\,ms$ with 66 MHz clock). In Software, this takes approximately 1 s on an Intel Core 2 Duo at 3.0 GHz [30].

We note that the optimization for large XOR sub-circuits described in Sect. 4.2 dramatically reduces the amount of communication between C and T: When using this optimization, the size of the AES circuit of Example 2 is $|C| = 1.1$ MB and the garbled AES circuit has size $|\widetilde{C}| = 1.1$ MB. Without this optimization, the circuit has no more 1-input gates ($G_1 = 0$), which results in a faster creation ($3,556,070$ clock cycles), evaluation, and size of the garbled circuit ($|\widetilde{C}| \approx 0.7$ MB). However, the size of the circuit is drastically larger ($|C| = 94.5$ MB), which might be a bottleneck if the communication between C and T is slow.

Acknowledgments We would like to thank Wilko Henecka for preparing test circuits and Ivan Damgård and reviewers of FC'10 for their helpful comments. The first author was supported by EU FP7 project CACE. The third and fourth authors were supported by EU FP6 project SPEED, EU FP7 project CACE, and ECRYPT II.

References

1. W. Aiello, Y. Ishai, O. Reingold, Priced oblivious transfer: How to sell digital goods. in *Advances in Cryptology – EUROCRYPT'01*. Lecture Notes in Computer Science, vol. 2045 (Springer-Verlag, Berlin, Heidelberg, New York, NY, 2001), pp. 119–135
2. M. Barni, P. Failla, V. Kolesnikov, R. Lazzeretti, A.R. Sadeghi, T. Schneider, in *Secure Evaluation of Private Linear Branching Programs with Medical Applications*. European Symposium on Research in Computer Security (ESORICS'09). Lecture Notes in Computer Science, vol. 5789 (Springer, Saint-Malo, France, 21–23 Sept 2009), pp. 424–439
3. C.L. Berman, Circuit width, register allocation, and ordered binary decision diagrams. IEEE Trans. CAD **10**(8), 1059–1066 (1991)

4. R. Canetti, Y. Lindell, R. Ostrovsky, A. Sahai, in *Universally Composable Two-party and Multi-party Secure Computation*. ACM Symposium on Theory of Computing (STOC'02), Montréal, Québec, Canada, 19–21 May 2002, pp. 494–503
5. D. Canright, in *A Very Compact S-box for AES*. Cryptographic Hardware and Embedded Systems (CHES'05), Edinburgh, UK, 29 Aug–1 Sept 2005. Lecture Notes in Computer Science, vol. 3659 (Springer, 2005), pp. 441–456
6. G.J. Chaitin, M.A. Auslander, A.K. Chandra, J. Cocke, M.E. Hopkins, P.W. Markstein, Register allocation via coloring. Comput. Lang. **6**(1), 47–57 (1981)
7. N. Chandran, V. Goyal, A. Sahai, New constructions for UC secure computation using tamper-proof hardware. in *Advances in Cryptology – EUROCRYPT'08*, Istanbul, Turkey, 13–17 Apr 2008. Lecture Notes in Computer Science, vol. 4965 (Springer, 2008), pp. 545–562
8. I. Damgård, J.B. Nielsen, D. Wichs, in *Universally Composable Multiparty Computation with Partially Isolated Parties*. Theory of Cryptography (TCC'09), San Francisco, CA, USA, 15–17 Mar 2009. Lecture Notes in Computer Science vol. 5444 (Springer, 2009), pp. 315–331
9. M. Feldhofer, J. Wolkerstorfer, in *Strong Crypto for RFID Tags — A Comparison of Low-Power Hardware Implementations*. International Symposium on Circuits and Systems (ISCAS'07) (IEEE Computer Society, 2007), pp. 1839–1842
10. M. Fort, F.C. Freiling, L.D. Penso, Z. Benenson, D. Kesdogan, in *Trustedpals: Secure Multiparty Computation Implemented with Smart Cards*. European Symposium on Research in Computer Security (ESORICS'06), Hamburg, Germany, 18–20 Sept 2006. Lecture Notes in Computer Science, vol. 4189 (Springer, 2006), pp. 34–48
11. Google Health (2009). `https://www.google.com/health`
12. V. Goyal, P. Mohassel, A. Smith, Efficient two party and multi party computation against covert adversaries. in *Advances in Cryptology – EUROCRYPT'08*, Istanbul, Turkey, 13–17 Apr 2008. Lecture Notes in Computer Science, vol. 4965 (Springer, 2008), pp. 289–306
13. V. Gunupudi, S. Tate, in *Generalized Non-interactive Oblivious Transfer Using Count-Limited Objects with Applications to Secure Mobile Agents*. Financial Cryptography and Data Security (FC'08), Cozumel, Mexico, 28–31 Jan 2008. Lecture Notes in Computer Science, vol. 5143 (Springer, 2008), pp. 98–112
14. C. Hazay, Y. Lindell, in *Constructions of Truly Practical Secure Protocols Using Standard Smartcards*. ACM Conference on Computer and Communications Security (CCS'08) (ACM, New York, NY, USA 2008), pp. 491–500
15. D. Hofheinz, J. Müller-Quade, D. Unruh, in *Universally Composable Zero-Knowledge Arguments and Commitments from Signature Cards*. Central European Conference on Cryptology (MoraviaCrypt'05), Brno, The Czech Republic, 15–17 June 2005
16. A. Iliev, S. Smith, *More Efficient Secure Function Evaluation Using Tiny Trusted Third Parties*. Technical Report TR2005-551, Dartmouth College, Computer Science, Hanover, NH (2005). URL `http://www.cs.dartmouth.edu/reports/TR2005-551.pdf`
17. Y. Ishai, J. Kilian, K. Nissim, E. Petrank, Extending oblivious transfers efficiently. in *Advances in Cryptology – CRYPTO'03*. Lecture Notes in Computer Science, vol. 2729 (Springer-Verlag, Berlin, Heidelberg, New York, NY 2003) pp. 145–161
18. K. Järvinen, V. Kolesnikov, A.-R. Sadeghi, T. Schneider, in *Embedded SFE: Offloading Server and Network Using Hardware Tokens*. In 14th International Conference on Financial Cryptography and Data Security (FC'10). Lecture Notes in Computer Science vol. 6052 (Springer, Jan 2010) pp. 207–221
19. J. Katz, Universally composable multi-party computation using tamper-proof hardware. in *Advances in Cryptology – EUROCRYPT'07*, Barcelona, Spain, 20–24 May 2007. Lecture Notes in Computer Science, vol. 4515 (Springer, 2007), pp. 115–128
20. V. Kolesnikov, T. Schneider, in *Improved Garbled Circuit: Free XOR Gates and Applications*. International Colloquium on Automata, Languages and Programming (ICALP'08), Reykjavik, Iceland, 6–13 July 2008. Lecture Notes in Computer Science, vol. 5126 (Springer, 2008), pp. 486–498
21. H. Krawczyk, M. Bellare, R. Canetti, HMAC: Keyed-hashing for message authentication. RFC 2104 (Informational), (1997). `http://tools.ietf.org/html/rfc2104`

22. Y. Lindell, B. Pinkas, An efficient protocol for secure two-party computation in the presence of malicious adversaries. in *Advances in Cryptology – EUROCRYPT'07* Barcelona, Spain, 20–24 May 2007. Lecture Notes in Computer Science, vol. 4515 (Springer, 2007), pp. 52–78

23. Y. Lindell, B. Pinkas, A proof of Yao's protocol for secure two-party computation. J. Cryptol. **22**(2), 161–188 (2009).   Cryptology ePrint Archive, Report 2004/175, http://eprint.iacr.org

24. Y. Lindell, B. Pinkas, N. Smart, in *Implementing Two-party Computation Efficiently with Security Against Malicious Adversaries*. Security and Cryptography for Networks (SCN'08), Amalfi, Italy, 10–12 Sept 2008. Lecture Notes in Computer Science, vol. 5229 (Springer, 2008), pp. 2–20

25. D. Malkhi, N. Nisan, B. Pinkas, Y. Sella, in *Fairplay — A Secure Two-party Computation System*. USENIX Security Symposium (Security'04), San Diego, CA, USA, 9–13 Aug 2004 (USENIX Association, 2004)

26. T. Moran, G. Segev, David and goliath commitments: UC computation for asymmetric parties using tamper-proof hardware. in *Advances in Cryptology – EUROCRYPT'08*, Istanbul, Turkey, 13–17 Apr 2008. Lecture Notes in Computer Science, vol. 4965 (Springer, 2008), pp. 527–544

27. M. Naor, B. Pinkas, in *Efficient Oblivious Transfer Protocols*. ACM-SIAM Symposium On Discrete Algorithms (SODA'01), Washington, DC, USA, 7–9 Jan 2001. (Society for Industrial and Applied Mathematics, 2001), pp. 448–457

28. NIST, U.S. National Institute of Standards and Technology: Federal information processing standards (FIPS 197). Advanced Encryption Standard (AES) (2001). http://csrc.nist.gov/publications/fips/fips197/fips-197.pdf

29. NIST, U.S. National Institute of Standards and Technology: Federal information processing standards (FIPS 180-2). Announcing the Secure Hash Standard (2002). http://csrc.nist.gov/publications/fips/fips180-2/fips-180-2.pdf

30. B. Pinkas, T. Schneider, N.P. Smart, S.C. Williams, Secure two-party computation is practical. in *Advances in Cryptology – ASIACRYPT 2009*, Tokyo, Japan, 6–10 Dec 2009. Lecture Notes in Computer Science, vol. 5912 (Springer, 2009), pp. 250–267

31. C.E. Shannon, The synthesis of two-terminal switching circuits. Bell Syst. Tech. J. **28**(1), 59–98 (1949)

32. J. Song, R. Poovendran, J. Lee, T. Iwata, The AES-CMAC Algorithm. RFC 4493 (Informational) (2006). http://tools.ietf.org/html/rfc4493

33. Y.N. Srikant, P. Shankar (eds.), *The Compiler Design Handbook: Optimizations and Machine Code Generation* (CRC Press, Boca Raton, FL, 2002)

34. S. Tate, R. Vishwanathan, in *Improving Cut-and-Choose in Verifiable Encryption and Fair Exchange Protocols Using Trusted Computing Technology*. Data and Applications Security (DBSec'09), Concordia University, Montreal, Canada, 12–15 July 2009. Lecture Notes in Computer Science, vol. 5645 (Springer, 2009), pp. 252–267

35. B.C.H. Turton, Extending Quine-McCluskey for exclusive-or logic synthesis. IEEE Trans. Educ. **39**, 81–85 (1996)

36. A.C. Yao, in *How to Generate and Exchange Secrets*. IEEE Symposium on Foundations of Computer Science (FOCS'86), Toronto, Canada, 27–29 Oct 1986 (IEEE, 1986), pp. 162–167

Towards Reliable Remote Healthcare Applications Using Combined Fuzzy Extraction

Jorge Guajardo, Muhammad Asim, and Milan Petković

1 Introduction

There are several important trends in healthcare that call for the deployment of remote healthcare applications. It is expected that people will live longer and that chronic diseases, such as hypertension and diabetes, will become more prevalent among older adults. That, in turn, will increase demand and cost of healthcare (in the United States already it is more than 17% of GDP). On the other hand, the healthcare sector is facing a decrease in resources (number of beds and qualified healthcare givers) relative to the increase in demand. Finally, an increasingly important trend in healthcare is consumerism. Patients demand more voice and choice in their healthcare. They are taking a more active role in their own health management.

As a consequence of the previously mentioned trends, the delivery of healthcare is gradually extending from acute institutional care to outpatient care and home care. Advances in information and communication technologies have enabled remote healthcare services (tele-health) including tele-medicine and remote patient monitoring. A number of services in the market already deploy tele-health infrastructures where the measurement devices are connected via hubs to remote backend servers. Healthcare providers use this architecture to remotely access the measurement data and help the patients to manage their conditions (Philips Motiva is an example of a disease management service). Next to that, a number of solutions [24, 25, 37] have been introduced in the market that allow patients to collect their own health-related information and to store them on portable devices, or PCs, and in online services. These solutions are often referred to as personal health record (PHR) services. Already a number of products in the market allow patients to automatically enter measurements and other medical data into their PHRs [22, 26]. For example, a weight scale sends its information via Bluetooth to a PC from which the data

J. Guajardo (✉)
Philips Research Eindhoven, Information and System Security Group, The Netherlands
e-mail: jorge.guajardo@philips.com

A.-R. Sadeghi, D. Naccache (eds.), *Towards Hardware-Intrinsic Security,* Information
Security and Cryptography, DOI 10.1007/978-3-642-14452-3_18,
© Springer-Verlag Berlin Heidelberg 2010

is uploaded to the person's PHR. This allows patients to collect and manage their health data but, more importantly, to share these data with the various healthcare professionals involved in their treatment.

This chapter addresses one of the basic security and safety problems in the domain of tele-health, which is the problem of user and device authentica- tion/identification (see Fig. 1). Namely, when data remotely measured by patients is used by tele-health services or in the medical professional world, healthcare providers need to place greater trust in the information that patients report. In par- ticular, they have to be ensured that a measurement is originating from the right patient and that an appropriate device was used to perform the measurement. For example, consider a blood pressure measurement taken at home. It is crucial to know that the blood pressure of a registered user is measured (not that of his friends or children) and that the measurement was taken by a certified device and not a cheap (potentially) fake device. This is very important because if this is not guaranteed there can be critical healthcare decisions made based on wrong data.

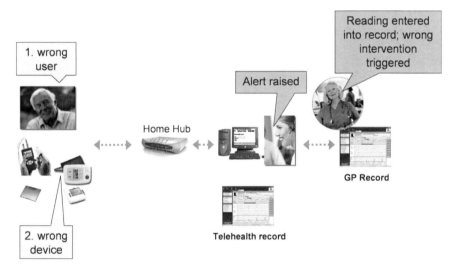

Fig. 1 Typical security and safety problems found in remote tele-monitoring health applications

Thus, user authenticity and device authenticity must be supported. Providing user and device authentication during the measurement acquisition process results in (i) increased patient safety (diagnosis and health decisions are based on reliable data), (ii) reduction of costs (patient provided data is reused in the professional healthcare domain), and (iii) convenience for the patient (patients take healthcare measure- ments at home).

In current practice, a device identifier (device ID) is used either as a user identifier (user ID) or together with some additional information (the position of an A/B user switch) as a means to derive a user ID (if multiple users are using the same device). For example, the Continua Health Alliance [8] requires each Continua-certified measurement device to send over personal area network interface its own unique device ID. The user ID is optional (and can be just as simple as 1, 2, or A, B). The valid user ID is obtained at the medical hub device (e.g., home PC) which can provide mapping between the simple user ID (associated with a device ID) and a valid user ID. There might also be measurement devices that can send a valid user ID next to the device ID. Then the mapping is not needed.

There are several problems with the current approach:

1. The current mapping approach does not quickly lock the user and device ID together, but it introduces room for mistakes. First of all, a user can make an unintended mistake. For example, the user sets the switch on the device in the wrong position (user A instead of user B), or he assigns his wife's measurement to his identity after all the measurements are transmitted from a blood pressure meter to the home PC. Furthermore, the system can mix the users. In particular, the application designer should take special care to provide data management in a way to reduce the potential for associating measurements with the wrong user.
2. A malicious user can introduce wrong measurements by impersonating the real user. Admittedly, malicious users in the cryptographic sense should be rare to come by in the home-monitoring scenario. However, it is very likely that children would be inclined to play with such devices and possibly send wrong measurements to the service provider as part of a practical joke, for example.
3. A device ID can be copied onto forged devices, which can be easily introduced in the ecosystem. Then, a user can use these devices to produce data that will look reliable but in fact will be unreliable. Here, the threat is users buying cheaper devices which appear to be certified but in fact are not. This is likely to occur as we move to tele-health services in which the consumer has more freedom to choose providers, services, and health-monitoring devices.

To address these drawbacks, we propose a method to bind the identity of a user and a device identifier as early as possible so as to certify that data coming from a specific device indeed originate from that particular device and a particular user. To ensure proper device and user authentication/identification the use of physically unclonable functions (PUFs) in combination with biometrics is proposed. Basically, we cover the three previously mentioned problems by providing

1. Close coupling of the user ID and the identification of the device used to take the measurement (the use of unregistered device/user is immediately detected),
2. Strong user authentication (biometrics), and
3. Anti-counterfeiting and strong device authentication.

Observe that in theory, strong user authentication could be provided via other more standard (and widely deployed) methods such as passwords. However, passwords have the drawback of being easily forgotten. This is even more likely in elderly patients, who, in turn, will be the most likely users of such services.

The reminder of the chapter is organized as follows. Section 2 describes in more detail remote healthcare services and their requirements. Biometrics, physically unclonable functions, and fuzzy extractors are explained in Sect. 3. Section 4 describes a new method we propose to bind the identity of the person and the measurement device to the measurement itself by combining fuzzy extraction processes for key extraction from PUFs and biometrics. Finally, Sect. 6 concludes the chapter.

2 Remote Patient Monitoring Services and Data Reliability Issues

Figure 2 illustrates a possible architecture for a remote patient monitoring solution. In particular, it illustrates the architecture that is envisioned by two standard organizations, i.e., Continua Health Alliance and Healthcare Information Technology Standards Panel (HITSP) in the domain of the remote patient monitoring solutions. The purpose of Continua is to establish a system of interoperable personal health solutions. Such solutions have the potential to foster independence by allowing home-monitoring medical services and empowering individuals by providing the opportunity for truly personalized health and wellness management. HITSP is a cooperative partnership between the public and private sectors in the United States.

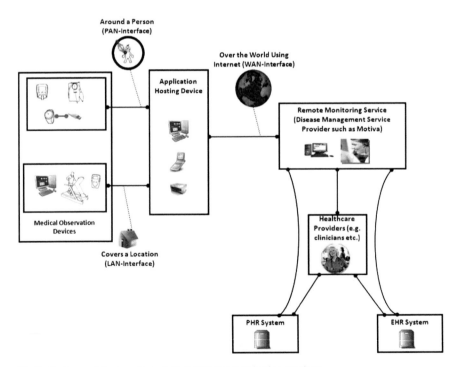

Fig. 2 System architecture for remote healthcare monitoring services

Similar to the Continual Health Alliance, its purpose is to harmonize and integrate standards that will meet clinical and business needs for sharing information among organizations and systems. Figure 2 identifies the following components and stake-holders:

- *The medical observation devices* measure vital signs such as weight, blood pressure, and glucose. These devices can be stationary, portable, body-worn, and have wired (e.g., USB) or wireless connectivity to transmit the measurements to the application hosting devices. The wireless connectivity to the application-hosting devices can be based on standards such as Bluetooth or low-power technologies such as ZigBee or Bluetooth-Low Energy.
- *The application-hosting device* serves as the direct interface to extract and store remote-monitoring information (patient's vital signs) from numerous medical observation devices inside the patient's home. Application-hosting devices can be stationary (e.g., PC, dedicated medical router) or portable (e.g., mobile phone, PDA). The application-hosting device forwards vital signs' observations from the devices under its vicinity to a remote-monitoring system in a secure fashion. In some cases, application-hosting devices can also be used for gathering subjective input from the patient, via short surveys.
- *The remote-monitoring service* collects data from all subscribed patients and makes it available for review by care-coordinators (e.g., nurse) who are the main stakeholders of the remote-monitoring system. If measurements fall outside an expected range then the nurse may forward patient-detailed vital-signs observations to the corresponding healthcare providers (e.g., patient's family physician).
- *Healthcare providers* interact with the remote-monitoring system and patient's EHR and PHR systems to obtain necessary data to provide health care. The main stakeholders of the healthcare provider system are clinicians.
- *The EHR system* "is a set of components that form the mechanism by which patient records are created, used, stored, and retrieved. A patient record system is usually located within a healthcare provider setting" [17]. The primary stake-holder of the EHR system are clinicians.
- *The PHR system* is also used to collect, store, and manage consumer health records, but, in contrast to EHR systems, PHRs are usually maintained by the patient himself. They may include any aspects of health conditions, medications, medical problems, allergies, vaccination history, visit history, or communications with healthcare providers. Examples of PHR systems include solutions such as the Microsoft HealthVault and Google health.

2.1 Data Reliability Issues

With the introduction of remote patient monitoring services professional healthcare, providers face a major dilemma with respect to the use of health data obtained from them and PHRs. The issue is whether they can trust and use the information that is being accumulated from patients using remote service providers. Remote monitoring takes place in an uncontrolled environment and without direct supervision of clinicians. In addition, remote-monitoring devices can be bought over the counter and are often shared by multiple users. For example, a blood pressure meter may be used by both husband and wife registered with a disease management service. Hence, due to these factors and with little knowledge about the context under which the measurements have been taken, clinicians have no way of knowing whether the data entered by the patients is correct and reliable. As a result, clinicians may be reluctant to access and use such remote-monitoring information unless they have sufficient assurance that the data is valid, accurate, and reliable. It is clear that clinicians want the measurements obtained from the remote-monitoring system to be for the right patient and device, right date/time, and, equally important, that the measurement is accurate or reliable (comparable to a measurement performed at the hospital under controlled conditions). Without the appropriate mechanisms to ensure the provenience and quality (reliability) of remote-monitoring information, full benefits of remote monitoring cannot be realized and its adoption by clinicians will be limited. Notice that in addition to other issues and (potential) obstacles such as confidentiality, security, and privacy of the patients' data, information reliability has been recognized by US Department of Health and Human Services (DHHS) as one of the main issues and obstacles with respect to the adoption of remote-monitoring technologies by clinicians [36]. Note that the issue of trust refers in our setting (mostly) to the reliability of the data and not necessarily to whether someone *trusts* a service provider.

3 Fuzzy Extractors, PUFs, and Biometrics

In this section, we review physical unclonable functions (PUFs), biometrics, and fuzzy extractors as well as their properties. The idea is to make the treatment of this chapter self-contained. Readers familiar with these concepts may skip it and jump to Sect. 4.

3.1 Preliminaries

We begin by briefly recalling some definitions, which will be used in the remainder of this chapter. Unless otherwise stated, we follow the presentation of [3, 9].

> *Hamming distance.* The Hamming distance between two vectors $x, y \in Q^n$, where Q is some field is denoted by $\mathrm{dis}(x, y)$ and it is defined to be the number of coordinates in which they differ. For our applications Q will be a finite field of characteristic p and often of characteristic two.

> *Error-correcting codes.* A q-ary block code $C = \{w_1, w_2, \ldots, w_k\}$ of length n is any non-empty subset of Q^n, where Q has cardinality q, i.e., Q has q distinct symbols. For example, if Q is the Galois field $\mathbb{F}_q$ then Q has q elements and

q is a prime power. The elements w_i of $\mathcal{C} \subseteq Q^n$ are called the code-words. Notice that the w_is are n-tuples of symbols taken from the alphabet Q. The minimum distance of the code $\mathcal{C}$, written $d_{\min}$, is defined to be

$$d_{\min} := \min\{\text{dis}(w_i, w_j) | w_i, w_j \in \mathcal{C}, w_j \neq w_i\}$$

For a given $d_{\min}$, the error-correcting capability or error-correcting distance e is

$$e := \left\lfloor \frac{d_{\min} - 1}{2} \right\rfloor$$

Geometrically, it can be seen as the radius e such that for every element $w \in Q^n$ there is at most one code-word in the ball of radius e centered on w.

Linear codes. A q-ary linear code $\mathcal{C}$ is a linear subspace of $\mathbb{F}_q$. If $\mathcal{C}$ is a k-dimensional linear code of length n and minimum distance d, we write it as an $[n, k, d]$-code. Thus, a q-ary $[n, k, d]$-code has cardinality q^k, i.e., it can encode up to q^k possible messages. For linear codes the minimum distance is equal to the minimum non-zero weight in $\mathcal{C}$.

Permutation groups. The set of all permutations of a set $\mathcal{M}$ is called the symmetric group on $\mathcal{M}$. Usually we take $\mathcal{M}$ to be the set $\{1, \ldots, n\}$, and denote the symmetric group by S_n, for some positive integer n. The order of S_n is $n!$. As it is well known, any permutation can be written as a product of disjoint cycles: we call this its cycle decomposition. For example, the permutation of $\{1, \ldots, 5\}$ which maps 1 to 4, 2 to 5, 3 to 1, 4 to 3, and 5 to 2 has cycle decomposition $(1, 4, 3)(2, 5)$. The cycle decomposition is unique up to writing the cycles in a different order and starting them at different points: for example, $(1, 4, 3)(2, 5) = (5, 2)(3, 1, 4)$. A permutation group $\mathcal{P}$ on a set $\mathcal{M}$ is a subgroup of the symmetric group on $\mathcal{M}$; that is, it is a set of permutations closed under composition and inversion and containing the identity permutation. The group operation is simply the action of the permutations π_i on the elements of the set $\mathcal{M}$. The permutation group $\mathcal{P} = \{\pi_i : \mathcal{M} \to \mathcal{M}\}$, indexed by i, is transitive on the set $\mathcal{M}$ if for any pair of points w, w' there exists a permutation $\pi_i \in \mathcal{P}$, such that $\pi_i[w] = w'$. The permutation group $\mathcal{P}$ is isometric with respect to the distance function dis in the set $\mathcal{M}$ (we assume the set $\mathcal{M}$ is a space with a distance function) if for all permutations $\pi_i \in \mathcal{P}$ and points $w, w' \in \mathcal{M}$, it holds that $\text{dis}(\pi_i[w], \pi_i[w']) = \text{dis}(w, w')$. These two last properties are used in the construction of fuzzy extractors based on permutations.

Universal hash functions [7]. A universal hash function is a map from a finite set $\mathcal{A}$ of size $|\mathcal{A}|$ to a finite set $\mathcal{B}$ of size $|\mathcal{B}|$. For a given hash function h and two strings x, x' with $x \neq x'$, we define the function $\delta_h(x, x')$ as equal to 1 if $h(x) = h(x')$ and 0 otherwise. For a finite set (or family) of hash functions $\mathcal{H}$, $\delta_{\mathcal{H}}(x, x')$ is defined to be $\sum_{h \in \mathcal{H}} \delta_h(x, x')$. In other words, $\delta_{\mathcal{H}}(x, x')$ counts the number of functions $h \in \mathcal{H}$ for which x and x' collide. For a random $h \in \mathcal{H}$ and any two distinct x, x', the probability that $h(x) = h(x')$ is $\delta_{\mathcal{H}}(x, x')/|\mathcal{H}|$, where $|\mathcal{H}|$ denotes the size of the set $\mathcal{H}$. There has been extensive research on universal hash functions (see for example [27, 30]). In the hardware domain, their implementation has been investigated in [20] and the work of [19].

Encryption schemes. Throughout this chapter we will refer to *encrypting* [2], denoted $\mathsf{Enc}(K, \cdot)$, meaning an encryption scheme using key K, which provides semantic security under chosen plaintext attacks [10, 13], commonly written IND-CPA. The corresponding decryption operation under key K is denoted by $\mathsf{Dec}(K,)$. There are stronger versions of security, such as semantic security under chosen ciphertext attacks (IND-CCA), however, common modes of operation (e.g., CBC) only provide IND-CPA. Although we focus on the symmetric key setting, the security notions themselves are general enough that apply to the public-key setting as well. Thus, our schemes can also be easily extended to public-key counterparts by deriving appropriate public-keys from the secret key information. Finally, we write $\mathsf{MAC}(K, \cdot)$ to indicate a message authenticating code (MAC) computed with the secret key K providing integrity of plaintexts (see [1]).

3.2 Physical Unclonable Functions

A function in mathematics is a relation which associates elements of a set $\mathcal{A}$, typically referred to as the domain, with elements of a set $\mathcal{B}$, known as the range or image. The relation which associates elements of set $\mathcal{A}$ to those of set $\mathcal{B}$ is defined via a mathematical formula, a graph, a table, etc. In 2001, [28, 29] introduced the concept of physical random functions or physical unclonable functions (PUFs). In this case the function is defined via a physical object or device. In particular, upon challenging such a PUF with a challenge C_i, a response R_i is generated. Physical Unclonable Functions have essentially two parts: (i) a physical part and (ii) an operational part. The physical part is a physical system that is very difficult to clone. It inherits its unclonability from uncontrollable process variations during manufacturing. In the case of PUFs on an IC such process variations are typically deep submicron variations such as doping variations in transistors. The operational part corresponds to the function. In order to turn the physical system into a *function* a set of challenges C_i (stimuli) has to be available to which the system responds with a set of sufficiently different responses R_i.

3.2.1 PUFs Types and Examples

In the literature, two different classes of PUFs have been introduced divided according to the number of challenge–response pairs that the particular PUF accepts. In particular, PUFs have been divided into strong and weak PUFs. A strong PUF accepts a large number of challenge–response pairs (C_i, R_i), $i = 1, \ldots, N$; i.e. the PUF has so many CRPs such that an attack (performed during a limited amount of time) based on exhaustively measuring the CRPs only has a negligible probability of success and, in particular, $1/N \approx 2^{-k}$ for large $k \approx 100$ [28, 35]. If the number of different CRPs N is rather small, we refer to it as a weak PUF. Notice that a weak PUF is usually used for secure key storage applications and thus, it is very similar to the concept of physically obfuscated keys (POKs) as introduced by [11]. We also emphasize that the *weakness* of a weak PUF does not refer to its unclonability or randomness properties but to the number of challenges it accepts.

Examples of PUFs include optical PUFs [28, 29], silicon PUFs [12], coating PUFs [33], and intrinsic PUFs [14]. Notice that an IPUF is a PUF inherently present in a device due to its deep submicron manufacturing process variations and no additional hardware has to be added for embedding the PUF. In [14], Guajardo et al. show that the start-up values of SRAM memory cells (present, for example, in an FPGA) are an IPUF. A variant of SRAM PUFs constructed by cross-coupling latches has been introduced by Kumar et al. [21]. We observe that while PUF structures can always be created from scratch using full-custom design of chips, intrinsic PUFs based on SRAM memory are omnipresent wherever SRAM memory is present in a device. In particular, SRAM PUFs are present in embedded processors as well, such as those available inside medical devices (see, for example, [16], for the typical size of SRAM in implantable medical devices).

3.2.2 PUF Security Properties

As in any security system, in order to evaluate the security of the system, it is necessary that we state the necessary assumptions for the system to be secure. Previous works [12, 14, 15, 28, 33] have either explicitly or implicitly made the following assumptions:

1. It is assumed that a response R_i (to a challenge C_i) gives only a small amount of information on another response R_j (to a different challenge C_j) with $i \neq j$.
2. Without having the corresponding PUF (i.e., the actual physical device or structure) at hand, it is impossible to come up with the response R_i corresponding to a challenge C_i, except with negligible probability.

In many cases, it is also reasonable to assume that PUFs are tamper evident. This implies that when an attacker tries to investigate the PUF to obtain detailed information about its structure, the PUF is damaged and the PUF's challenge–response behavior is changed substantially. In this chapter, this property is not used explicitly.

As noticed previously, the above assumptions are guaranteed based on the hardness of copying the actual device (or structure) used as a PUF. This hardness is due to the unfeasibility to copy the structure and it is not due to some physically impossible process (as the cloning of quantum bits, for example). We observe, however, that in our particular application scenario, our requirements can be relaxed. In many situations, we just need that the string derived from the PUF be unclonable and not necessarily secret. We will describe explicitly the situations in which this is the case in the next sections.

3.3 Biometrics

Biometric measurements have been proposed as a way to authenticate individuals. They are very appealing because they are inherently unclonable and cannot be easily reproduce except by the person whose biometric is being authenticated. Thus, one can think of a biometric as a PUF (or POK) embedded in a person (the opposite analogy is often used in the PUF literature). Unlike the PUF case, biometric measurements are highly privacy invasive. Thus, lots of work has been devoted to studying methods to store and process biometrics in a privacy friendly manner (see, e.g., [6, 9, 23, 34]).

Following recent work [5, 31] on the (in)security of the techniques described in [6, 9, 23], it is apparent that templates based on helper data techniques need to comply with at least three properties. Informally, these properties are

- negligible information leakage (from a single helper data): given helper data W for biometric X, W should leak negligible information about X. This property was guaranteed by the constructions in [6, 9, 23];
- non-reversibility: given (public) helper data $\{W_i\}_{i=1}^{K}$ corresponding to possibly several (related) biometric measurements $\{X_i\}_{i=1}^{K}$, it should be unfeasible to recover any X_i from information leaked by the (public) $\{W_i\}_{i=1}^{K}$;
- indistinguishability: given (public) helper data $\{W_i\}_{i=1}^{K}$ corresponding to possibly several (related) biometric measurements $\{X_i\}_{i=1}^{K}$, it should be unfeasible to tell if any of the W_i originate from the same biometric measurement X_i.

Interestingly, it is shown in [5, 31] that for most constructions and choices of parameters the techniques in [6, 9, 23] only provide the first guarantee, failing to provide non-reversibility and indistinguishability when lists of potentially related helper data are taken into consideration when mounting an attack.

3.4 The Need for Fuzzy Extractors

Observe that a common property of PUFs and biometric measurements is that they are noisy. In addition, PUF responses and biometric measurements are not fully uniformly distributed, which is undesirable for security applications (e.g., key derivation functions or authentication applications). As a result, a fuzzy extractor or helper data algorithm is required to extract secure keys from the PUF responses (respectively, biometric measurements). For formal definitions of fuzzy extractors and helper data algorithms we refer to [6, 9, 23]. Informally, we need to implement two basic primitives: (i) *information reconciliation* or error correction and (ii) *privacy amplification* or randomness extraction. In order to implement those two primitives, helper data W is generated during the *enrollment phase*. During this phase, carried out in a trusted environment, a probabilistic procedure called Gen is run. Later, during the *key reconstruction* or authentication phase, the key is reconstructed based on a noisy measurement R_i' and the helper data W. During this phase, a procedure called Rep is performed. We now present several constructions for such procedures previously described in [6, 9, 18, 23]. Constructions for other metrics can also be found in [9].

3.4.1 Construction Based on Code Offset (Discrete Distributions)

In order to implement the procedures Gen and Rep an error correction code C and a set $\mathcal{H}$ of universal hash functions [7] are required. The parameters $[n, k, d]$ of the code C are determined by the length of the responses R and the number of errors t that have to be corrected. The distance d of the code is chosen such that t errors can be corrected with high probability. The definition of what is an acceptable "high" probability is directly dependent on the application requirements.

The Gen procedure takes as input a response(s) (from a PUF or a biometric measurement) R and produces as output a key K and helper data $W = (W_1, W_2)$. This is achieved as follows. First, a code word $C_S \leftarrow C$ is chosen at random from C. Then, a first helper data vector equal to $W_1 = C_S \oplus R$ is generated. Furthermore, a hash function h_i is chosen at random from $\mathcal{H}$ and the key K is defined as $K \leftarrow h_i(R)$. The helper data W_2 is set to i. Summarizing the procedure Gen is defined as follows, $(K; W) = (K; (W_1, W_2)) \leftarrow \text{Gen}(R)$.

During the key reconstruction phase the procedure Rep is run. It takes as input a noisy response R' from the same PUF (or person) and helper data W and reconstructs the key K, i.e., $K \leftarrow \text{Rep}(R', W)$. This is accomplished according to the following steps: (1) *information reconciliation*: using the helper data W_1, $W_1 \oplus R'$ is computed. Then, the decoding algorithm of C is used to obtain C_S. From C_S, R is reconstructed as $R = W_1 \oplus C_S$ and (2) *privacy amplification*: the helper data W_2 is used to choose the correct hash function $h_i \in \mathcal{H}$ and to reconstruct the key as $K = h_i(R)$. Notice that we have implicitly assumed the use of a binary code. This construction is a variant of [18] where the focus was on biometric applications. We observe that in the biometric literature, it is often the case that instead of using a universal hash function to compute the key K, a hash function such as SHA-2 is used. This has the effect of providing computational guarantees as opposed to information theoretic security as in the former case (i.e., using universal hash functions).

3.4.2 Construction Based on Permutations (Discrete Distributions)

The permutation-based construction is due to [9]. As in the code-offset construction, we choose a code $C \subseteq \mathcal{M}$ and, in addition, a corresponding permutation group $\mathcal{P}$ that is both transitive and isometric. The $(K, W) \leftarrow \text{Gen}(R)$ then computes K and W from input R by first selecting a random code word $C_S \leftarrow C$ and corresponding $\pi_P \in \mathcal{P}$, such that $\pi_P[R] = C_S$. Notice that the transitivity property of $\mathcal{P}$

guarantees that such π_P will exist. Then as before, we randomly choose a universal hash function $h_i \in \mathcal{H}$ and we output $(K; (W_1, W_2)) = (h_i(R); (P, i)) \leftarrow \mathsf{Gen}(R)$.

During the key reconstruction phase a procedure called Rep is run according to the following steps: (1) *information reconciliation*: using the helper data $W_1 = P$, we compute $\pi_P[R'] = C'_S$. Because of the isometric property of π, C'_S should be sufficiently close to C_S that, after applying the decoding algorithm of $\mathcal{C}$, we will obtain C_S. From C_S, R is reconstructed as $R = \pi_P^{-1}[C_S]$ and (2) *privacy amplification*: the helper data W_2 is used to choose the correct hash function $h_i \in \mathcal{H}$ and to reconstruct the key as $K = h_i(R)$.

3.4.3 Constructions for Continuous Distributions

Until now, we have assumed that the system obtains a binary vector which can be used as input to the fuzzy extractor $\mathsf{Gen}(\cdot)$ and $\mathsf{Rep}(\cdot)$ procedures. In practice, there needs to be a procedure that binarizes real-valued (biometric) measurements and converts them into a (noisy) binary vector. Linnartz and Tuyls are the first to consider this problem in [23], later generalized in [6]. Here we follow the presentation of [23].

As in [6, 23], we model biometric measurements as m-dimensional feature vectors $X \in \mathbb{R}^m$ with zero mean and independently and identically distributed (i.i.d.) components. Since we assume all components x_i for $i = 1, 2, \ldots m$ of X to be i.i.d., it is sufficient to consider the processing of one of them. During enrollment, a random bit $s_i \in_R \{0, 1\}$ is chosen. Then, helper data w_i is computed corresponding to x_i as

$$
w_i = \begin{cases} \left(2n + \frac{1}{2}\right)q - x_i & \text{if } s_i = 1 \\[2mm] \left(2n - \frac{1}{2}\right)q - x_i & \text{if } s_i = 0 \end{cases}
$$

where q is the quantization step and $n = \ldots, -1, 0, 1, 2, \ldots$ is chosen such that $-q < w_i < q$. The helper data is then published together with a hash of the value $S = s_1||s_2|| \cdots ||s_m$, where $||$ means concatenation. During the verification procedure a new (noisy) vector $X' \in \mathbb{R}^m$ is measured and for each component x'_i of X', the value s'_i is computed as

$$
s'_i = \begin{cases} 1 & \text{if } 2nq \le x'_i + w_i < (2n+1)q, \text{ for any } n = \ldots, -1, 0, 1, \ldots \\[2mm] 0 & \text{if } (2n-1)q \le x'_i + w_i < 2nq, \text{ for any } n = \ldots, -1, 0, 1, \ldots \end{cases}
$$

if the hash of $S' = s'_1||s'_2|| \cdots ||s'_m$ is the same as the one of S, then the verification is accepted and otherwise not. Notice that the above procedure guarantees that the verification will succeed as long as $x'_i = x_i + \delta$ and $|\delta| < q/2$. Similarly, the amount of information leaked by the w_i's can be made negligible as shown in [23].

3.4.4 Security

The security of the above constructions has been established in [3, 4, 9, 18, 23]. By security, here we mean two complementary things. First, [9, 23] provide a bound on the number of bits of entropy left after the fuzzy extractor operates on the source bits of the PUF. In other words, given a number of bits with certain entropy, we know from [9, 23], how many "secure" bits we are left with after processing with the fuzzy extractor. Second, [3, 4, 18] show that given the public helper data information, negligible information is learned about the derived secret. Finally, [3, 4] show how to protect the helper data against tampering and modification.

We end this section by noticing that there is recent work which indicates that additional security measures need to be taken if privacy against matching across different databases is desired. In particular, [5, 31] have shown that the helper data derived from fuzzy extractors for both discrete and continuous distributions are susceptible to attacks, in the sense that one can find out whether two helper data correspond to the same user (assuming the user has enrolled in two different databases). This seems to be a hard problem to solve if the only available extra information is the biometric itself (no additional secret key to encrypt the biometric helper data information). The solution we describe in this chapter can prevent these types of attacks by introducing a key derived from the device where the biometric is stored.

4 Combining PUFs and Biometrics

In this section, we propose a method to bind the identity of a user and a device identifier as early as possible so as to certify that data originating from the device originate from the particular device and the particular user. To ensure proper device and user authentication/identification the use of Physically Unclonable Functions in combination with biometrics is proposed. By combining these two authentication methods we closely couple the user and the device used to perform the measurements, achieving strong user-device combined authentication. Furthermore, PUFs allow us to realize in health care a very important anti-counterfeiting feature.

As previously mentioned, the main problem is to link a measurement to both a device ID and the particular user. Clearly a stable device ID can be derived from a PUF response and associated helper data. Traditionally, the helper data is chosen randomly from the code words of an error-correcting code. Here we propose to encode the helper data from the device using a string derived from the biometric. We then derive a key from the device information. In this manner, only the right combination of user and device will be able to produce the correct key. This key is, in turn, used to authenticate the patient data. In the next section, we describe several possible constructions.

In our presentation, we assume implicitly that the following procedures are available on the device that is being used:

- A PUF such that when challenged with C_i produces a response R_i, we write $R_i \leftarrow \mathsf{PUF}(Ci)$. Notice that this can be easily expected from any device containing SRAM memory, such as any microprocessor or microcontroller.
- GenPUF algorithm which upon getting a PUF response R_i outputs (Ki, Wi). We write $(Ki, Wi) \leftarrow \mathsf{GenPUF}(R_i)$.
- A RepPUF algorithm which upon getting a PUF response R_i' and helper data W_i outputs key K_i if R_i and R_i' are sufficiently close according to some appropriate distance measure. We write $K_i \leftarrow \mathsf{RepPUF}(R_i', W_i)$.
- A GenBio algorithm which upon getting a biometric measurement X_u from user U outputs (K_u, W_u). We write $(K_u, W_u) \leftarrow \mathsf{GenBio}(X_u)$. Observe that in practice that helper data can be computed according to any of the constructions described in Sect. 3.4 or a combination of these procedures (see, e.g., [32, 34]).
- A RepBio algorithm which upon getting a biometric measurement X_u from user U and helper data W_u outputs the key K_u if X_u and X_u' are sufficiently close. We write $K_u \leftarrow \mathsf{RepBio}(X_u', W_u)$.

As it is traditional in biometrics and PUFs, the system works in two stages: (i) registration or enrollment and (ii) authentication. During registration, the user(s) and device(s) biometrics and properties are measured and sent in a secure manner to the service provider. In the second stage, the service provider verifies that the user measurements correspond to the correct user and have been measured with a certified device. The registration procedure is fully specified in Algorithm 2. Notice that this procedure is defined per device and user pair. In particular, a device is expected to be used by a small number of different users in the home. Similarly, a user will use several devices at the same time.

Algorithm 2 Registration (Enrollment) Procedure for Device d_k and user U_j

---

Input: A device d_k and corresponding PUF challenge $R_{i,k}$, a biometric measurement $X_{U_j} \in \mathbb{R}^m$ corresponding to user U_j, an encryption scheme $\mathsf{Enc}(\cdot, \cdot)$.

Output: Key K_{i,k,U_j} derived from device d_K information and user's U_j biometric information, device encrypted helper data $\overline{W}_{i,k,U_j}$ with biometric derived key K_{U_j}, user's U_j helper data W_{U_j}, and hashes of K_{U_j} and $K_{i,k}$.

1: Run procedure $(K_{i,k}, W_{i,k}) \leftarrow \mathsf{GenPUF}(R_{i,k})$ on device d_k. This procedure does not need to be run by d_k. In particular, this procedure can be run by a separate entity. The only thing needed by the entity to run GenPUF is the response $R_{i,k}$.
2: User U_j runs procedure $(K_{U_j}, W_{U_j}) \leftarrow \mathsf{GenBio}(X_j)$ on his/her biometric X_j.
3: The helper data $W_{i,k}$ is encrypted using K_{U_j} as $\overline{W}_{i,k,U_j} \leftarrow \mathsf{Enc}(K_{U_j}, W_{i,k})$. This value is stored in the device's memory.
4: For user U_j, a key K_{i,k,U_j} is computed as $K_{i,k,U_j} \leftarrow \mathsf{hash}(W_{i,k}||W_{U_j}||K_{i,k}||K_{U_j})$. This key is transmitted in a secure manner to the health service provider. The hash of K_{U_j} is also computed and stored in the local user-device database.
5: **return** $(K_{i,k,U_j}, \overline{W}_{i,k,U_j}, R_i, W_{U_j}, \mathsf{hash}(K_{U_j}), \mathsf{hash}(K_{i,k}))$

Before describing the authentication procedure, it is important to describe how the registration procedure will be integrated in the overall system. Algorithm 2 outputs a key K_{i,k,U_j}, which is to be transmitted to the remote-monitoring health service provider. This key can then be used to authenticate a biomedical signal measured from the patient by, for example, computing a MAC over the measurement data. Observe, that in principle this could already be done using a key stored in the measurement device or in the application-hosting device. The main idea in this chapter is to derive this key from both device intrinsic data and patient biomedical data, in such a way that only the right combination of user and device can produce the key that correctly authenticates the data to the service provider.

It is expected that Algorithm 2 will be run for a few users and device combinations. The result is a small database of tuples of the form shown in Table 1, which will be stored either at the device or (more likely) at the application-hosting device.

The database in Table 1 is depicted as having user IDs in its first column. This can improve slightly the efficiency of the overall solution because it would mean that the device does not need to search for helper data W_{U_j} for the corresponding user. In practice, such user ID can be easily implemented with an up-down switch. This is clearly at the cost of user convenience (i.e., the user needs to move a switch or choose a user profile). A more user-friendly solution is to have the device search through the database until either it gets a matching $\text{hash}(K_{U_j})$ value (in which case it continues to determine which device is in use) or it rejects the user and asks him/her to register by running Algorithm 2. Notice that such linear search in practice does not affect the performance of the solution significantly as the number of users of the device in any given household will be rather small in practice. Similar comments are applicable to the device data. In particular, in Table 1, it is assumed that a device identifier (d_{k_t} for $t = 1, 2, \ldots, s$) is available. As in the biometric data case, a hash of the device dependent key $K_{i,k}$ is also stored in the database. The idea is to be able to detect failure before the actual measurement has been sent to the service provider. Clearly, the alternative is not to check the correctness of the device

Table 1 Example of home user-device database

User identifier	User data	Device identifier	Device data
U_1	$W_{U_1}, \text{hash}(K_{U_1})$	d_{k_1}	$\overline{W}_{i,k_1,U_1}, R_i, \text{hash}(K_{i,k_1})$
		d_{k_2}	$\overline{W}_{i,k_2,U_1}, R_i, \text{hash}(K_{i,k_2})$
		$\vdots$	$\vdots$
		d_{k_s}	$\overline{W}_{i,k_s,U_1}, R_i, \text{hash}(K_{i,k_s})$
U_2	$W_{U_2}, \text{hash}(K_{U_2})$	d_{k_1}	$\overline{W}_{i',k_1,U_2}, R_{i'}, \text{hash}(K_{i',k_1})$
		d_{k_2}	$\overline{W}_{i',k_2,U_2}, R_{i'}, \text{hash}(K_{i',k_2})$
		$\vdots$	$\vdots$
		d_{k_s}	$\overline{W}_{i',k_s,U_2}, R_{i'}, \text{hash}(K_{i',k_s})$
$\vdots$	$\vdots$	$\vdots$	$\vdots$
U_n	$W_{U_n}, \text{hash}(K_{U_n})$	d_{k_1}	$\overline{W}_{i'',k_1,U_n}, R_{i''}, \text{hash}(K_{i'',k_1})$
		d_{k_2}	$\overline{W}_{i'',k_2,U_n}, R_{i''}, \text{hash}(K_{i'',k_2})$
		$\vdots$	$\vdots$
		d_{k_s}	$\overline{W}_{i'',k_s,U_n}, R_{i''}, \text{hash}(K_{i'',k_s})$

dependent key $K_{i,k}$ at the patient home, but rather wait for an acceptance notification when the service provider verifies the message authentication code computed on the measurement data with the key K_{i,k,U_j} (see Algorithm 3). This, however, can become costly, in terms of re-transmissions if the amount of (measurement) data sent to the service provider is significant. What should become apparent from the previous discussions is that the system described is flexible enough that it allows for many trade-offs between user comfort and quality of service.

Having described the enrollment procedure and the database created as a result of it, we are now prepared to introduce the authentication or verification procedure in Algorithm 3. For ease of presentation, we assume in the description of Algorithm 3 that both the user and the device can provide an ID which will indicate what helper data to use from the database for purposes of biometric and device identification, respectively. The search variants can be easily derived from the current algorithms.

Algorithm 3 Authentication (Verification) Procedure for Device d_k and user U_j

---

Input: A device d_k and corresponding PUF challenge $R_{i,k}$, a biometric measurement $X_{U_j} \in \mathbb{R}^m$ corresponding to user U_j, an encryption scheme $\mathsf{Enc}(\cdot, \cdot)$, a database as shown in Table 1.

Output: Key K_{i,k,U_j} derived from device d_K information and user's U_j biometric information, device encrypted helper data $\overline{W}_{i,k,U_j}$ with biometric derived key K_{U_j}, user's U_j helper data W_{U_j}, and hash of K_{U_j}.

1: User U_j recovers his helper data W_{U_j} from the database, runs procedure $K'_{U_j} \leftarrow \mathsf{RepBio}(X'_j, W_{U_j})$ on his/her biometric measurement X'_j and obtains key K'_{U_j}.
2: **if** $\mathrm{hash}(K_{U_j}) \neq \mathrm{hash}(K'_{U_j})$ **then**
3:   Communicate to the user that he is not recognized by the system and that he needs to register by running Algorithm 2.
4: **end if**
5: The device then recovers the encrypted helper data $\overline{W}_{i,k,U_j}$ and decrypts it using K_{U_j} to obtain $W_{i,k} \leftarrow \mathsf{Dec}(K_{U_j}, \overline{W}_{i,k,U_j})$.
6: The device runs procedure RepPUF to obtain $K'_{i,k} \leftarrow \mathsf{RepPUF}(R_{i,k}, W_{i,k})$.
7: **if** $\mathrm{hash}(K_{i,k}) \neq \mathrm{hash}(K'_{i,k})$ **then**
8:   Communicate to the user that the device is not recognized by the system and that he needs to register by running Algorithm 2 with device d_k.
9: **end if**
10: Key K'_{i,k,U_j} is computed as $K'_{i,k,U_j} \leftarrow \mathrm{hash}(W_{i,k}\|W_{U_j}\|K_{i,k}\|K_{U_j})$.
11: Measure biomedical signal M with device d_k from user U_j.
12: The biomedical measurement M measured with device d_k, $\tau' \leftarrow \mathsf{MAC}(K'_{i,k,U_j}, M)$, the device ID d_k and the user ID U_j are sent to the service provider.
13: The service provider receives M, τ, d_k, U_j, retrieves K_{i,k,U_j} from its local database, and computes $\tau \leftarrow \mathsf{MAC}(K_{i,k,U_j}, M)$ (with its locally stored version of K_{i,k,U_j}).
14: **if** $\tau \neq \tau'$ **then**
15:   **return** (Verification failed)
16: **else**
17:   **return** (Verification passed)
18: **end if**

Algorithm 3 assumes that the service provider maintains a database, which includes tuples of the form $\{U_i, \{d_{k_1}, d_{k_2}, \ldots, d_{k_s}\}, \{K_{i,k_1}, K_{i,k_2}, \ldots, K_{i,k_s}\}\}$ (i.e., user ID, device ID's associated with the user, and a number of keys that have been derived from the device and the user biometric information). Observe that the number of devices and keys does not necessarily have to be the same for all users. These seems reasonable to assume as the service provider needs to associate the data that it receives with a corresponding patient.

4.1 A Practical Simplification

In the algorithms previously described, combined user and device authentication is achieved on the basis of combining PUFs and biometrics. This method assumes implicitly that PUFs are available on all medical devices. A unique and stable device ID can then be derived from the PUF response and its associated helper data. Although algorithms based on this assumption provide strong security guarantees since PUFs are unclonable, a disadvantage of this approach is that not all (current) devices have PUFs built into them. Thus, it is interesting to consider simplifications, which would allow for algorithms that link device and user, yet do not require an unclonable structure embedded in the device.

Therefore, in the sequel we consider the device permanent ID as a source of device identifier which is not necessarily secret or unclonable. Then, the device ID can represent a stable PUF which has no noise. This means that the databases (both at home and at the service provider) would have unique device identifiers in them. It is then straightforward to combine device identifiers and biometric information and link them both into a key that can be used to compute a MAC on the patient data. Examples include

- Using the device ID as the device-dependent key $K_{i,k}$ derived in Algorithm 2.
- Concatenating the device ID and the biometric into a single string, which is used as an "extended" measurement $Xext$ to be input to the procedures GenBio and RepBio in Algorithms 2 and 3, respectively.
- Using the device ID as part of the randomness to be used during the computation of the biometric helper data W_{U_j}.

An intuitive idea to design a method to combine device and user authentication based on the device ID and the user's biometric would then be to map together the biometric measurement and the device ID to a randomly selected error-correcting code word from the code word space. Notice that usually *only* the biometric measurement is directly mapped to a randomly selected error-correcting code-word and the helper data is then generated. The proposed approach will authenticate both the device and user at once and has the same (safety) advantages as the method based on the PUFs and biometrics. However, it does not provide the same anti-counterfeiting and strong device authentication guarantees that the solution based on PUFs does.

4.2 Other Variations

In our discussion so far, we have described the different procedures in terms of symmetric key primitives. It should be clear, however, that these constructions can be easily extended to the public-key setting. In particular, the encryption operation $\mathsf{Enc}(K_{U_j}, \cdot)$ can be performed with either a symmetric-key or asymmetric key cipher. If performed with an asymmetric key cipher then it is understood that a corresponding key derivation function has to be performed so as to create private-public key pairs. Similarly, in Algorithm 3 one could replace the **MAC** operation with a corresponding signature and verification operations. This implies that the corresponding public-key has been transmitted to the service provider during the enrollment procedure.

4.3 Security and Safety

Clearly, the algorithms described for enrollment and authentication are based on the **Gen** and **Rep** procedures as described in Sect. 3.4. Thus, they inherit all their security properties from them. It should also be clear that if we choose an encryption scheme, which is secure against chosen plaintext attacks (IND-CPA) then the encrypted value $\overline{W}_{i,k,U_j}$ is indistinguishable from any other string encrypted using the same secret key. Furthermore, assuming the security of the underlying cryptographic primitive (e.g., AES or RSA with OEAP) it is unfeasible to recover the plain value W_{i,k,U_j} without knowledge of the corresponding secret key (corresp. private key). As a result, it is not possible to reconstruct the device key $K_{i,k}$. It follows that it is not possible to reconstruct the user-device dependent key K_{i,k,U_j} and therefore the **MAC** operation in step 13 would fail.

Notice that no user information or helper data corresponding to the user's biometric is stored at the service provider. This essentially prevents any cross-correlation attacks as described in [5, 31]. We assume, however, that it is hard to obtain this data from either the patient devices or the application-hosting device since they are at home and presumably, either hard to reach or hard to attack. Interestingly enough, the techniques described here could also be used to protect against transferring biometric data between different devices by encrypting the biometric helper data and not the device helper data. The algorithms would be very similar with the roles of the device and the user data exchanged. This "non-transferability" property could be used to deter cross-matching attacks as this would imply that to perform the cross-matching you would have to first be able to decrypt the helper data using a device-specific key.

As a final remark, we would like to emphasize that our aim is to prevent *impersonation of measurements*. To this end, we stipulate that we need to generate device and user-dependent keys, which in turn are combined to certify the origin of the measurement data. In this sense, Algorithm 3 provides very strong guarantees. In particular, the algorithm derives a key every time that a measurement needs to be performed (as opposed to retrieving a key from secure storage). This simple observation guarantees that the measurement comes from the right person and the right device. Notice that the simplification of Sect. 4.1 could leak some biometric data if the device identifier is made public. This is not really our concern in the home setting, where it is assumed that people in your surrounding are "trusted."

5 Conclusions

Home healthcare solutions are expected to be widely available in the near future. As a result, it is of the utmost importance to provide solutions that can guarantee that a (physiological) measurement corresponds to the patient, who claims it to be and that it originates from the particular device. In this chapter, we describe a solution, which provides very strong (safety) guarantees. To this end, we combine techniques from the PUF and biometric literature and derive identifiers that uniquely link patient, device, and (potentially) the actual measurement.

Acknowledgments This work has been funded in part by the European Community's Sixth Framework Programme under grant number 034238, SPEED project – Signal Processing in the Encrypted Domain. The work reported reflects only the authors views; the European Community is not liable for any use that may be made of the information contained herein.

References

1. M. Bellare, C. Namprempre, Authenticated encryption: Relations among notions and analysis of the generic composition paradigm. in *Advances in Cryptology — ASIACRYPT 2000*, ed. by T. Okamoto. Lecture Notes in Computer Science, vol. 1976 (Springer-Verlag, Berlin, Heidelberg, New York, NY, 3–7 Dec 2000), pp. 531–545
2. M. Bellare, P. Rogaway, Encode-then-encipher encryption: How to exploit nonces or redundancy in plaintexts for efficient cryptography. in *Advances in Cryptology — ASIACRYPT 2000*, ed. by T. Okamoto. Lecture Notes in Computer Science, vol. 1976 (Springer-Verlag, Berlin, Heidelberg, New York, NY, 3–7 Dec 2000), pp. 317–330
3. X. Boyen, Reusable cryptographic fuzzy extractors. in *ACM Conference on Computer and Communications Security — ACM CCS 2004*, ed. by V. Atluri, B. Pfitzmann, P.D. McDaniel. (ACM, New York, NY, 25–29 Oct 2004), pp. 82–91
4. X. Boyen, Y. Dodis, J. Katz, R. Ostrovsky, A. Smith, Secure remote authentication using biometric data. in *Advances in Cryptology — Eurocrypt 2005*, ed. by R. Cramer. Lecture Notes in Computer Science, vol. 3494 (Springer-Verlag, Berlin, Heidelberg, New York, NY, 2005), pp. 147–163
5. I. Buhan, J. Breebart, J. Guajardo, E. Kelkboom, K. de Groot, T. Akkermans, A quantitative analysis of indistinguishability for a continuous domain biometric cryptosystem. in *Data Privacy Management and Autonomous Spontaneous Security — DPM 2009*, ed. by J. Garcia-Alfaro, G. Navarro-Arribas, N. Cuppens-Boulahia, Y. Roudier. Lecture Notes in Computer Science, vol. 5939 (Springer, St. Malo, France, 24–25 Sept 2009), pp. 78–92. Revised Papers.
6. I. Buhan, J. Doumen, P. H. Hartel, R.N.J. Veldhuis, Fuzzy extractors for continuous distributions. in *ACM Symposium on Information, Computer and Communications Security — ASIACCS 2007*, ed. by F. Bao, S. Miller. (ACM, New York, NY, 20–22 Mar 2007), pp. 353–355
7. J. Lawrence Carter, M.N. Wegman, Universal classes of hash functions. J. Computer Syst. Sci. **18**(2), 143–154 (1979)

8. Continua health alliance. Accessed December 2009. Available at `http://www.continuaalliance.org`
9. Y. Dodis, M. Reyzin, A. Smith, Fuzzy extractors: How to generate strong keys from biometrics and other noisy data. in *Advances in Cryptology — EUROCRYPT 2004*, ed. by C. Cachin, J. Camenisch. Lecture Notes in Computer Science, vol. 3027 (Springer, Heidelberg, 2004), pp. 523–540
10. D. Dolev, C. Dwork, M. Naor, in *Non-Malleable Cryptography (Extended Abstract)*. ACM Symposium on Theory of Computing — STOC'91 (ACM, New York, NY, 6–8 May 1991), pp. 542–552
11. B. Gassend, *Physical Random Functions*, Master's thesis, Computer Science and Artificial Intelligence Laboratory, MIT, February 2003. Computation Structures Group Memo 458
12. B. Gassend, D.E. Clarke, M. van Dijk, S. Devadas, Silicon physical unknown functions. in *ACM Conference on Computer and Communications Security — CCS 2002*, ed. by V. Atluri. (ACM, New York, NY, Nov 2002), pp. 148–160
13. S. Goldwasser, S. Micali, Probabilistic encryption. J. Comput. Syst. Sci. **28**(2), 270–299 (1984)
14. J. Guajardo, S.S. Kumar, G.-J. Schrijen, P. Tuyls, FPGA intrinsic PUFs and their use for IP protection. in *Cryptographic Hardware and Embedded Systems — CHES 2007*, ed. by P. Paillier, I. Verbauwhede. Lecture Notes in Computer Science, vol. 4727 (Springer, Berlin, Heidelberg, 10–13 Sept 2007), pp. 63–80
15. J. Guajardo, S. S. Kumar, G.-J. Schrijen, P. Tuyls, in *Physical Unclonable Functions and Public Key Crypto for FPGA IP Protection*. International Conference on Field Programmable Logic and Applications — FPL 2007 (IEEE Computer Security, 27–30 Aug 2007), pp. 189–195
16. D. Halperin, T.S. Heydt-Benjamin, K. Fu, T. Kohno, W.H. Maisel, Security and privacy for implantable medical devices. IEEE Pervasive Comput. **7**(1), 30–39 (2008)
17. HL7 EHR System Functional Model: A Major Development Towards Consensus on Electronic Health Record System Functionality, 2004. Available at `http://www.sanita.forumpa.it/documenti/0/100/140/148/EHR-SWhitePaper.pdf`. Accessed Nov 2009
18. A. Juels, M. Wattenberg, A fuzzy commitment scheme. in *ACM Conference on Computer and Communications Security — ACM CCS '99*, ed. by J. Motiwalla, G. Tsudik. (ACM, New York, NY, 1–4 Nov 1999), pp. 28–36
19. J.-P. Kaps, K. Yüksel, B. Sunar, Energy scalable universal hashing. IEEE Trans. Comput. **54**(12), 1484–1495 (2005)
20. H. Krawczyk, LFSR-based hashing and authentication. in *Advances in Cryptology – CRYPTO '94*, ed. by Y. Desmedt. Lecture Notes in Computer Science, vol. 839 (Springer, London, 21–25 Aug 1994), pp. 129–139
21. S.S. Kumar, J. Guajardo, R. Maes, G.-J. Schrijen, P. Tuyls, in *The Butterfly PUF: Protecting IP on every FPGA*. ed. by M. Tehranipoor, J. Plusquellic. IEEE International Workshop on Hardware-Oriented Security and Trust, HOST 2008, Anaheim, CA, USA, 9 June 2008. Proceedings. (IEEE Computer Society, Washington, DC, 2008), pp. 67–70..
22. Lifesensor.Accessed Dec 2009. Available at `https://www.lifesensor.com/en/us/`
23. J.-P.M.G. Linnartz, P. Tuyls, New shielding functions to enhance privacy and prevent misuse of biometric templates. in *Audio-and Video-Based Biometrie Person Authentication — AVBPA 2003*, ed. by J. Kittler, M.S. Nixon. Lecture Notes in Computer Science, vol. 2688 (Springer, Heidelberg, 9–11 June 2003), pp. 393–402
24. Medkey personal health records system. Accessed Dec 2009. Available at `http://www.medkey.com/`
25. Metavante's healthmanager. Accessed Dec 2009. Available at `http://www.phrforme.com/index.asp`
26. Microsoft, healthvault. Accessed Dec 2009. Available at `http://search.healthvault.com/`
27. W. Nevelsteen, B. Preneel, Software performance of universal hash functions. in *Advances in Cryptology – EUROCRYPT'99*, ed. by J. Stern. Lecture Notes in Computer Science, vol. 1592 (Springer, Berlin, 2–6 May 1999), pp. 24–41

28. R. S. Pappu, *Physical One-Way Functions*. Ph.D. thesis, Massachusetts Institute of Technology, Mar 2001. Available at `http://pubs.media.mit.edu/pubs/papers/01.03.pappuphd.powf.pdf`

29. R. S. Pappu, B. Recht, J. Taylor, N. Gershenfeld, Physical one-way functions. Science **297**(6), 2026–2030 (2002) Available at `http://web.media.mit.edu/~brecht/papers/02.PapEA.powf.pdf`

30. V. Shoup, On fast and provably secure message authentication based on universal hashing. in *Advances in Cryptology – CRYPTO '96*, ed. by N. Koblitz. Lecture Notes in Computer Science, vol. 1109 (Springer, Berlin, Heidelberg, 18–22 Aug 1996), pp. 313–328

31. K. Simoens, P. Tuyls, and B. Preneel. in *Privacy Weaknesses in Biometric Sketches*. IEEE Symposium on Security and Privacy — S&P 2009, (IEEE Computer Society, Washington, DC, 17–20 May 2009), pp. 188–203

32. P. Tuyls, A.H.M. Akkermans, T.A.M. Kevenaar, G.-J. Schrijen, A.M. Bazen, R.N.J. Veld-huis, Practical biometric authentication with template protection. in *Audio- and Video-Based Biometric Person Authentication — AVBPA 2005*, ed. by T. Kanade, A.K. Jain, N.K. Ratha. Lecture Notes in Computer Science, vol. 3546 (Springer, Heidelberg, 20–22 July 2005), pp. 436–446

33. P. Tuyls, G.-J. Schrijen, B. Škorić, J. van Geloven, N. Verhaegh, R. Wolters, Read-proof hardware from protective coatings. in *Cryptographic Hardware and Embedded Systems — CHES 2006*, ed. by L. Goubin, M. Matsui. Lecture Notes in Computer Science, vol. 4249 (Springer, Heidelberg, 10–13 Oct 2006), pp. 369–383

34. P. Tuyls, B. Škorić, T. Kevenaar, (eds.), *Security with Noisy Data: On Private Biometrics, Secure Key Storage and Anti-Counterfeiting* (Springer-Verlag New York, Inc., Secaucus, NJ, 2007)

35. B. Škorić, P. Tuyls, W. Ophey, Robust key extraction from physical uncloneable functions. in *Applied Cryptography and Network Security — ACNS 2005*, ed. by J. Ioannidis, A.D. Keromytis, M. Yung. Lecture Notes in Computer Science, vol. 3531 (Springer, Heidelberg, 7–10 June 2005), pp. 407–422

36. U.S. Department of Health and Human Services,Remote Monitoring Detailed Use Case, March 2008. Available at `http://www.himss.org/content/files/RMON_Use_Case.pdf`. Accessed Nov 2009

37. Webmd. Accessed Dec 2009. Available at `http://www.webmd.com/`

Printed by Books on Demand, Germany